Modern Potable Rainwater Harvesting
Second Edition

System Design, Construction, and Maintenance

Daniel M. Brown

ISBN-13: 978-1983497650
ISBN-10: 1983497657

Disclaimer

This publication is for informational and educational purposes only. Nothing written here should be taken as a warranty or guarantee of performance of your own rainwater system. Although the author has made every attempt to provide correct and accurate information on a specific water system that provides safe potable water for his own private residence, there is no guarantee the same system will work similarly for you under your specific circumstances, nor that it meets all the state and local regulations of your locality. You and you alone are responsible for ensuring the safety and proper working condition of your rainwater harvesting system and its compliance with state and local regulations. Neither the author nor the publisher are responsible for any mishaps or consequences that result from the reader's use of the information in this book since they have no control over how the reader chooses to utilize the information, nor do they have any capability of determining the reader's level of proficiency or physical condition.

Dedicated to my children and grandchildren,
as they bow before the Giver of all wisdom and knowledge,
and labor to be faithful stewards of the King's garden (Psalm 24).

For He "sends rain on the just and on the unjust" (Matthew 5:45).

Acknowledgments

I am thankful to the many engineers, scientists, and water technicians that reviewed the draft of this book and provided helpful feedback to make this book more readable and informative. Some of them are listed on the back cover of this book. Special thanks goes to my neighbor and friend, Andy Weaver, who edited my grammar and punctuation. Andy became my first customer and motivated further R&D of rainwater hardware. He maintains his own potable rainwater system and has first-hand experience with this book's subject matter. Special thanks also goes to Brandon Mantel, president of Donamarc Water Systems in Ohio, who greatly encouraged me in this effort. Brandon installs several rainwater systems every year for customers. I am always grateful to my lifelong friend, Larry Gelder of South Carolina, a solid fellow engineer and theologian. Larry always thoroughly reads my manuscripts and returns them with much helpful red ink! Most of all, I am thankful for my bride, the wife of my youth, who continues to patiently tolerate all my R&D experiments, including the ones that fail. She remains my greatest encourager in life.

Safety Warning

The system described in this book uses dangerous chemicals that can cause injury or death if not properly handled. These chemicals include liquid sodium hypochlorite (bleach), solid calcium hypochlorite, and concentrated hydrogen peroxide (35%). All these chemicals are strong oxidants of organic matter, including skin and eyes. Others include liquid epoxies, plastics, and solvents. Always wear eye protection and rubber gloves when handling these chemicals. Do not breathe dusts or vapors of chemicals. Never let calcium hypochlorite come in contact with any organic solvent because a violent exothermic reaction could occur that results in fire or explosion. Only mix disinfectants with water, and never with each other in concentrated form. Make sure you read and understand the Material Safety Data Sheet (MSDS) of all chemicals that you use.

Be aware that 120 volt AC electricity can kill. Always de-energize a 120 VAC circuit and check it with a voltage meter to make sure you turned off the correct switch before touching any bare wires. If for some reason you must measure voltages on an energized 120 VAC circuit, hold the voltmeter probes in one hand like chop sticks and put the other hand in your pocket. This reduces the potential for electricity to travel through your heart if you are accidentally shocked.

Table of Contents

Abbreviations

AC	alternating current
AOC	assimilable organic carbon
AOP	advanced oxidation process
BCD	binary-coded decimal
°C	degrees Celsius
CHARM	Cooperative Huntsville Area Rainfall Measurements
cm	centimeters
CMOS	complementary metal-oxide-semiconductor
CT	product of concentration and time
DC	direct current
DIP	dual in-line package
DPD	N,N-diethyl-p-phenylenediamine, chemical for testing chlorine level
DPDT	double pole double throw switch
DP6T	double pole six throw rotary switch
EPA	U.S. Environmental Protection Agency
ESR	equivalent series resistance of a capacitor
°F	degrees Fahrenheit
FIP	female iron pipe thread
ft	feet
GAC	granular activated charcoal
gal	gallons
GFCI	ground-fault circuit interrupter
GPH	gallons per hour
GPM	gallons per minute
GUI	graphical user interface
HOCl	hypochlorous acid
HP	horsepower
hr	hours
Hz	hertz
IC	integrated circuit
ID	inside diameter
in	inches
in^3	cubic inches
ISR	interrupt service routine
K	kilo, or kilo-ohms
°K	degrees Kelvin
LCD	liquid crystal display
LED	light-emitting diode
mA	milliamps
min	minutes
MIP	male iron pipe thread
ml	milliliters
mm	millimeters
m/s	meters per second
MSDS	Material Safety Data Sheet
MUX	multiplexer
mV	millivolts
N.C. or NC	normally closed contacts
N.O. or NO	normally open contacts
NOM	natural organic matter

NSF	NSF International (formerly National Sanitation Foundation)
NTC	negative temperature coefficient
OCl⁻	hypochlorite ion
OD	outside diameter
ORP	oxidation reduction potential
Ω	ohms
PC	printed circuit
PCB	printed circuit board
pF	picofarad
pH	a measure of acidity or alkalinity, 7 being neutral
ppm	parts per million, identical to milligrams per liter (mg/L) or $\% \times 10^4$
psi	pounds per square inch
PVC	polyvinyl chloride plastic
°R	degrees Rankine
RC	product of resistance and capacitance values
RO	reverse osmosis
RV	recreational vehicle
SPST	single pole single throw switch
SPDT	single pole double throw switch
SP3T	single pole three throw rotary switch
TBR	tipping bucket rain gauge
TDS	total dissolved solids
TempCo	temperature coefficient in parts per million per degree of temperature
THHN	thermoplastic high heat-resistant nylon-coated wire
THMs	trihalomethanes
uF	microfarads
UPC	Uniform Plumbing Code
UV	ultraviolet
VAC	volts alternating current
VDC	volts direct current
W	watts

Preface to Second Edition

Why a second edition? I continued to look for ways to improve potable rainwater harvesting after publication of the first edition of <u>Modern Potable Rainwater Harvesting</u>. Modifications to my own system proved necessary after encountering several unexpected challenges with my original design. These challenges included; consistent chlorine metering, prefilter clogging, accurate first flush control, accurate storage level measurement, and water consumption monitoring during long droughts. Implementing full system control with an Arduino microcontroller helped solve these challenges. Furthermore, an Arduino eliminated the design and fabrication cost of complex digital control circuits. Subsequent research and development of new digital rain gauges and wireless control with a Raspberry Pi made it obvious that a second edition of this book was long overdue.

Consistently metering chlorine with my original Mazzei injector proved to be illusive and unstable; I was constantly adjusting it. Chlorine is only injected when water is transferred from raw storage to clean storage (see Fig. 1-1 in Chapter 1). However, chlorine metering was highly dependent on transfer pump flow rate, which depended on filter condition, pump condition, and amount of ozone bubbles in the raw water circulation loop. When my transfer pump finally wore out (needing replacement) chlorine injection completely stopped. I replaced the Mazzei injector with a Stenner peristaltic metering pump, making chlorine metering completely independent of other hardware conditions. Nevertheless, we do propose a potential simple solution for venturi injectors in Chapter 7.

The first edition discussed a mechanical AND gate to control a first flush valve. In order to achieve a proper balancing of forces in the mechanical AND gate, my prefilter was originally located ahead of the first flush valve. This meant that all roof water (including first flush water) went through the prefilter, which somewhat defeats the purpose of a first flush valve in the first place. This required prefilter washing after practically every rainfall, particularly during warmer periods with high mold growth in gutters and drain lines. I removed the mechanical AND gate and replaced it with an Arduino microcontroller connected to a digital rain gauge and float switch on the raw water tank. The old mechanical first flush valve was replaced with a motor-driven valve incorporating a 12 VDC linear actuator. The prefilter was moved to a position after the first flush valve, where it belongs, and turned upside down to provide some self-backwashing capability after every rainfall. It is now a simple single-stage filter. This has worked much better; it has hugely reduced prefilter bag washing. I have removed the mechanical AND gate discussion from this second edition and replaced it with the much superior Arduino control.

We successfully survived an extremely severe drought in 2016 without needing to truck in water and refill our cisterns. According to the U.S. Drought Monitor map (http://droughtmonitor.unl.edu) the Southeast land area with Drought Monitor's most severe drought category (D4) actually exceeded the D4 area of southern California! Spring rains of that year were below normal and I began implementing household water conservation measures in May 2016, well before anyone was even talking about drought. Water conservation measures became more rigorous as the drought continued. Near the drought's end our household was consuming an average of only 18.2 gallons per day for over a month, in spite of still taking showers every day, flushing toilets, washing dishes, and having plenty of drinking water! We did wash clothes elsewhere since our clothes washer consumes nearly 50 gallons per load. Before the rains returned and topped off our storage tanks our water storage level was only down slightly under 50% and we could have survived another 77 days without rain. The degree to which we were able to conserve water while still meeting ordinary needs of life surprised both my wife and I.

At the end of 2016 I tallied up total amount of water captured by our roof and total amount of water consumed by our household, using my daily spreadsheet of rainfall amounts and water meter readings. To our surprise, in spite of one of the most severe droughts we have seen in many years, our rainwater system still dumped 56% of collected rainfall to the ground, either due to first flushing or due to full tanks that could not hold any more water! An additional 1500 gallons of storage would have made weathering the drought much easier without significantly reducing the amount of water dumped to ground.

The following year, 2017, was a much wetter year. In that year we dumped 69% of collected rainfall to ground. In other words, our roof area of only 1366 square feet could have supported a family three times our size during that year, assuming all persons conserved water as we did and water storage was adequate. Obviously, those who argue that rainwater harvesting robs the ground of needed water are either completely uniformed or are intentionally lying.

Many people mistakenly believe that well water is more reliable than rainwater harvesting. Our 2016 drought showed the very opposite is true. A common underground aquifer generally serves multiple wells and nobody knows exactly how much water is in that aquifer. Nor can anyone control how much water their neighbors draw from that same aquifer. Many private wells in northern Alabama went suddenly dry without any warning during the drought of 2016. My household suffered through the same drought but my rainwater storage tanks never went dry. I knew exactly how much water remained in storage and when our tanks would run dry if we did not get rain. We controlled water usage and successfully weathered that drought. Information on our water resource, which helped us conserve water, was far superior to that of groundwater resources for well owners.

This drought also showed me that locating a flow meter in the transfer pump line between raw and clean water tanks, as shown in the first edition, was not an ideal position. In the drought's severest part, when knowing one's daily water consumption became most crucial, we were unable to obtain that information because my flow meter completely stopped once the raw water tank became empty and automatic water transfer ceased. I relocated the flow meter to a position between the pressure tank and house to solve this problem. We can now immediately measure exact water consumption for every single water-consuming activity in our household.

The pressure-based water level sensors discussed in the first edition proved to be problematic and unreliable. Due to their delicate construction these sensors must be isolated from stored water with a flexible membrane and immersed in either a small pocket of air or distilled water. Maintaining a constant small volume over long periods of time was exceedingly difficult, even when no discernible leaks existed. After a sensor would stop working I would find its membrane stretched to maximum deflection and all air squeezed out of the cavity. Furthermore, acoustic vibrations from circulation pumps were superimposed on the pressure signals, seriously degrading signal-to-noise ratio. Pressure noise could be removed with low pass filtering, but this also made signal measurement less reliable. I finally abandoned the pressure sensors altogether and fabricated capacitive tank level sensors with much superior results. These sensors are robust, low cost, easy to build, insensitive to acoustic vibrations, and very accurate. Discussion of pressure-based tank level sensors is removed and replaced with capacitive tank level sensors in this second edition.

I still use the original custom digital runtime counter discussed in the first edition on my own system since it continues to operate and do its job. And it once again saved me from a pipe rupture by shutting down pressure pump power after it had wasted about 100 gallons of water to my basement floor while I was away from home. This rupture was due to my own plumbing error in moving the flow meter. I had tightened a PVC fitting down too tight causing it to eventually rupture. We humans make mistakes! Prudence dictates that we install safety mechanisms to protect us and our hardware systems from both our own errors and other causes. However, I have removed discussion of the old custom digital runtime counter and replaced it with an Arduino microcontroller in this second edition. The incredible low cost of microcontroller hardware such as an Arduino or Cypress PSoC simply does not justify the effort and expense of building custom electronic control circuits. Practically everything that needs to be done in automating a rainwater harvesting system can be done with low-cost microcontrollers.

No doubt improvements and cost reductions will continue to be made as more people install potable rainwater systems. In spite of the fact that rainwater harvesting has been practiced for thousands of years, much more research and development is still needed in potable rainwater harvesting technology. Since the vast majority of our nation depends on large centralized public water supplies rather than decentralized rainwater harvesting, very little R&D has been done on alternative roofing and gutter materials or on low-power advanced disinfection processes suitable for rainwater harvesting. Furthermore, architects remain unconcerned about how their roof line designs affect quality of captured rainwater or gutter maintenance. Most gutter installers do not install gutters and gutter screens suitable for rainwater harvesting, but rather for appearance and moving storm drainage away from foundations. Other potential research areas are mentioned throughout this book. I hope this book helps stimulate new innovations in potable rainwater harvesting technology.

As of this writing, the state of Colorado still makes private rainwater harvesting nearly impossible, in spite of the fact that rainwater harvesting is obviously the most effective means of water conservation. Stupid excuses from Colorado politicians for outlawing rainwater harvesting insult the intelligence of people who understand that rainwater harvesting conserves water and benefits all of society, as proven by vast scientific experimental facts. Colorado politicians even ignore scientific studies from their own engineers and scientists! But politicians and bureaucrats are not really interested in water conservation. Their primary goal is control, job security, and maximizing taxes extracted from productive society. And centralized water supply facilitates that primary goal.

Political excuses that rainwater harvesting deprives others of available water from public water supplies simply holds no water (pun intended). Where does all drinking water ultimately come from? Does a rainwater harvester somehow magically reduce average rainfall? How does a rainwater harvester's removing himself from public water supply and reducing his consumption of water deprive others of needed water? Think about this: our ancestors manually drew or pumped a bucket of water from a well, which lasted almost an entire day for their household. Drawing water from a well was real work. Now we have teenagers wasting 400 gallons of hot water every day down a sewer drain just taking long hot showers because they don't have to do any work for it! The real culprit for water waste in America is public water supply, not rainwater harvesting. And politicians are ultimately guilty of water shortages when they outlaw rainwater harvesting and mandate purchasing public water supply.

To drive home the silliness of political arguments against rainwater harvesting, consider the fact that nobody living in Boulder, Colorado gets their drinking water from their neighbor's storm drainage runoff. Drinking water for Boulder comes from reservoirs high in the mountains west of Boulder, such as Silver Lake and Barker Reservoir. Everyone (except politicians) knows that water does not flow uphill! Therefore, a rainwater harvester in Boulder has **zero** effect on water flowing **into** Boulder's water reserves. However, by removing himself from public water supply and capturing a tiny amount of rainwater that would otherwise go into storm drainage and sink into the desert plains east of Boulder, a rainwater harvester would actually be helping his neighbors have a little more available water from public water reservoirs. Furthermore, every drop of captured rainwater still ultimately either goes down a sewer drain or is immediately dumped to ground. So those dependent on water wells east of Boulder remain completely unaffected by rainwater harvesters.

If you are reading this book you probably already know how rainwater harvesting is incredibly conserving of water. Rainwater harvesting forces humans to conserve water like nothing else can possibly do (other than wilderness backpacking or drawing a bucket of water from a well), especially if rainwater harvesting is your only source of potable water. When properly performed, as promoted in this book, rainwater harvesting provides water that is orders of magnitude cleaner than any public water supply. Rain is essentially distilled water, except for a tiny amount of pollutants picked up from the atmosphere. But those pollutants are benign compared to the dangerous chemicals and pharmaceuticals currently plaguing our nation's ground water reserves. If politicians and government bureaucrats were truly interested in water conservation and the health of human societies they supposedly serve, they would strongly promote rainwater harvesting. But they are not; their primary goal is job security in a cushy government position that requires no real work.

Some people still promote a myth that drinking distilled water or rainwater is unhealthy. They imply public water supplies are healthier! I would not be surprised if anti-rainwater harvesting politicians are the source of such nonsense. Rainwater is obviously void of essential minerals our bodies need. However, we don't drink water to get our needed minerals, but rather to dissolve minerals in foods and essential supplements we consume. We also drink water to properly flush our bodies of normal bodily pollutants. I would rather drink pure water and know exactly what I'm putting into my body than to take a risk of drinking polluted ground water while hoping I might get some essential minerals!

There are many conscientious government scientists, engineers, and regulators who strive to do right for the public they serve. My criticism is not against them, but against their political bosses who prevent them from doing right because of taking bribes from lobbyists and big campaign donors who use the sword of government to eliminate their competition! If you are among those conscientious government engineers trying to do the right thing, don't be offended by my criticism of politicians and bureaucrats. This book's primary audience are readers like you, engineers and technical types who want to make potable water safe and healthy for everyone. Besides, liberal politicians are not going to read this book. They would prefer a book on hair grooming for TV appearances!

Reluctantly, I have converted this second edition to an all black and white printing (except for the cover). The few color figures in the first edition do not justify the significantly increased cost of full color printing. Where necessary figures have been reworked so that there is no loss of information in this conversion to black and white printing. Some readers may be disappointed by this change, but I want to keep this book's cost as low as possible to maximize accessibility to its information. In order to compensate for the loss of color in this edition, I am also publishing a separate color photo optional supplement for those who want to see more hardware details.

No doubt there will be some SI units snobs who read this book and complain that I have not used kilograms, meters, liters, and degrees Celsius throughout this book. Such complaints are frequently lodged against authors of American technical books. I have seen these people on the Internet even condemn technical explanations in anything other than SI units as "pseudo science," which is utterly ridiculous. Real scientists and engineers can comfortably work with any set of units, including cubits, rods, and parsecs! For an engineer or scientist it does not matter what units he works in, as long as those units are clearly defined. However, we write technical books to increase the understanding of the community in which we live, not to satisfy some international units committee trying to impose their political agenda upon the earth. And the community in which I live is most comfortable with feet, inches, gallons, and degrees Fahrenheit. Therefore, I have primarily used the United States customary units in this book, with brief references to SI units for those who wish to work in other units.

Potable rainwater harvesting is not difficult. Anyone with average intelligence and a willingness to learn a few new skills can build and maintain a system that meets all of one's household water needs. Through your own efforts to supply clean water to your household you will learn to conserve water like you never thought possible. And you will come to appreciate who the true Giver of pure water really is. Hint, it's neither the public water supplier nor the government! Rainwater is orders of magnitude cleaner than any public water supply. The trick is to cleanly capture it and keep it clean during storage. That is what this book is all about.

Dan Brown

Chapter 1.
Introduction and Background History

The goal of this book is to provide technical instructions on designing, building, and maintaining a potable rainwater harvesting system for private residences or small businesses. It is not meant for those with only a passing curiosity about rainwater harvesting, nor those who only want to use rainwater harvesting for non-potable purposes (e.g. garden or animal watering). Indeed, numerous other books cover these rainwater harvesting basics, many of which are referenced at the end of this book. My personal favorites are the Texas rainwater harvesting documents.[1][2][3] Rather, this book is for the technical engineering-minded person that has read all the basic books and still has questions on design details, construction, and maintenance of a practical rainwater harvesting system that serves as one's only or primary means of potable water supply. If rainwater is not your primary or only source of drinking water, or you harvest rainwater only for watering gardens, then other systems are less costly to implement and operate than the one described in this book. I assume the reader has read the various state rainwater harvesting manuals and understands rainwater harvesting basics for non-potable purposes. This book takes a technical reader's knowledge to the next level and provides confidence for building a practical modern potable rainwater harvesting system.

Being a physicist and engineer, I find writing on any subject without mathematical equations to be very difficult. Physical processes and the invariant laws of creation are described with mathematical equations. Rainwater harvesting is definitely a physical process, governed by mathematics and physical laws. One cannot do an adequate job of system design without referring to appropriate governing equations. Although I have endeavored to keep mathematical equations at the high school graduate level or first year college level, this book necessarily contains many equations. Examples of how to use them in your design process are also included.

Rainwater collection systems discussed in most basic books, including many state rainwater harvesting documents, are adequate for garden watering and possibly a temporary emergency source of drinking water if public utilities fail. However, I believe that some methods taught in those books are inadequate for providing a reliable daily supply of drinking water. Specific details and my reasons for disagreement are discussed later, particularly in Chapter 2. We distinguish between

---

1   Dr. Hari J. Krishna, et. al. <u>The Texas Manual on Rainwater Harvesting</u>, 3rd Edition, Texas Water Development Board, 2005.

2   <u>Harvesting, Storing, and Treating Rainwater for Domestic Indoor Use</u>, Texas Commission on Environmental Quality (TCEQ), January 2007.

3   <u>Rainwater Harvesting: System Planning</u>, AgriLIFE Extension Texas A&M System, 2009.

rainwater harvesting methods suitable only for garden watering and those that are suitable for primary potable (drinking) uses. Our goal here is the latter. Furthermore, the rainwater harvesting method described here is not suitable for large public water suppliers because a change in consumer mindset and lifestyle is required that centralized suppliers of cheap abundant water cannot possibly achieve.[4] This will become much more apparent when we discuss management of limited resources. Successful potable rainwater harvesting can only be achieved by private ownership of a system. We discuss briefly in the last chapter how larger private systems could be shared among a few households living in close proximity. Other than this, potable rainwater harvesting is not a solution for large centralized public utilities: it is a decentralized private solution.

My own household water system evolved over time as I tried different solutions to provide clean potable water for our personal residence. We often hear that necessity is the mother of invention and that has surely been the case here. This brief background history shows why we chose rainwater harvesting, and hopefully, will help you avoid the costly mistakes I made along the way. My wife and I built a house on a mountainside that had no access to municipal water. Bringing electricity to the site was costly, but bringing municipal water to the site was cost-prohibitive. Our county water supplier quoted us $300,000 to pipe water only 0.7 miles – significantly more than the combined cost of both land and house!

We first tried drilling a well near our house. Our driller bored through three caves in a little over 100 feet which caused drill tailings to stop blowing back to the surface as normally occurs. Fearing that he might lose his bit from tailing buildup down in the hole, he abandoned his efforts and informed us he could not drill in our area. Our local area consists of a deep limestone karst geology with low probability of finding a better drilling site. Even if we did hit water with a drilled well it would most likely produce only surface water in need of heavy purification since limestone karst provides little to no natural filtering. Limestone layers are hundreds of feet thick in northern Alabama and riddled with thousands of vertical and horizontal caves, allowing surface water to fall hundreds of feet to the water table with virtually no filtration. So we gave up on the idea of drilling a water well.

Since we have a cave on our property from which water flows all year long, even during droughts, we next tried to use this as our water source. The cave is located 230 vertical feet below our house and produces only a trickle of water during dry summer months. The cave has a shallow pool of water in an area that sees little visitor traffic due to a tight crawl on hands and knees required to reach it, making this pool a somewhat protected place to locate a water pump foot valve. Laboratory testing confirmed our cave water would need some heavy duty purification. In addition to

---

4   Notwithstanding that taking water from rivers, lakes, and wells can be considered "rainwater harvesting," we limit our definition of rainwater harvesting to the immediate catchment and storage of rainfall using man-made structures rather than natural structures.

being biologically unsafe for drinking, this water contained lead, iron, and several other metals.[5] Due to low flow into the cave pool during dry periods, a typical deep well water pump could not be used because it would suck the pool dry and lose pump prime. Furthermore, sufficiently purifying water at high flow rates typical of well pumps would be costly. Lead-removal filters typically allow only very low flow rates due to their small porosity (e.g. 0.5 micron activated charcoal filter). I installed a high-pressure low-flow diaphragm pump (150 psi, 0.7 GPM) near the cave and passed the water through a 0.5 micron lead removal filter, then through an ultraviolet (UV) lamp purifier, and into a 1650-gallon water cistern near our house. A small circulation pump on the cistern continuously circulated cistern water through a filter and UV purifier to maintain stored water purity. A shallow well pump and pressure tank in our basement then re-pressurized cistern water for household use.

Another reason for using a low flow diaphragm pump at our cave, rather than a standard well pump, was its low power consumption of 120 watts, which allowed running 14-gauge wire across the 600-foot distance from house to pump. I designed and built an electronic controller that monitored water level in both cistern and cave pool to automate cistern filling and prevent the cave foot valve from sucking air. This system sufficiently purified the water but filters had to be changed about every two weeks due to heavy silt and minerals in the cave water. Fine silica silt caused pumps to wear out quickly, requiring frequent replacement. I added two more filters, including a back-washable filter, to extend filter life. This helped somewhat, but heavy silt, iron, and other cave water pollutants still meant frequent and costly filter changes. Increasing porosity micron size of the final charcoal filter extended its life but also increased recirculation flow rate to more than specified capacity of my UV purifier. Fine silt that did get through the filters also coated the UV lamp quartz tube rendering it useless. So I began using chlorine instead of a UV lamp to disinfect our water. However, my activated charcoal filter, needed to remove lead, also removed chlorine as cistern water circulated through it! Chlorine had to be added two or three times each week. Additionally, chlorine reacts with organic matter, producing trihalomethanes (THMs), which are carcinogenic. This was not an acceptable solution.

Final motivation to abandon our cave water and go with rainwater harvesting actually came from my wife. Practically every time we had unusually heavy rains our cave flooded and washed the foot valve out of the cave pool in spite of burying it under a piles of rocks. Access to this cave pool was difficult; a wet crawl on hands and knees. Frequent trips into the cave to repair or reset the foot valve and re-prime the pump did

---

5   Local legend has it that a highly productive lead mine in our area was once used by Confederates in the Civil War. When Union troops advanced into northern Alabama the Confederates collapsed the mine entrance with explosives to prevent it from falling into Yankee hands. The exact location of the entrance has since been lost. The well-known presence of lead and radon in northern Alabama prompted a search for uranium several decades ago, with local hardware stores selling Geiger counters, but no significant uranium deposits were ever found.

not constitute a low-maintenance water system! One of the last times my wife and I were resetting the cave foot valve, shivering from crawling through cold cave water in the middle of winter, my wife said to me; "Honey, we're getting too old to keep doing this. You need to come up with a better way to get water to our house!" I decided right then to abandon cave water and install a potable rainwater harvesting system. Rainwater is orders of magnitude cleaner than surface water and should provide us with a high-quality, low-maintenance water supply, if our roof area and average rainfall are sufficient to meet daily water demands. Rainwater had to be much easier to filter than cave water. Fortunately, I had been thinking about rainwater harvesting for quite some time and had installed a metal roof on our house just in case we had to resort to rainwater harvesting when all other options failed. Looking back in hindsight, I should have made rainwater harvesting my first choice rather than last choice.

This experience taught us that using a strong oxidant and disinfectant like chlorine is unavoidable if potable water is derived from surface or ground water. Chlorine is needed to disinfect water of pathogenic bacteria, viruses, and protozoa. It also eliminates harmless iron bacteria common in most soils that build rust-colored biofilms in plumbing. Chlorine helps protect against more harmful micro-organisms that can thrive inside biofilms, and it oxidizes unpleasant odors from natural organic matter in water. Depending on local geological conditions, deep wells and municipal water may not be much cleaner than surface water. Most of my neighbors on wells wrestle with high levels of iron, iron bacteria biofilm, and other pollutants in their water. One of my neighbors, whose home is connected to chlorinated municipal water, installed a cistern due to very low water pressure. Municipal water trickles into his cistern which he then re-pressurizes for household use. He showed me his cistern one day and complained that after only a few years of operation there was more than an inch of sediment and biofilm in it, all from "clean" public chlorinated water.

If done properly, rainwater harvesting can provide much cleaner water than any of the traditional sources, although chlorine or other residual disinfectant is still needed. Primary sources of pollution in rainwater come from roof and gutters, but this is a tiny fraction of the amount of pollutants in typical surface or ground water supplies used by public water suppliers! Due to very low total dissolved solids (TDS) and absence of silt in rainwater, a UV lamp purifier can be used quite effectively. UV lamp purifiers are useless when its quartz tube becomes coated with dirt and silt, which frequently happened with our cave water. Now, with rainwater harvesting, I only check the UV lamp quartz tube when I change the bulb (once each year) and this tube has always been crystal clear with zero deposits.

Another advantage of rainwater harvesting is that it is generally much less expensive than drilling a well. Drilling a well may cost tens of thousands of dollars (depending on depth), regardless of whether or not you hit water; a homeowner bears the cost of dry holes. Average "do-it-yourself" (DIY) persons with basic tools can easily install potable rainwater systems and save on labor costs. This is not possible with well drilling, which requires very expensive drilling equipment. The most expensive hardware items in a rainwater system are your storage tanks, and these are usually under $1000 for above-ground plastic tanks. I provide detailed cost estimates of hardware items throughout this book.

Groundwater pollution is a very serious problem that most people are oblivious to. Pollutants may be diluted below EPA limits during certain periods of the year, but they always seem to wildly fluctuate with rainfall amount. I personally observed this with our cave water; sometimes lead was detectable and at other times it was not. High levels of toxic metals and radioactive compounds are showing up in groundwater aquifers all across the nation due to injection well uranium mining and remaining in groundwater long after mining operations have been shut down. Pharmaceutical drugs are likewise showing up in groundwater along with herbicides and pesticides, some of which are very difficult to remove without advanced oxidation processes.

Due to extensive groundwater pollution today, the probability of drilling a clean water producing well is actually quite low. Most well water today, both private and public, requires heavy purification. My father had a well on his west Texas ranch over 400 feet deep that produced water contaminated with E.coli along with bad smell and taste. Practically every state, county, and public utility practices roadside herbicidal spraying to reduce cost of mowing along roadways and power lines. These powerful and long-lasting herbicides get into groundwater during rainfalls. Some farmers mindlessly dispose of empty pesticide and herbicide canisters into a sinkhole on their property. Pesticide and herbicide residues from rusting canisters may then show up in a municipal water well many miles away.[6] Massive quantities of chemical weapons have been buried throughout the United States by our own government since WWII. And the government has essentially forgotten exactly where those weapons are buried! Spelunkers are well aware of the fact that subterranean rivers and creeks can be just as lengthy as those on the surface. The karst subsurface of northern Alabama and other states with deep limestone deposits is extremely porous, riddled with thousands of horizontal and vertical caves. In some places surface water can free fall in mid-air over 400 feet underground without ever touching a rock (e.g. Surprise Pit in Fern Cave).

Unfortunately, state and county governments and public utilities are often major groundwater polluters through their roadside herbicidal spraying programs. In an effort to save money maintaining state and county roads, local governments often resort to spraying herbicides rather than mowing along roadsides. A strong argument can be made that roadside herbicidal spraying is actually more detrimental to the environment and groundwater resources than agricultural spraying. Farmers try to minimize runoff to prevent erosion of precious topsoil and force rainwater to soak into topsoil. By contrast, roadsides are specifically designed to maximize runoff and divert rainwater to creeks and rivers as quickly as possible to prevent road flooding. These

---

6   This actually happened in northern Alabama.

rivers serve as raw water supplies for municipal water companies. Herbicides and pesticides frequently show up in annual comprehensive water tests performed by public water suppliers, although usually below EPA limits. I am amazed that EPA does not shut down herbicidal spraying along roadsides and electric utility right-of-ways. However, EPA is a government agency subject to political manipulation. Abundant research exists on the Internet showing that herbicidal spraying along roadsides, railroads, and electric utility right-of-ways is a serious groundwater pollution problem, which currently seems to have no cost-effective solution. This is another reason for choosing potable rainwater harvesting rather than trusting government-regulated public water utilities to supply clean potable water.

Centralized supply of basic resources needed for survival makes societies vulnerable to human error, natural disasters, political corruption, and terrorism. The April 2011 series of tornadoes in northern Alabama shut down electrical power and municipal water supplies for a week. Many people had no clue how to supply their families with clean drinking water. In 1993 a Cryptosporidium outbreak in a Milwaukee public water supply caused over 400,000 people to get sick and resulted in over 100 deaths. In January 2014 a chemical spill of 4-methylcyclohexane methanol into the Elk River a mile upstream of the single intake point for the West Virginia American Water company caused 300,000 residents to lose drinking water for several days. In May 2014 an E.coli outbreak in the Portland, Oregon public water supply prompted officials to warn residents to boil water before drinking it. Escherichia coli is one of the easiest pathogens to kill with chlorine and E.coli outbreaks indicate a failure of water utilities to maintain proper chlorine levels. Ohio has one of the most highly regulated public water supplies in our nation. Yet their source waters frequently suffer from harmful algal blooms (HABs) that generate cyanotoxins and cause people to become sick. Although U.S. public water supplies are usually quite safe, waterborne disease outbreaks in public water supplies occur frequently enough across the U.S. that the Centers for Disease Control (CDC) continually tracks these outbreaks.[7]

Centralized public water suppliers often draw source water from single points on large rivers (which are essentially nature's sewer systems) making a large population segment vulnerable to water safety at that single intake point. Decentralized water supplies (i.e. private supply) both improves water safety and minimizes impact of human error or terrorism. Decentralization makes water system operators more accountable to users and less tempted to take short cuts with system maintenance. Decentralization of essential utility resources truly is a national security issue.

Each year usually sees several news reports of municipal water authorities in various states recommending that people not drink municipal water due to high levels of biological or chemical pollutants. Biofilm buildup in distribution systems is often the main culprit for disease outbreaks, but removing biofilms is nearly impossible

---

7   See http://www.cdc.gov/healthywater/statistics/wbdoss/surveillance.html

without a massive disruption of public water supplies. Unlike public supplies, private water supplies can be temporarily shut down in order to flush a distribution system with strong cleaning agents to remove biofilms. Users prepare for it and tolerate the temporary inconvenience for the sake of cleaner and safer water.

Both public and private wells are no longer immune from the intentional "legal" pollution of groundwater reserves. Injection well uranium leach mining and oil shale fracking are primary examples of this. Uranium mining companies have discovered that injecting an oxidizing solution (usually called a lixiviant) into uranium rich groundwater sands through an injection well causes a huge spike in dissolved uranium compounds in the groundwater. Groundwater is then pumped from a nearby well and processed to remove the uranium. Hydrology engineers usually assume certain conditions about subsurface groundwater flow to determine impact on nearby potable water wells, but can make no guarantee about the future. Mining companies try to pump sufficient water from aquifers during mining operations to cause local depressions in water tables, causing groundwater flows inward toward mines and reducing risk of lixiviant contaminating nearby water wells. But subsurface groundwater levels also wax and wane like surface bodies of water from rainfall amounts. The U.S. Nuclear Regulatory Commission requires that mining companies sweep all lixiviant out of aquifers and restore their original conditions after mining operations cease, but this practically never happens.[8] The cost of groundwater restoration can amount to 40% or more of total cost of the mining operations. With such a huge cost for groundwater restoration, huge temptations to cut corners or "fudge" data obviously exist.

In a recent report prepared by Davis and Curtis for the U.S. Nuclear Regulatory Commission[9] the authors state that restoring groundwater to its original condition is extremely difficult and usually far more costly than originally estimated. Their data show that dissolved uranium levels in groundwater at a Wyoming mine at the end of "restoration," eight years after mining operations ceased, was still 70 times higher than initial levels before mining operations started! Dissolved uranium in groundwater at a Nebraska mine was 25 times higher after restoration than before mining operations began. At another Wyoming mine it was 41 times higher after groundwater restoration. Restoration after mining usually leaves high levels of uranium, radium, selenium, arsenic, and vanadium in groundwater aquifers. Exemptions to originally agreed-to terms are almost always applied for and granted. Huge regions of aquifer groundwater along Texas coastal plains, from which many people draw their drinking

---

8   Susan Hall, <u>Groundwater Restoration at Uranium In-Situ Recovery Mines, South Texas Coastal Plain</u>, Open-File Report 2009-1143, U.S. Geological Survey, Central Energy Resources Science Center, Denver, CO, 2009.

9   J.A. Davis and G.P. Curtis, <u>Consideration of Geochemical Issues in Groundwater Restoration at Uranium In-Situ Leach Mining Facilities</u>, NUREG/CR-6870, U.S. Geological Survey, Menlo Park, CA, January 2007.

water, have been polluted by soluble radioactive compounds and toxic metals. The Navajo and Sioux Nations in western states have been devastated by radiation poisoning in their drinking water due to mining waste from abandoned uranium mines. Big energy companies make massive campaign donations to both Republican and Democrat politicians (euphemistically called "greasing the skids") to encourage them to make rulings favorable to mining companies. Where there is money to be made there is also money to keep the public deceived.

Lead was once considered to be a useful metal and was profitably mined throughout the United States. Today we know it to be a highly toxic metal with devastating effects upon brain function, particularly in children. Modern societies have exerted tremendous efforts to remove it from our environment and minimize its use. However, we forget that lead deposits often show up in limestone layers. Indeed, many abandoned lead mines exist in limestone karsts. Today, mining the limestone for road and building construction is more profitable than lead mining. However, these limestone quarries unwittingly put groundwater resources at risk of lead poisoning when they crush limestone and rinse it with water to reduce dust. Lead veins in limestone are not easy to recognize, particularly if a miner is not looking for it. And limestone miners are definitely not interested in finding lead!

Deep groundwater resources are rapidly becoming polluted with all kinds of chemical, biological, and mining wastes from normal productive human activities. Do an internet search on groundwater remediation in any state in our nation and you will quickly realize how huge this problem is! Karst areas with thick exposed limestone layers are particularly susceptible. Large tilted limestone formations, such as the Edwards Plateau in Texas, allow surface pollution to infiltrate groundwater and travel hundreds of miles to an unsuspecting deep water well. I am surprised that well drillers are still in business. Unfortunately, most people (including myself) tend to first think of well drilling when public water supplies are unavailable. Rainwater harvesting is just not on most people's "radar." Having personally attempted both well drilling and pulling water from a cave spring, I now realize that rainwater harvesting should have been my first choice for a potable water supply. In fact, I would rather absorb the cost of replacing an asphalt shingle roof with a metal roof in order to pursue rainwater harvesting than to bear the cost and health risks of drilling a well.

It is easy to make potable water safe (free of pathogens) by simply adding sufficient disinfectant. But making it healthy to drink (free of chemical toxins) is very difficult without distillation. The more polluted initial source waters are, the higher the risk of generating unhealthy potable water. One of the stupidest ideas currently circulating among some public water professionals is the idea of using black water for a potable water source! Only distillation (and its accompanying huge energy cost) could have any hope of turning black water into safe and healthy potable water. Rainwater harvesters obviously use distillation as their initial water purification step, but the energy comes from the sun and is free of cost.

Rainwater harvesting provides confidence that you do not need to depend on government or anyone else to provide your most basic and essential resource needed for survival. And it is not subject to groundwater pollution. Both my own personal experience and that of all my neighbors who either own water wells or have attempted drilling wells, prove that potable rainwater harvesting is far cheaper and much more reliable than drilling a water well. I have neighbors that have drilled multiple wells, costing them tens of thousands of dollars, with all those wells producing inadequate water quality or quantity. If electricity fails a modern well is as useless as a dry hole, but one can still dip water out of a rainwater tank without electricity. Even in dry areas like west Texas, the cost of installing an additional shed roof and a few more plastic storage tanks is far cheaper than drilling a well, and will no doubt provide a cleaner and more reliable supply of water. Rainwater harvesting is simply a low-cost, common sense solution to practically all water problems in America.

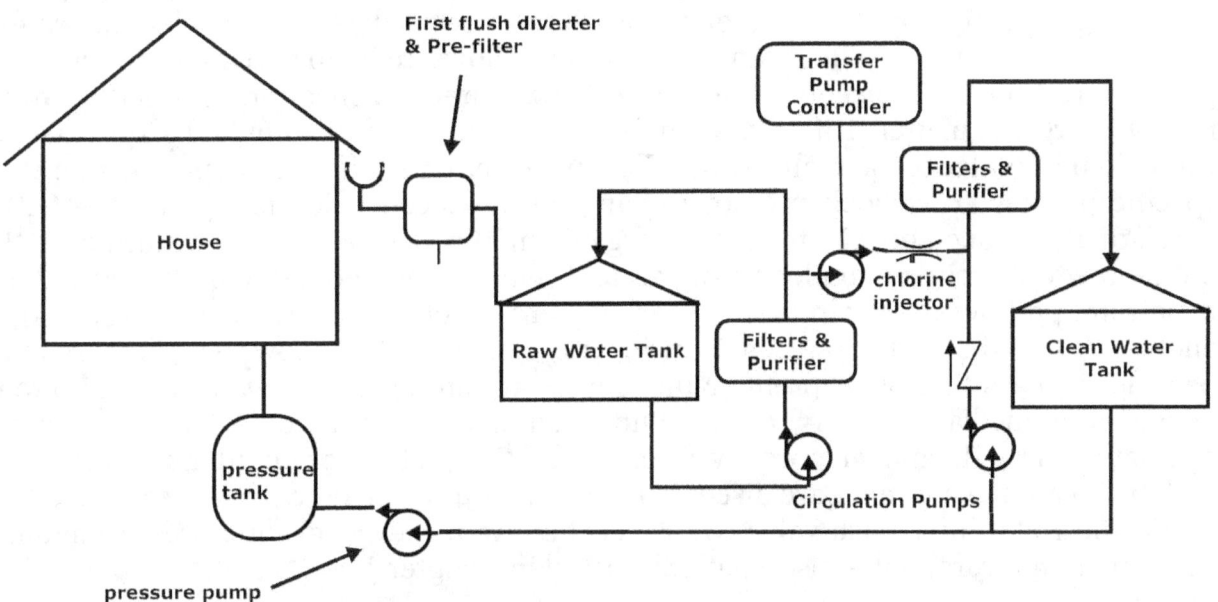

Fig. 1-1. Basic dual-tank rainwater harvesting system.

The rainwater system discussed in this book is a double-tank design, illustrated in Fig. 1-1, with separate continuous-circulation purification loops on each tank. A raw water tank receives immediate runoff from roof after passing through a prefilter. A portion of initial rainfall is dumped to ground through an automatically controlled first-flush valve just ahead of the prefilter. This rinses roof and gutters of dirt and contaminants before sending rainwater to storage. A clean water tank, with its own separate purifier circulation loop, is kept topped off at all times to maximize available space in raw water storage to receive new rainfall. When clean water level drops by about 100 gallons a transfer pump automatically turns on and moves filtered raw water to clean water storage, forcing it through the clean water filter. Note that transferred raw water comes from downstream of raw water filters and is injected upstream of clean water filters in the clean water circulation loop. This helps insure that only the cleanest water is transferred to clean water storage. A check valve just above the clean water circulation pump prevents transferred water from going backwards through the circulation pump and bypassing the clean water filter. Chlorine is also added during this transfer process. Household water is drawn from clean water storage by a pressure pump as needed.

This basic design approach has worked well for us and has not changed. Although many single-tank rainwater systems are currently providing private potable water supplies, potable rainwater harvesting is far more effective and less costly with two separated tanks or storage systems. Rainwater quality can vary throughout the year due to seasonal variation of pollen, leaf matter, and mold in gutters and drain pipes. Splitting water storage into raw water storage and clean water storage allows a stable high quality supply of water to our house, regardless of variable raw water quality from the roof. A dual-tank design also allows reducing THM formation by postponing chlorine injection until after raw water purification has sufficiently removed organic matter. Different disinfection processes can be applied sequentially. Both circulation loops have filters, but filter types may be different, depending on the specific process. This book promotes using an advanced oxidation process (AOP), consisting of ozone and UV, to remove organic matter before injecting chlorine. AOP on the raw water circulation loop oxidizes and removes small organic particulates that penetrate prefiltering during rainwater capture. Prefilter porosity must be large enough to handle high flow rates during rainfall, but this allows smaller organic particles to pass through a prefilter into raw water storage. Raw water quality often degrades immediately after rainfall and then slowly improves as raw water is circulated through smaller porosity filters and AOP. Details of circulation plumbing and purification equipment are discussed in Chapter 4. Electronic controllers (transfer pump controller, first-flush valve controller, tank water level sensors, pressure pump runtime monitor, rainfall sensors) are discussed in Chapter 5.

Single-tank systems are more suitable for non-potable irrigation purposes or an emergency water supply when public water supplies fail. Potable rainwater collection using only a single tank requires more expensive on-demand filtration and disinfection systems. However, filtration and disinfection are always more effective and less costly when done slowly at low flow rates, which a dual-tank system permits. Continuous circulation in both tanks prevents stagnation and keeps water crystal clear. Anyone that has ever maintained a swimming pool or spa understands that continuous circulation through filters is essential to keep exposed bodies of water clean and biologically safe. Although exposure of stored rainwater is less than that of a typical swimming pool, rainwater is still exposed to contaminants from roof, gutters, and atmosphere. Even with superior filtration, potable water still contains some amount of natural organic matter (NOM) that can serve as food for bacteria and biofilm. NOM is a primary cause of high levels of disinfection by-products, particularly THMs, when chlorine is used as a disinfectant.[10] All these issues are discussed in more detail in following chapters.

My rainwater system evolved over time as I modified it to solve various disinfection problems, increase automation, and reduce maintenance. Indeed it has evolved since the first edition of this book. Initially, my rainwater system used only a combination of UV and chlorine for disinfection. Under normal operation, chlorine (sodium hypochlorite or calcium hypochlorite) was primarily added to raw water storage. A low level of chlorine was maintained in clean water storage to provide a residual disinfectant to fight biofilm growth inside distribution plumbing. Higher levels of chlorine were maintained in raw water since it was being consumed by oxidizing organic matter. THMs are produced when chlorine oxidizes organic matter. In spite of continuous turbulent circulation to help dissipate THMs to the atmosphere, water testing performed by National Testing Laboratories still confirmed presence of THMs in excess of EPA limits. Using chlorine for disinfection and its associated THM production remained a problem, particularly during periods of high organic matter shedding from trees. When I initially investigated ozone to reduce THM formation I found that practically all ozone generators required dry air to work properly. Many use a desiccant to dry input air. However, rainwater harvesting households usually don't live in dry areas. High humidity means high maintenance of an ozone generator, including frequent replacement or drying out of a desiccant.

Ozone is a strong oxidizer on contact, but it does not provide any residual disinfection capability like chlorine. Ozone does not protect against biofilm buildup in hard-to-reach places like inside distribution plumbing; only a residual disinfectant such as chlorine can do this. I assumed that using ozone in conjunction with chlorine might help reduce chlorine demand and reduce THM generation. Running a reduced chlorine system was attractive to me, even if periodic shock chlorination of the entire system was still required to combat biofilm buildup.

After discovering a hybrid ozone generator that does not require dry air, manufactured by Prozone Water Products in Huntsville, AL, I added ozone generators to both tank loops hoping to reduce system consumption of chlorine. However, that did not happen. If anything, chlorine demand increased and THMs remained a problem. Further research indicated that both ozone and UV light break down hypochlorite, requiring more hypochlorite to be added in order to maintain proper free chlorine levels. This prompted an investigation into alternative disinfectants and a series of modifications to reduce NOM in raw water and to move the point of chlorine injection further downstream in the purification process. My system currently uses an ozone/UV advanced oxidation process (AOP) in the raw water tank with chlorine

---

10 In this book "chlorine" primarily refers to sodium hypochlorite or calcium hypochlorite. But the problems of disinfection by-products are also associated with chlorine gas, chlorine dioxide, and chloramine.

injected when water is transferred to the clean water tank. Residual chlorine levels are much more stable with this method and THMs have been essentially eliminated. Disinfection and oxidation are discussed in more detail in Chapter 2.

One of my earlier attempts to eliminate chlorine, using a copper-silver ionizer designed for swimming pools and spas, did not work at all. Due to all the added pool chemicals, swimming pool water conductivity is much higher than rainwater conductivity. Low cost copper-silver ionizers designed for pools and spas require high conductivity water in order to work properly. All the copper-silver ionizers for potable water systems that I found were expensive high-volume units designed for hospitals and hotels, used to supplement normal chlorination. These systems contain large electrode plate areas and high voltages to overcome low conductivity of potable water. Copper-silver ionizers have proven to be effective against Legionnaires disease, but testing against other pathogens is limited. Using copper-silver ionizers on potable water systems also requires proper monitoring of heavy metal ions to ensure their concentrations stay below safe levels for human consumption, while also maintaining high enough levels to be effective against harmful organisms. Since chlorine is a proven disinfectant for potable water sources and practically all copper-silver ionizers in operation today on potable water systems are only additions to chlorine disinfection, rather than total replacements of chlorine disinfection, I abandoned the copper-silver ionizer concept and continued to use chlorine as my primary disinfectant, focusing on other ways to reduce generated THMs.

Trying to minimize raw water NOM and reduce THM generation proved to be a difficult problem. I initially installed a first-flush diverter consisting of the typical standpipe and floating ball illustrated in most rainwater harvesting manuals, but quickly found it to be totally inadequate. Floating organic matter (pollen and leaf matter) still got past the diverter and into the tank, particularly during spring and fall times of the year. I then installed a two-stage prefilter (discussed in the first edition of this book) which stopped practically all large particulates, but did not stop fine organic particulates such as pollen. During spring rains clumps of yellow pollen floated on the surface of raw water, discoloring water and requiring large amounts of chlorine to restore clarity. Furthermore, heavy rainfall during spring often overflowed raw water storage and made organic matter removal even more difficult.

To overcome these problems I needed a first-flush diverter that completely dumped first flush water to ground, rather than relying on a floating ball to isolate it from subsequent cleaner water. First-flush amount should be adjustable to allow compensating for seasonal changes in rainwater quality. In my area rainwater quality is lower during spring (pollen) and fall (leaf litter) and higher during summer and winter. Additionally, this diverter must divert all roof water to ground when the tank becomes full, preventing organic matter from unnecessarily entering the tank. Overflowing a tank wastes all the effort spent in cleaning stored water. These problems motivated a first flush diverter design (discussed in first edition) which was essentially a mechanical logic AND gate. Its design allowed rainwater to only enter raw water storage when the tank was not full AND a sufficient amount of first-flush water has been diverted to ground. Once full all additional rainwater was diverted to ground. This second edition shows a superior electronic version of this same diverter concept.

Raw water quality significantly improved when I modified my gutters and gutter screens (discussed in section 4.1). This reduced contact time of fresh rainwater with decaying leaves and decay by-products and reduced NOM in raw water. I do not know which of these modifications had the greatest effect on eliminating THMs – advanced oxidation process and delayed chlorination, automatic rainwater diverter, or micro-mesh gutter screens – but subsequent laboratory testing proved these combined modifications practically eliminated all THM problems. Raw water quality is now much higher and more stable throughout the year.

A few minor problems, mentioned in the preface of this second edition, still persisted. These were solved after publishing the first edition to this book and are discussed in detail in this second edition. Notably, the mechanical AND gate was replaced with an electronic AND gate that does the same thing, only better (section 4.3). It provides more reliable and accurate adjustment of first flush amounts. The prefilter was redesigned to make fabrication simpler and provide some self-backwashing capability (section 4.4). This significantly reduced prefilter clogging. Electronic system control and status monitoring has greatly improved with installing Arduino microcontrollers (Chapter 5). This second edition provides cost details of various system parts along with complete example system designs.

I'm often asked about using gray water for toilets, since gray water systems are advertised to save water. From personal experience, I believe that gray water should not be used for anything other than watering plants, and that only with an immediate discharge to a garden (i.e. no storage of gray water). Gray water systems for toilet flushing require some amount of gray water storage, which is problematic. Dissolved NOM, soap, and skin debris in gray water is difficult to remove and can lead to ugly biofilms in toilets, requiring frequent scrubbing with cleansers. Modern dual flush toilets consume very little water. Showering and clothes washing are the major water consumers and gray water cannot be used for these activities. The extra plumbing, filtering, and cost required to outfit a house with a gray water system do not seem to me to be worth the benefit.[11] Furthermore, if you use clean potable water to flush toilets then toilet bowls become initial indicators of potential biofilm problems in your water distribution system. If you see the typical reddish slime of iron bacteria biofilm forming in your toilet bowl or toilet tank then you know it is most likely also forming inside your potable water pipes. Iron bacteria biofilms are not harmful to humans, but they can become breeding grounds for other more harmful microorganisms.

---

11   The one exception to this might be a desert area where water is extremely scarce. In such cases a cost tradeoff between additional plumbing and larger storage capacity might make sense. Alternatively, solar distillation of gray water would make it more useful than simply flushing toilets.

Storing gray water for plant watering is simply a bad idea as it quickly becomes foul smelling black water with high levels of dangerous pathogens. I tried storing gray water for plant watering almost 40 years ago when we lived in California under water-rationing conditions and found it to be seriously problematic. Web sites promoting gray water use also strongly discourage gray water storage. Under most circumstances, I believe the only reasonable use for gray water is watering plants with an immediate discharge from a gray water source to your garden. However, if you select plant species that naturally grow in your region they should not need much extra watering. If plant watering for a vegetable garden or non-native plants is a major concern, my recommendation is to install a larger raw water storage tank and keep stored water free of organic matter build-up.

Gardens usually need extra watering during dry conditions, right at the same time that you are trying to cut back on water usage. If you do install a gray water system for immediate discharge to garden plants, I recommend using only your shower or tub and possibly a washing machine (using only bio-friendly laundry detergents) for gray water discharge. However, a shower is not a huge source of gray water during droughts if sea showers are implemented. A kitchen sink and dishwasher usually have too much oil, grease, and detergents in their water which can result in other plant diseases. A bathroom sink also contains high concentrations of plant-unfriendly chemicals (toothpaste, mouthwash, shaving cream, etc.) with only very small flows of water to dilute these, especially with rainwater harvesting. Make sure the gray water discharge line in your yard or garden has a screen on its end to prevent rodents, reptiles, and insects from finding their way into your bathroom! You also need to frequently move the discharge line as too much soap and detergent will kill plants. A good website for gray water use is www.oasisdesign.net.

This book contains unique information on potable rainwater harvesting not found elsewhere. Nevertheless, I recommend you read the several rainwater harvesting manuals and guidelines written by various State and Federal governments, if you have not already done so. These publications contain excellent information on basics of rainwater harvesting generally not repeated here. These publications can be downloaded from the Internet free of charge. However, be aware that most of the systems discussed in these manuals are for non-potable water use. My own experience shows that some methods recommended in these manuals can be problematic in terms of biofilm growth and harmful microorganisms in a potable water system.

For example, I do not believe so-called "calming inlets" and devices that promote sedimentation are helpful for systems designed primarily for potable water use. Sediment and sludge in bottoms of tanks are food sources for microorganisms that quickly become breeding grounds for biofilms and harmful pathogens. I believe that filtering foreign matter from storage tanks as quickly as possible, rather than letting it settle to tank bottoms, is far superior. Maintaining a swimming pool for seventeen years taught me this. Water in a swimming pool must stay turbulent and circulating through filters to discourage growth of biofilms and sediment. Periodic sweeping and scrubbing of walls are needed to stir up sediment and loosen biofilms so that filters can remove these contaminants. If you would not swim in a pool loaded with biofilm and sediment, why would you even consider drinking such water? Regular filter backwashing or replacement is essential. Regular water testing and maintaining proper disinfectant level are absolutely critical. Every pool owner knows what happens when regular maintenance is ignored: a stagnant green pond will form! Maintaining a potable rainwater harvesting system is very similar to maintaining a swimming pool, although the work is generally less. Like a swimming pool, you want crystal clear water continuously circulating inside totally clean tanks.

Before jumping into design details of potable rainwater harvesting, permit me to make one final introductory point. Rainwater harvesting is most definitely a so-called "green" technology that encourages water conservation like nothing else. Every person that moves from cheap abundant public water supply to potable rainwater harvesting experiences huge reductions of household water consumption. One of my neighbors on public county water consumes about 400 gallons of water per day whereas my house consumes about 50 gallons per day. During dry periods of rigorous conservation we drop this down to less than 20 gallons per day. According to Huntsville Utilities the average customer in northern Alabama consumes approximately 8000 gallons of water per month, over five times the consumption of my household. My brother and his wife live "off-grid" in the dry hill country of western Texas with rainwater harvesting as their only water source. They consume less than 20 gallons per day. Anyone that has ever spent time wilderness backpacking understands how little water is truly needed for drinking and personal hygiene. A backpacker learns how to bathe with just a cup of water!

People learn how to conserve scarce resources when their life depends upon it. Simple things like not leaving the faucet running while brushing teeth or shaving, using a small trickle to rinse dishes or vegetables, using a modern water-conserving clothes washer, and not taking long showers can greatly conserve water. A household dependent on rainwater harvesting is highly motivated to find all kinds of ways to conserve water, whereas a household on cheap public water supply is not. Nothing can stimulate water conservation like being dependent on rainwater harvesting. In hindsight, I am actually grateful that our county gave us a cost-prohibitive price to bring public water to our home. They forced me to investigate a technology that I otherwise might never have considered. That investigation made us far more conserving of water than we ever expected to become.

Almost all 50 states have legalized rainwater harvesting in some form or fashion, although some (most notably Colorado) have so heavily restricted it that it's practically illegal. Old laws from long-forgotten water-rights battles are now used to effectively outlaw modern rainwater harvesting in Colorado. These old laws are slowly being modified as people with common sense put pressure on state legislatures to change these counterproductive laws. Arguments against rainwater harvesting are nonsensical since everyone, including both public water utilities and private individuals, are actually harvesting rainfall when they draw water from lakes, rivers, and wells! The only difference between these rainwater harvesters and what we normally think of as rainwater harvesting is the catchment surface. A public utility "rainwater harvester" uses the ground as a natural catchment surface, removes pollutants leached into rainwater by the ground, and then sells the potable water for profit. A small private rainwater harvester uses a man-made catchment surface and storage system, which he keeps as clean as possible, and purifies his water for personal use rather than for profit. Private rainwater harvesting is a non-profit business that benefits everyone.

Scientific studies examining rainwater harvesting in light of Colorado water rights law clearly demonstrate that rainwater harvesting produces no injury to previous water rights owners. One such study performed over ten years ago in Colorado by three engineering and legal firms and stamped with the seals of four State of Colorado professional engineers concluded that rainwater harvesting would benefit their entire state. Seventeen persons employed with public water utilities and county government served on the advisory and peer review committees for this study. This document, entitled, Holistic Approach to Sustainable Water Management in Northwest Douglas County,[12] demonstrated that since only 2% to 3% of rainfall on undeveloped land becomes runoff into local streams and rivers, on average, laying claim to an additional 98% runoff from another person's roof is without warrant. Everyone knows that urbanization significantly increases runoff. When a house is built on land that previously had only 3% runoff, the entire land footprint covered by that house becomes impermeable and 100% of the rainfall on it becomes runoff. Therefore, no injury is done to local rivers and previous water rights owners when a homeowner captures and uses a portion of that increased runoff from his roof. No rainwater system captures 100% of rainfall on a roof; a large fraction (much larger than 3%) is spilled to ground as first flush divert water and tank overflow water.

To drive home my point above, consider the fact that 2016 was an exceptionally dry year for northern Alabama in which many water wells ran dry. My roof captured 39,564 gallons of rainfall that year, but only 17,321 gallons of that made it into my storage tanks either due to first-flush divert or tank overflow. In other words, 56% of the rainfall hitting my roof became immediate runoff to ground or a nearby creek, and that was during an exceptionally dry year reaching a "D4" category on the U.S. Drought Monitor map (www.droughtmonitor.unl.edu)! And that does not even include the portion of stored water I used to water my garden. The argument that rainwater

---

12 Holistic Approach to Sustainable Water Management in Northwest Douglas County, prepared for: Colorado Water Conservation Board, Dominion Water and Sanitation District, Castle Ines North Metropolitan District, Douglas County, Thunderbird Water and Sanitation District, and Plum Valley Heights HOA. Prepared by: Leonard Rice Engineers, Inc., Meurer & Associates, and Ryley Carlock & Applewhite Professional Association. January 2007.

harvesters steal water from rivers or other water rights owners is just pure nonsense! But Colorado politicians are not interested in scientific studies and facts, only in maximizing their tyrannous extraction of wealth from productive society.

The problem with public water is public perception. Although the catchment surface for any public utility is in fact limited, it is exceedingly large (near infinity) in the minds of practically all utility customers, and is therefore perceived to be capable of supplying an unlimited amount of potable water. However, a small private rainwater harvester knows exactly the limits of his catchment surface and storage capacity, and is therefore highly motivated to conserve water. There are some stupid people arguing that rainwater harvesters hoard rainwater and rob the recharging of groundwater resources. This is an incredible myth (as the above Colorado scientific study demonstrates) and truly amazing that anyone even believes this. Every drop of rainwater captured by a rainwater harvesting household is returned to the ground surface very near where that rain originally fell. In fact, if a rainwater harvesting household is on a septic tank then they actually provide captured rainwater a 4-foot head start in recharging groundwater reserves! Most unrestrained rainwater runoff finds its way back to the oceans through a multitude of rivers and creeks. Slowing down rainwater runoff via man-made lakes and rainwater storage tanks is a proven way to best recharge groundwater reserves. The State of Texas clearly understood this and brought in the Army Corp of Engineers many decades ago to dam up practically every creek in their state and build lakes!

County and city governments are slowly relaxing laws on rainwater harvesting as they realize how counterproductive and illogical these old laws are (or enough pressure is put on them by intelligent people). Government is usually the last entity to adopt new technology since it is the ultimate consensus organization; politicians usually only do what they think will buy them the most votes. So it will take time for governments to realize that encouraging rainwater harvesting is one of the best ways to stimulate water conservation, protect our environment, and increase national security. Water does not disappear when it is consumed by a rainwater harvesting household. The same amount of water is returned to the ground, either directly through roof gutters, garden hoses, septic systems, or public sewers, regardless of whether or not that household harvests a small portion of rainwater for personal use. Surface water, which started out as rainwater, will eventually find its way to a municipal water purification plant where it will be treated and resold, regardless of whether or not it saw prior use in a rainwater harvesting household. All water in the earth is recycled water!

Municipal water supplies are increasingly strained by normal population buildup, water-wasteful practices, and by agricultural and industrial pollution. Sludge removed at municipal water purification plants and sewage processing plants is in fact an environmental pollutant. Unfortunately, much of this sludge is being shipped to other states and dumped on agricultural fields! Some man-made pollutants cannot be removed from drinking water without very expensive advanced oxidation processes or energy-consuming distillation. Herbicides, pesticides, pharmaceuticals, and toxic wastes from injection well mining are showing up in deep drinking water aquifers all across our nation. Deep ground waters are no longer clean as they once were a hundred years ago. Voters must pressure local governments to admit that private rainwater harvesting is one of the best ways to protect our environment, conserve water, and protect national health and security. Potable rainwater harvesting is a "no-brainer" common sense approach to our nation's drinking water problems.

Chapter 2.
Rainwater Filtration and Disinfection

Good filtration and disinfection are absolutely essential for any potable water system, including potable rainwater harvesting. Rainwater starts out as pure distilled water in the sky, but picks up contaminants along the way to your storage tanks, including contaminants in the air, on your roof, gutters, and drain pipes. Mold, algae, biofilms, and other products of organic matter decay in gutters and drain pipes are primary sources of contamination. You cannot stop birds from roosting on your gutters, nor trees from shedding organic matter, nor wind from blowing dirt and pollen on your roof. Harmful microorganisms, including E.coli, Giardia, Cryptosporidium, etc. can enter your potable water system from all these sources without proper filtration and disinfection. Just because water came from the sky and looks clear does not mean that it is safe to drink. Although rainwater has far less pollutants than typical surface waters from rivers and creeks, it still needs to be treated as surface water in need of purification.

In addition to a well-designed filtration and disinfection system, you should periodically test your water for proper disinfectant levels and total dissolved solids (TDS). TDS is a potential food source for bacteria. Proper residual disinfectant levels must be diligently maintain at all times. Water test kits, electronic TDS meters, and oxidation reduction potential (ORP) meters can be purchased online or from local suppliers of water purification equipment. More comprehensive water testing of all contaminants in your water must be done by a certified water testing laboratory. They have the expensive equipment needed to measure small traces of toxic metals and chemicals in water. A well-designed and well-maintained rainwater harvesting system will provide you water that is safer and cleaner than practically any other public water supply. The reasons for that surprising statement will become obvious in this chapter.

No person is more motivated to provide your household with clean potable water than you are! Testing has repeatedly shown that my rainwater system consistently provides much cleaner water than either our cave water or our local public water supply. Laboratory tests on both my water supply and public water supplies are too extensive to list here to prove that statement. However, the reason that this is true is simply due to the fact that rainwater is essentially distilled water, except for a small amount of pollutants imparted by a man-made collection system, which is orders of magnitude cleaner than the pollutants in rivers, lakes, and ground water sources used by public water suppliers. There are no radiological emitters in rainwater to be concerned about, nor volatile organic compounds (other than disinfection by-products), no lead, fluoride, nitrates, nitrites, or sulfates. Public water supplies practically always contain these pollutants simply because their source water comes from the ground, although pollutants are generally below EPA requirements.

Testing with a simple TDS meter has always shown total dissolved solids in my rainwater are at least an order of magnitude lower than TDS in local public water supplies. Rainwater leaves no scale deposits in a coffee pot or tea kettle, unlike public supplies, since it has no dissolved calcium or magnesium in it from the ground. Water softeners, which exchange calcium and magnesium with sodium ions, are not needed with rainwater. Research has shown that scale build-up inside distribution plumbing from naturally occurring calcium, magnesium, and other minerals in ground water promote biofilm growth. Public utilities constantly fight biofilm growth, usually by raising chlorine levels, which is why public water often smells like a chlorinated swimming pool. Rainwater avoids this problem: it is naturally distilled water. Therefore, it should come as no surprise that a properly maintained potable rainwater supply is orders of magnitude cleaner than typical public water supplies.

My system incorporates redundancy in both filtration and disinfection to ensure our potable water is chemically and biologically safe to drink. Some might argue my system design is overkill, but I am not running it to make a profit; I am running it to provide clean healthy water for my family. I test our water weekly for residual free chlorine, pH, ORP, and TDS (or conductivity). Kitchen tap water is periodically sent to a certified water testing laboratory for thorough chemical analysis. Other inspections (discussed in more detail in Chapter 6) are included in my regular weekly maintenance to ensure our household water is biologically and chemically safe to drink. All tests, inspections, and maintenance are recorded in a log book to help spot trends and provide a stable residual disinfectant level. If I become incapacitated my log book becomes the primary resource for others to take over system maintenance.

There are essentially only three practical processes (excluding distillation) for small-scale whole-house water purification. They include filtration, disinfection, and oxidation. Filtration removes large particulates, including some pathogens, that are larger than about 1 micron (depending on type of filters used). Harmful micro-organisms smaller than this can pass through filters and must be inactivated with a disinfectant. Some chemicals (mostly organics in our case) that discolor water or give it bad odor or taste can also pass through filters. These are removed by oxidation, sometimes referred to as bleaching. Fortunately, most common disinfectants are also good oxidants. So we discuss these together in section 2.1, often using disinfectant and oxidant interchangeably, even though these two processes are technically different. Filtration is discussed in section 2.2.

2.1 Disinfection and Oxidation

Potable water does not necessarily imply sterile water. Public water supplies are sufficiently clean to drink but that water is not sterile. Numerous species of bacteria and viruses are airborne and any exposure of water to air, including filtered air, exposes water to potentially harmful microorganisms. If a stored body of potable water does not have sufficient residual disinfectant to continuously kill off all airborne microorganisms, and that water has enough nutrients to support bacterial growth, microorganisms will multiply and eventually render that water unsafe to drink. Practically everyone has seen a good example of this when "clean" potable water has been allowed to stagnate in household plumbing for days or weeks. A homeowner returns from a long vacation and sees stinky rust-colored water flowing from the faucet when it is turned on. Plumbing biofilms had depleted residual chlorine to the point where iron bacteria and other microorganisms could thrive in stagnant water. Flushing out plumbing with freshly chlorinated water from a water utility restores water potability. Bacteria still thrive in biofilms and tiny cracks in plumbing walls, but their numbers are low enough that the water is safe to drink.

The same thing can happen in a rainwater harvesting system, which is why we keep our water circulating. Rain falls through the air and picks up all kinds of bacteria along the way, and picks up nutrients (assimilable organic carbon) from roof and gutter surfaces. If stored rainwater is then allowed to stagnate without sufficient residual disinfectant, it will quickly become unsafe to drink. The best way to maintain potability of stored rainwater is to keep it continuously circulating through filters and maintain proper residual disinfectant levels.

Designing redundancy in disinfection/filtration systems is wise. Redundancy of sequential stages of disinfection/filtration will greatly reduce risk of harmful pathogens surviving inactivation. Chlorine inactivates a broad range of common harmful bacteria and viruses, but is not very effective on Giardia and Cryptosporidium protozoa. Fortunately, these pathogens are large enough that filtration can usually remove them. As the EPA's Alternative Disinfectants and Oxidants Guidance Manual[13] shows, using combinations of different disinfectants can be more effective at pathogen inactivation than just a single disinfectant. However, one cannot arbitrarily mix different disinfectants together since some are not compatible with each other. For example, mixing hydrogen peroxide with sodium hypochlorite will neutralize both disinfectants, leaving the water with no residual disinfectant. UV light and ozone also tend to consume hypochlorite. I found UV to be a more powerful neutralizer of free chlorine than ozone on my system.

There are many ways to design an effective system with multiple disinfectants. You essentially need a primary disinfectant for the initial kill of all pathogens in the water and a secondary residual disinfectant to protect water in the distribution system from reinfection. You also need an oxidant and filtering system to remove as much organic matter as possible from raw water and remove the food source for bacterial growth, without generating excessive carcinogenic disinfection by-products such as THMs.

---

13 Alternative Disinfectants and Oxidants Guidance Manual, EPA 815-R-99-014, United States Environmental Protection Agency, April 1999.

Normal potable water chlorination levels are not very effective against biofilms. Biofilms are nature's way of protecting micro-organisms against chemical disinfectants! Certain microorganisms produce biofilms, which are slimy organic polymeric substances that attach to objects in a water system and protect microorganisms growing under it from a residual disinfectant. One of the most common producers of biofilm is iron or rust bacteria. This bacteria is harmless to humans and often shows up as a redish or rust-colored film in toilet tanks and around sink drains. However, rust bacteria biofilm can also shield harmful pathogens living symbiotically in its biofilm. Biofilms are one of the primary causes of disease outbreaks in public water supplies, so they should not be ignored. However, the only way to remove them is with either mechanical scrubbing or with a strong oxidant, like chlorine or hydrogen peroxide, at elevated "shock" levels that are unsuitable for drinking. Removing biofilms without interrupting public water supplies is a major on-going research topic in the water industry. We discuss how to temporarily shut down your system and remove biofilms in Chapter 6. In spite of all the negatives of chlorine, it is a proven disinfectant for potable water with a long history of effectiveness. Although chlorine does not stop biofilms, it does slow down their progress. Chlorine is safe to drink at the low residual levels normally used for water treatment, but its chlorination by-products (THMs) are considered carcinogenic. Our dual-tank system provides a means of overcoming these chlorine-related problems.

Common disinfectants/oxidants in use today for water purification include ozone, hydrogen peroxide, sodium hypochlorite, calcium hypochlorite, chlorine gas, and chlorine dioxide gas. UV is also used in combination with ozone to produce highly oxidizing hydroxyl radicals. Many public water utilities are moving away from chlorine gas and using hypochlorites due to safety and environmental concerns. So we do not consider chlorine gas or chlorine dioxide gas to be useful for our purposes either. A rainwater harvester will primarily be interested in either ozone, UV, hydrogen peroxide, and the two hypochlorites, or combination of these, for disinfection and oxidation. Of these four only hydrogen peroxide and the hypochlorites provide residual disinfection capability (disinfection inside plumbing). UV and ozone do not provide adequate residual disinfection capability, but can be used very effectively in oxidizing and removing organic matter. The hypochlorites are the least expensive, so we'll discuss these first.

Use sodium hypochlorite or calcium hypochlorite to produce free (uncombined) chlorine in water. Liquid sodium hypochlorite (common household bleach) has a limited shelf life of only about three months, particularly at higher concentrations of 12% or more. This is usually not a concern for water utilities using large amounts of chlorine on a continual basis. However, it is a concern for a small private rainwater harvester. Do not buy 5-gallon containers of 12% sodium hypochlorite because it will decompose, mostly into salt water, long before you consume it. Find a reliable local supplier of small quantities (1 gallon or less) of 5% to 8% solution sodium hypochlorite and try to use it up within a month or two. Make sure it is NSF-approved[14] for potable water purification. Many household bleaches have added chemicals to improve odor and/or laundry whitening, but these are not approved for drinking water purification. Use only pure sodium hypochlorite in potable water. Non-scented Clorox regular bleach, 6%, is NSF-approved for drinking water purification and is readily available at most local stores. A quarter cup of 6% sodium hypochlorite will usually raise free chlorine level in 1600 gallons of water approximately 1 ppm in my tank, depending on amount of dissolved organic matter in the water.

When using hypochlorite as my primary oxidant/disinfectant, I usually maintained at least 2 ppm free chlorine in raw water storage and slightly less than 0.5 ppm chlorine in clean water storage. I raised these levels under special circumstances, such as tank cleaning or battling a biofilm problem in distribution plumbing. I usually only added chlorine to raw water storage. Make-up feed from highly chlorinated raw water was sufficient to maintain a low residual level of chlorine in clean water storage. A UV lamp in the clean water circulation loop was turned on periodically to help regulate free chlorine if it got too high. This allowed a high chlorine level in raw water storage, where oxidation is needed most, and a low residual chlorine level in clean water storage with sufficient time to dissipate THMs before household use (assuming turbulent circulation and air mixing). A more complex method, discussed in greater detail below, incorporates an advanced oxidation process (AOP) in the raw water loop and introduces chlorine later in the process in order to reduce THM production. This is the method I currently use on my system and recommend to others.

Alternatively, you can use solid calcium hypochlorite, which has a shelf-life of well over a year, and dissolve a small amount in water as needed. I have used both liquid sodium hypochlorite and solid calcium hypochlorite in my system. I like the long shelf life of calcium hypochlorite but prefer the convenience of sodium hypochlorite. Liquid bleach is readily available and eliminates any need for pre-mixing. Whether you use sodium hypochlorite or calcium hypochlorite, make sure it meets the requirements of NSF/ANSI Standard 60 for water sanitation. Calcium hypochlorite is manufactured in both tablet and granular form. The granular form has much higher shipping costs due to its potential fire hazard.[15] I use Pro Chlor 2.6-inch tablets manufactured by Global Water Treatment Chemicals of Weatherford, TX (www.gwtcinc.com). These tablets are easily crushed inside a couple of plastic Ziploc bags with a gentle tap of a hammer. One level tablespoon of calcium hypochlorite powder will raise free chlorine levels approximately 1 ppm in 1600 gallons of water, depending on organic matter levels in the water.

---

14 NSF International, http://www.nsf.org/certified-products-systems NSF/ANSI 61 for drinking water system components and NSF/ANSI 60 for drinking water treatment chemicals.

15 Granular or powdered calcium hypochlorite can decompose violently and unpredictably when it comes in contact with a number of different substances. Do not store large amounts of powdered calcium hypochlorite and protect it from coming in contact with other organic liquids and chemicals.

Pre-mixing is easily done by adding calcium hypochlorite powder to several cups of warm water in a gallon jug and shaking. Let residual calcium solids settle to the bottom of the jug before pouring clear liquid concentrate into your water storage tank. Wear rubber gloves and eye protection when handling calcium hypochlorite as it is a powerful oxidant of organic materials like skin. (The same applies to 35% hydrogen peroxide.) Crush and mix outdoors or wear a dust mask so as not to breath the dust. If you are unsure about sufficiency of chlorine added, wait several hours for chlorine to become thoroughly mixed in your storage tank and then retest for free chlorine levels. Keep accurate records of test results, amount of chlorine added, and date that chlorine was added. Good records help you determine proper amounts of chlorine to add to stored water and when to add it. Another method I have used in the past is to put a half tablet in a floating swimming pool chlorine tablet dispenser with all ports closed off except for one small hole in the bottom of the dispenser. This is less accurate than the above measuring method, but accuracy is not all that critical when chlorine is only added to raw water storage.

Manually adding chlorine (hypochlorite) to stored water is easy and the least expensive way to maintain disinfectant levels. Testing for free chlorine levels and adding chlorine as needed takes very little time. The disadvantage of manual disinfectant maintenance is that you must frequently perform this disinfectant maintenance, sometimes every other day, in order to maintain stable chlorine levels. Manual disinfectant maintenance also slightly increases exposure of water to outside contaminants when the tank lid is opened and increases operator exposure to potentially dangerous chemicals. Automatic chlorine injection is preferable for these reasons, but more expensive.

Stable automatic chlorine injection is easily achieved on dual-tank systems if it is injected at the point of water transfer from raw water storage to clean water storage. Transferred water equals the amount of water consumed by the household and chlorination at this point allows sufficient time for chlorine to do its job. The large clean water storage volume provides buffering against random injection errors. Maintaining stable chlorine levels is harder if it is injected as water exits clean water storage and enters the distribution system. Another common method is to electronically monitor chlorine levels and automatically inject disinfectant as needed using a peristaltic metering pump, but this method has higher hardware costs.

Two common devices for liquid chlorine injection include low-cost venturi injectors, such as a Mazzei injector (www.mazzei.net), and more expensive peristaltic metering pumps, such as a Stenner metering pump (www.stenner.com). Although venturi injectors are much less expensive, I found them to be more troublesome to adjust than peristaltic metering pumps. After using a low-cost venturi injector for a few years I finally broke down and bought a Stenner Econ FP metering pump for my system. A venturi injector depends on a pressure differential to operate properly, but this changes with transfer pump flow rate, which is dependent on pump condition and filter condition. I found myself constantly readjusting the venturi injector and often having to manually add chlorine anyway. A venturi injects chlorine over the entire length of time water is being transferred through it. So injecting a measured amount of chlorine with a short run of the transfer pump is really not feasible. However, a peristaltic metering pump injects a measured amount of chlorine over a short length of time, usually under one minute, regardless of how long it takes a transfer pump to move water. So if chlorine levels are low they can be easily raised with a peristaltic metering pump by switching it on and off a couple of times.

With either type injector, venturi or peristaltic, chlorine can be injected into the water either as it leaves the clean water tank and enters the distribution system or as it is transferred from raw water storage to clean water storage. I prefer the latter because this leaves a residual disinfectant in clean water storage that can be easily tested. Venturi injectors have no moving parts and require no electrical input; they derive their effect from water flowing through them. They can be metered with a needle valve, although not as accurately as peristaltic pumps. They do need to be properly sized for specific flow rates and pressures. Mazzei provides tables to properly size injectors according to expected flow rates and pressure drops across the injector. Stenner peristaltic pumps can meter chlorine either by length of time the pump runs or by running speed of the pump. An input signal or dry contact closure initiates the pumping cycle. I currently have my Stenner set up to turn on and run for a fixed amount of time when it senses transfer pump activation via a relay switch.

Venturi injectors require flowing water in order to work. However, you cannot put a venturi injector in a circulation loop unless you have an electronic valve to completely turn off metering, otherwise you will over-chlorinate your water. The two positions for easily injecting chlorine with a venturi as water is consumed are 1) at the transfer pump when water is moved from raw water storage to clean water storage, and 2) on the suction line between clean water storage and house pressure pump, ahead of distribution plumbing. Water flows at both these two points only as it is used by the house, thus permitting chlorine to be properly metered. A peristaltic pump injecting at the second position would need to be triggered with the pressure pump or an inline flow sensor. Note that ozone, due to its short lifetime, can be injected continuously in circulation loops without risk of too much disinfectant. But chlorine cannot be continuously injected without over-chlorinating your water. Chlorine must be injected as water is consumed by the house, or as determined by an electronic chlorine sensor.

If chlorine is injected at the exit port of clean water storage, the contact time of chlorine with water will be much lower than if it is injected into clean water during transfer from raw water storage. However, if another primary disinfectant is being used in both tanks and chlorine only serves as a secondary residual then this would be an appropriate position for chlorine injection. Contact time can be estimated by volume of stored water after point of injection (including water in pipe and pressure tank) divided by your daily usage rate. For example, if you inject chlorine after water leaves clean water storage, and total volume of water in your supply line and pressure tank is 30 gallons, and you use 50 gallons per day, then approximate contact time is 30 gallons divided by 50 GPD, or 864 minutes. Minimum chlorine levels can then be determined from standard chlorine concentration-time (CT) charts.

Minimum chlorine level is usually between 0.2 ppm to 2 ppm. CT values are published for various disinfectants and pathogens. CT value is the product of disinfectant concentration in mg/L (or ppm) and minimum required contact time in minutes to achieve a 99% inactivation of specified pathogens. For Giardia the CT value is set for 99.9% inactivation and is around 150 ppm-minutes, depending on concentration and temperature. So 0.2 ppm times 864 minutes exceeds minimum CT for Giardia and shows that, for the above example, water volume after point of injection and flow rate provide sufficient time to inactivate Giardia to the 99.9% level. What the CT value means is that it takes time to kill pathogens with a disinfectant, and the lower the concentration of disinfectant the longer it takes. E.coli has a much lower CT value that Giardia, meaning that E.coli is more easily killed with free chlorine. Free available chlorine level (in ppm) should be monitored at the kitchen tap or some other faucet in your house and it must remain above a minimum specified concentration for your system. The longer free chlorine remains in water before being consumed the lower its concentration can be, but it should stay above 0.2 ppm at any rate. One advantage of chlorine injection at the transfer pump, rather than ahead of the distribution system, is that it gives you two different positions in your water system for measuring chlorine residuals. A significant reduction in residual chlorine level after water has traveled through your distribution plumbing indicates a potential biofilm problem.

The two chemical components that contribute to free chlorine measurement include hypochlorous acid (HOCl) and the hypochlorite ion (OCl⁻) from partial disassociation of hypochlorous acid. Of the two, hypochlorous acid is 80 to 300 times more effective at killing pathogens than the hypochlorite ion.[16] The ratio of HOCl to OCl⁻ is a strong function of pH. At 7.5 pH free chlorine is approximately evenly divided between HOCl and OCl⁻. At a pH of 6.5 approximately 90% of free chlorine is HOCl and significantly more effective as a bactericide. Therefore, pH must also be measured along with ppm free chlorine levels in order to assess disinfection effectiveness. Several studies have shown that ORP (oxidation reduction potential) is a better measurement of sanitizing capability of chemically treated potable water than just free chlorine measurement.[17,18,19] ORP is a more direct measurement of hypochlorous acid concentration. ORP readings should be maintained above 650 mV for adequate

16 Jacques M. Steininger, PPM or ORP: Which Should Be Used? Swimming Pool Age & Spa Merchandiser, November 1985.

17 Jacques M. Steininger, Catherine Pareja, ORP Sensor Response in Chlorinated Water, NSPI Water Chemistry Symposium, Phoenix, AZ (1996).

disinfection potency. ORP sensors are frequently used with peristaltic metering pumps for automatic disinfectant maintenance on commercial swimming pools and can also be used on rainwater systems.

When either chlorine or hydrogen peroxide is used as a primary disinfectant and oxidant in raw water storage, automatic injection can be achieved with a peristaltic metering pump and an ORP sensor, or with a microprocessor-controlled system (discussed in more detail in Chapter 5). A microcontroller, such as an Arduino, could determine amount of new water added using a tank level sensor and then turn on a peristaltic disinfectant metering pump for an appropriate length of time. However, a venturi injector on the drain line would be problematic due to rainfall rate variability.

Chlorine consumption can potentially be reduced with UV and/or ozone injection, but it cannot be completely eliminated because it is the only practical residual disinfectant approved by NSF and EPA for potable water. Hydrogen peroxide is generally not considered to be a broadly effective residual disinfectant, which is why NSF and EPA recommend following up hydrogen peroxide treatment with chlorine. Chlorine has a long history (nearly a century) of proven effectiveness as a residual disinfectant to control bacteria and biofilm growth in potable water systems. Nevertheless, other oxidants, including hydrogen peroxide, can be used as primary disinfectants/oxidants and delay chlorine injection until most of the natural organic matter has been removed. Hydrogen peroxide is very effective at eliminating chlorine-resistant biofilms in distribution systems.

As previously mentioned, oxidizing natural organic matter with chlorine has an undesirable side-effect of producing THMs which are carcinogenic and regulated by EPA to be no greater than 80 parts per billion in drinking water. THMs include bromodichloromethane, bromoform, chloroform, and dibromochloromethane. Chloroform is the primary by-product on rainwater systems since bromine is not normally present in rainwater. Chlorine also combines with ammonia and nitrogen compounds in natural organic matter to form chloramines (also referred to as combined chlorine) which is far less effective as a disinfectant than free chlorine. Chloramines have a strong chlorine-like smell that can fool an operator into thinking disinfectant levels are adequate when in fact they are not. One's nose is not an appropriate sensor of adequate chlorine levels.

Completely eliminating all undesirable water disinfection by-products is nearly impossible. However, several studies have shown that combining different disinfectants in an appropriate manner can have a synergistic effect on increasing pathogen inactivation while reducing undesirable disinfection by-products. An

---

18 Trevor V. Suslow, <u>Oxidation-Reduction Potential (ORP) for Water Disinfection Monitoring, Control, and Documentation</u>, Publication 8149, Agriculture and Natural Resources, University of California at Davis.

19 Richard James Spahl, <u>FC^E: Groundbreaking Measurement of Free Chlorine Disinfecting Power in a Handheld Instrument</u>, Myron L Company White Paper, January 2012.

excellent source of information for this is <u>Alternative Disinfectants and Oxidants Guidance Manual</u> published by the U.S. Environmental Protection Agency.[20] Ozone and UV can work together as an advanced oxidation process (AOP), forming hydroxyl radicals ($^-$OH) which break down organics more effectively than ozone or chlorine alone, but hydroxyl radicals can also consume free chlorine and increase hypochlorite usage rate. These disinfectants must be combined sequentially rather than simultaneously. Hydrogen peroxide can also be used to oxidize organic matter and biofilms in plumbing and distribution lines more effectively than chlorine, without producing THMs, but it must be followed up with residual chlorine injection according to EPA. This prevents using hydrogen peroxide in public water distribution systems.

Hydrogen peroxide and hypochlorite cannot be used simultaneously because they neutralize each other. However, ozone combined with hydrogen peroxide (also called peroxone) is an effective disinfectant/oxidant included in several advanced oxidation processes. Hydrogen peroxide is about four times as expensive as chlorine, but may be needed occasionally for stubborn chlorine-resistant biofilm problems. The synergistic effect of combining different disinfectants in appropriate ways has been observed by other researchers, but experimentation with advanced oxidation and disinfection processes is rarely practiced by municipal water suppliers due to cost, unless they have a serious problem meeting EPA requirements. Experimentation in water disinfection technology almost always comes from small private water suppliers who are motivated more by health concerns than by cost. Our dual-tank design provides flexibility to combine different disinfectants/oxidants sequentially or simultaneously to disinfect water without producing significant levels of disinfection by-products.

EPA acknowledges the usefulness of hydrogen peroxide as an oxidant for water purification, but not as a stand-alone disinfectant. According to EPA's web site on drinking water treatment:

> "Hydrogen peroxide (H_2O_2) is rarely used in drinking water treatment as a stand-alone treatment process. H_2O_2 is a weak mirobiocide compared to chlorine, ozone, and other commonly used disinfectants. Consequently, it is not approved by regulatory agencies as a stand-alone disinfection treatment process. However, there are a number of technologies where H_2O_2 is used as part of the treatment program. The advanced oxidation process (AOP) uses H_2O_2 in conjunction of O_3 and/or UV light to produce hydroxyl radicals ($\cdot$OH), which are very effective in removing taste and odor (T&O) compounds, and inorganic and organic micropollutants."

Hydrogen peroxide has some significant advantages over chlorine, particularly with distribution system biofilm removal. It is also one of main components of AOP. So hydrogen peroxide should be considered as a useful tool in your oxidation/disinfection process. Mark Krueger of South Texas Groundwater Solutions

---

20 <u>Alternative Disinfectants and Oxidants Guidance Manual</u>, EPA 815-R-99-014, United States Environmental Protection Agency, April 1999.

has had tremendous success using 35% hydrogen peroxide to clean up contaminated well water in situations where chlorine proved inadequate (www.southtxgs.com). He has a few hundred very satisfied customers. I had a stubborn biofilm problem in my distribution system that elevated chlorine levels could not remove even after several weeks. Mark encouraged me to try hydrogen peroxide to solve my problem. A single treatment with hydrogen peroxide immediately removed the biofilm. Others also find hydrogen peroxide very useful for water purification. For example, Oxytech Solutions in Germany (www.oxytechsolutions.com) combines hydrogen peroxide with silver in a product called D50/500 to produce a water disinfectant alternative to chlorine. So even if hydrogen peroxide is not used as a stand-alone disinfectant, it should be a part of your "tool box" for solving problems that chlorine cannot.

NSF's web site (http://info.nsf.org/Certified/PwsChemicals/) lists a number of suppliers of NSF-approved food-grade hydrogen peroxide for drinking water treatment. If you have a biofilm buildup problem that normal chlorination cannot eliminate, or chlorine-generated THM levels are too high, you might consider cleaning your system and distribution lines with hydrogen peroxide. We discuss how to do this in section 6.6. An unpleasant taste or musty odor in your water, along with increased chlorine demand, is a good indicator of biofilm buildup in plumbing that needs immediate attention. Pure rainwater is tasteless and odorless. You want water coming from your kitchen tap that is completely tasteless and odorless.

Another chemical that appears to be more effective at biofilm removal than hydrogen peroxide alone is peracetic acid. Peracetic acid (also referred to as peroxyacetic acid or PAA) readily decomposes into hydrogen peroxide and acetic acid (vinegar) and has been used as a cleanser and disinfectant in the food processing industry since the early 1950s. However, it is currently not NSF-approved for water disinfection. Its oxidation potential is higher than sodium hypochlorite or chlorine dioxide and does not produce any of the halogenated disinfection by-products of chlorine. It is also much more environmentally friendly than chlorine. However, I am not aware of any research on using it with potable rainwater harvesting systems.

Advanced oxidation processes (AOPs) are usually implemented where stubborn difficult-to-remove organics are found in ground water, including pesticides, herbicides, and pharmaceuticals. A rainwater harvester usually does not have to deal with these pollutants, but AOP is still useful for oxidizing natural organic matter from decaying leaves, bird droppings, mold, and algae without producing THMs. AOP can be implemented by continuously injecting ozone immediately ahead of UV exposure in the raw water circulation loop. This eliminates any need to meter chlorine in raw water storage and allows moving chlorine injection further downstream in the disinfection process, which helps reduce THM generation. Prozone Water Products, Inc. (www.prozoneint.com) in Huntsville, AL has recently combined UV and ozone into a single AOP unit, called the Eco Master. I currently recommend the medium sized unit for my rainwater harvesting customers.

Chlorine, ozone, UV, and hydrogen peroxide can be combined in many different ways on a rainwater system. Ozone can be injected ahead of filters, using biofilm buildup in a granular activated charcoal (GAC) filter to help remove oxidized by-products. A UV lamp immediately following the filters keeps biofilms confined to the filters. Backwashing or changing filters controls biofilm buildup. Another method, the one I use, is placing an ozone injector after the filter immediately followed by a UV lamp to generate hydroxyl radicals to oxidize organics that penetrate the filter. To me this seems preferable for a number of reasons. In addition to producing strongly oxidizing hydroxyl radicals, ozone bubbles help keep the UV lamp quartz tube clean. Ozone and UV following filters ensure biofilms remained confined to filters. Also, an ozone injector is less affected by back-pressure as filters age. Hydrogen peroxide can be used in an AOP, but properly metering hydrogen peroxide has the same problems as metering chlorine. It would require sensors and a peristaltic metering pump. Manually adding hydrogen peroxide when needed is less costly.

AOP is implemented on raw water storage primarily to reduce THM production and reduce biofilm buildup in plumbing by eliminating organic matter. It also allows a fully automated disinfection/oxidation process with chlorine injection occurring at either the transfer pump or at bottom outlet of clean water storage. Installing an injector on suction line from clean water storage is wise, even if you do not normally use it. If you need to inject hydrogen peroxide to eliminate a distribution system biofilm problem, the least costly way is to inject it ahead of distribution plumbing rather than dumping gallons of it into clean water storage. If you use a peristaltic pump for injection, you can easily provide this alternative injection capability by simply adding a ¼-inch port immediately after the main cutoff valve and putting a threaded plug in it. If you ever need to inject hydrogen peroxide you simply remove the plug and screw in the peristaltic pump injector. If you use a venturi injector for hydrogen peroxide injection, Mazzei's web site shows how to install an injector with a bypass loop so that it can either be placed in operation or completely bypassed.

AOP can be implemented in various ways with rainwater harvesting and much more research is needed on this specific application. All AOP processes involve ozone, UV, hydrogen peroxide, or combinations of these to generate strongly oxidizing hydroxyl radicals. Chlorine must be used sequentially, not simultaneously with AOP since it consumes chlorine. AOP should be used as a primary disinfectant/oxidant, with chlorine being injected later downstream as a secondary residual disinfectant. If significant industrial or agricultural air pollution (e.g. crop dusting, fertilizing with sewage treatment waste, incineration, etc.) exists in your area, you should have laboratory testing performed on raw rainwater to measure industrial pollutants. If testing shows presence of industrial or agricultural chemicals then AOP should definitely be considered. Nevertheless, removing atmospheric pollutants in rainwater should be much easier than removing soil pollutants leached into ground water.

Table 2-1. Practical disinfectants for potable rainwater harvesting.

Primary Oxidant & Disinfectant	Secondary Residual Disinfectant	Advantages	Disadvantages
chlorine	chlorine	low cost	potential THM problem
ozone	chlorine	reduced THM, broader inactivation	more costly hardware and electricity
ozone & UV (AOP)	chlorine	very effective oxidation, biological inactivation	more costly hardware and electricity
ozone & UV & hydrogen peroxide (AOP)	ozone & hydrogen peroxide (peroxone)	very effective oxidation and biofilm removal	hydrogen peroxide is expensive, may not be EPA-approved

Table 2-1 shows common disinfectant combinations that can be used with modern potable rainwater harvesting. The primary disinfectant and oxidant is applied in raw water storage. The secondary residual disinfectant exists in the distribution system and is applied either at clean water storage or ahead of distribution plumbing. Activated charcoal filters cannot be used in circulation loops where residual chlorine is being maintained, but can be used in a raw water AOP loop. Activated charcoal filters can also be used as point-of-use filters under a kitchen sink to remove all residual chlorine and disinfection by-products immediately prior to drinking.

Just before publication of this second edition I discovered a new ozone generator that has a built-in air dryer and an incredibly long service life. The device, called "Oxidize It," was designed to eliminate the need for soap in washing machines. It generates 0.3 ppm of ozone in the water, which is more than adequate for water disinfection, along with hydroxyl radicals. After talking with a company representative I believe this device could easily serve our purposes in potable rainwater harvesting, but I have not yet had an opportunity to test it. According to the company some customers successfully use it on swimming pools. The company's web site is: www.oxidizeit.com.

In summary, chlorine may be used for both primary and secondary disinfectants as the simplest and least costly method. I used this method for a couple of years with very good results, except for a THM problem. THMs were removed with a point-of-use activated charcoal filter under our kitchen sink. Implementing more expensive AOP has many advantages, including automatic chlorine injection and reduced THM generation. I currently use the third method in Table 2-1 (AOP in raw water tank, chlorine in clean water tank). For biofilm removal I will temporarily switch over to the fourth method in Table 2-1, with elevated levels of hydrogen peroxide in distribution plumbing. Safe drinking levels of hydrogen peroxide are around 25-30 ppm. Activated charcoal filters do not adequately remove hydrogen peroxide, but boiling the water does. Safe levels of free or available chlorine for drinking are 0.2-2 ppm and activated charcoal filters do easily remove all residual chlorine.

Required amounts of disinfectant/oxidant depends on age of disinfectant and organic matter levels in the water. Testing and experience are your best guidance in this. Assuming no loss of disinfectant through oxidation of organic matter, the following simple equation provides an initial quick calculation of disinfectant amount to add to stored water in order to achieve a desired ppm change in disinfectant level:

$$C_1 \times V_1 = C_2 \times V_2 \qquad\qquad (2\text{-}1)$$

where C_1 is the ppm level of the disinfectant concentrate, V_1 is the amount of concentrate to add to the water, C_2 is the desired ppm change of disinfectant in the stored water, and V_2 is the volume of stored water. A concentrate of 6% liquid sodium hypochlorite is approximately 60,000 ppm free available chlorine. A concentrate of 35% hydrogen peroxide is 350,000 ppm of hydrogen peroxide. For solid calcium hypochlorite with 65% free available chlorine by weight, 2 ounces dry weight of calcium hypochlorite mixed in 10,000 gallon of water gives a 1 ppm rise in free chlorine. The conversion factor for calcium hypochlorite 65% is 0.0002 oz/(ppm-gal). For example, if I wish to raise free chlorine in 1600 gallons of water by 1 ppm, using 6% liquid sodium hypochlorite, Eq. (2-1) can be used to calculate volume, V_1, of concentrate using 60,000 for C_1, 1.0 for C_2, and 1600 for V_2. Solving for volume gives 0.0267 gal or 3.4 fluid ounces of Clorox bleach. Alternatively, if I want to raise the chlorine level in 1600 gallons of water by 1.6 ppm, using calcium hypochlorite (65% chlorine), I simply find the product of the conversion factor, volume of water, and change in ppm: (0.0002 oz/(ppm-gal))×(1600 gal)×(1.6 ppm) = 0.51 ounces. This is approximately equal to 14.5 grams or 1 level tablespoon of calcium hypochlorite powder.

Automatic disinfectant injection increases stability of water quality, but the operator must take care that automation does not tempt him/her to reduce frequency of inspections, and possibly overlook an unexpected problem or hardware failure. Regular testing of residual disinfectant levels must be diligently maintained regardless of whether disinfectant is added manually or automatically. Automation should assist an operator, reducing his work and increasing reliability, but not eliminate his system oversight. A potable rainwater system is a mechanical, electrical, chemical, and biological system that requires knowledgeable human oversight. THMs (primarily chloroform) are primary concerns for a rainwater harvester using chlorine as a disinfectant. But this depends greatly on amount of organic matter in the water. No doubt dryer climates will have less organic matter in rainwater and THMs may not be a concern. Only laboratory testing can determine if a THM problem exists, requiring a more complex disinfection process such as AOP. But if you implement a dual-tank design from the start, making an adjustment later to include AOP is much easier.

In addition to potability, there is another seldom appreciated reason for carefully monitoring chlorine levels in your household water supply and not allowing it to get unnecessarily high. If your house uses a septic tank, the "health" of that septic tank depends on maintaining a thriving anaerobic bacteria community of biofilms in the septic tank that can easily handle residual chlorine flushed into it from the house. Highly chlorinated water can weaken or kill those beneficial bacteria colonies. An unhealthy septic tank that is not properly consuming waste will belch huge quantities of stinky gas, but a healthy tank will have virtually no smell to it (at least when the lid is on). The end of the inlet waste line into a septic tank is supposed to terminate just below the water surface to minimize septic tank odors. Sometimes, either from tank settling or from improper installation, the inlet may terminate above the surface allowing septic tank gasses to be vented out the roof stack vent. This happens to be the case with my septic tank and it allows me to quickly know when my septic tank is having "health" problems! I discovered that practically every time I raised chlorine levels above normal for shock treatment during maintenance, I could smell the septic tank downwind of my house. No amount of commercial septic additives would solve this problem either. Septic tanks do not need additives that are often promoted by marketers, and in fact our local county health department discourages their use. Septic tanks simply require that you not kill off its naturally-occurring anaerobic bacteria with highly chlorinated waste water. This is the one and only tank where you want a thriving and healthy biofilm community!

2.2 Filtration

Practically all filter cartridges have some small amount of leakage around their end seals or through their filter media. One should not depend on just a single filter cartridge in a system. My system incorporates a series of multiple filters, including microfiltration at gutters, prefiltering immediately before raw water storage, a 25-micron back-washable filter, a whole-house 5-micron cartridge filter, a 1-micron cartridge filter, and a 0.5-micron cartridge filter under the kitchen sink. This gradual reduction in filter pore size helps maximize filter life and reduces risk of pathogens penetrating filtration. It also helps keep flow rates strong in the continuous circulation loops, which is required for ozone injectors to work properly. Water is purified in stages, continuously circulating through two separate filtration/purification loops, a raw water loop and a clean water loop. On my system, raw water continuously circulates through the 25-micron back-washable filter and the 5-micron cartridge filter. Ozone injection immediately follows these filters, followed by UV radiation, to produce hydroxyl radicals to oxidize dissolved organics that penetrate filters. Backwashing the 25-micron filter was required more often before modifying gutter screens. After adding micro-mesh gutter screens this filter rarely needs backwashing, except during periods of high pollen levels. Gutters and gutter screens are discussed in more detail in Chapter 4.

I should note that this 25-micron backwashable filter is left over from my original cave water system. If I were building a system afresh I would not include this backwashable filter. It requires 25 gallons of water to backwash it, whereas cleaning the prefilter bag requires only one or two gallons from a hose nozzle. Another problem with both filters on my raw water loop is that I cannot visually inspect filter condition because they do not have clear housings like my clean water filter has. I now recommend my customers incorporate clear filter housings for both raw and clean water filters, such as the large Pentek 20-inch clear housings (part no. 150568). Pleated filter cartridges can be sprayed down with a hose nozzle using far less water than my backwashable filter.

Clean tank water continuously circulates through a 1-micron cartridge filter with a clear housing. Since a residual chlorine level is being maintained in the clean water tank this 1-micron filter cannot be an activated charcoal type. I use a white polypropylene depth filter for this clean water filter. A 0.5-micron activated charcoal filter resides under the kitchen sink and removes residual chlorine and any remaining THMs. If I suspect biofilm buildup in the line between the charcoal filter and the faucet (usually by taste) it is an easy matter to remove the filter cartridge, pour in some concentrated hypochlorite or hydrogen peroxide, and flush out the line.

Filters on both circulation loops should have flow rate capacities much greater than circulation pump flow rates. As a filter accumulates dirt and biofilms, its flow capacity decreases and pressure drop increases. This causes the pump to work harder and consume more electricity. Low wattage circulator pumps do not have much pressure capacity to begin with. So if a filter robs too much head pressure from the pump, flow will be insufficient to keep tanks clean. Keeping filter pressure drops low also reduces risk of water leakage around end seals. A larger filter increases system electrical efficiency and extends filter life. End of life for a filter is indicated when flow no longer produces a good stream of ozone bubbles from the ozone injector. Ozone injectors provide a very easy way to monitor filter condition, other than visual appearance. A clear plastic tube section immediately after an ozone injector allows visual inspection of flow rate. The cartridge filter on my clean water loop has a clear plastic housing, facilitating easy visual inspection of its white polypropylene filter. This filter is replaced when it becomes gray or black, regardless of flow rate.

As previously mentioned, activated charcoal filters must not be used in circulation loops with chlorinated water since they remove the chlorine. However, I have not found activated charcoal filters to be very effective at removing hydrogen peroxide. So using activated charcoal filters with hydrogen peroxide disinfectant in an AOP loop is possible, but more research is needed on this. Granulated activated charcoal (GAC) can supposedly be used effectively with ozone as it helps remove oxidized by-products of ozonation. The charcoal filters that I have experimented with tend to have low flow rates, which hinders proper operation of the ozone injector. For water circulation loops with residual chlorine disinfectant I recommend sticking with pleated or polypropylene filters. Make sure they are NSF-approved for drinking water.

Micro-mesh gutter screens are essentially back-washable filters that get washed every rainfall. For that reason they do slightly reduce roof collection efficiency. However, if you live in a dry area where every drop of rainfall is precious you probably will not have many trees around your house clogging up your gutters anyway. For wetter climates with lots of trees, micro-mesh gutter screens are absolutely essential for rainwater collection. They are your first-stage filters that immediately separate leaves and other organic matter from rainwater. Unlike other typical gutter guards, micro-mesh gutter screens do not allow leaves to sit on top of gutters and begin rotting.

Filtration actually starts at the roof gutters, one of the most important but usually most overlooked part of a rainwater harvesting system. This topic is so important that I have dedicated an entire section to it (section 4.5). Properly installed gutters with micro-mesh gutter screens are essential to minimizing dissolved organic matter in rainwater, reducing biofilm buildup, and reducing THM production, especially if you live in a high rainfall area with lots of trees. Assimilable organic carbon (AOC) and NOM are bacterial food sources that get into rainwater from decaying leaves on top of gutter screens and from mold and algae in gutters and drain lines. Properly installed micro-mesh gutter screens significantly reduce mold and algae buildup in gutters and drain lines, but they do not completely eliminate it. Periodic cleaning with bleach and a pressure washer is still needed.

The next important place for stopping AOC from entering water storage is the first-flush diverter. This automatic rainwater diverter (discussed in section 4.3) diverts the initial several gallons of roof water to ground, rinsing the entire catchment system. First-flush diverters are essential in potable rainwater systems, but typical diverters shown in state rainwater harvesting manuals are not very effective, as previously mentioned. Following the diverter valve is a raw water tank prefilter. This filter is the last line of defense just before rainwater enters raw water storage. It contains a high flow capacity 100-micron bag filter. Like strainers on swimming pools, this bag filter must be washed periodically with a garden hose nozzle. During high pollen season it may need washing after every rainfall. If the bag prefilter frequently gets clogged after every rainfall during periods of low pollen levels, this may indicate a serious mold and algae problem in gutters and/or drain lines. Pollen is small enough to pass through micro-mesh gutter screens and also through high capacity filter bags. Although most of the pollen will be dumped to ground by the diverter valve, you may find floating globs of yellow pollen in your raw water tank during high pollen season. If so, scoop these globs out with a swimming pool net to minimize water discoloration and odor.

Some people ask me about reverse osmosis filtration for rainwater collection. The primary reason I chose not to incorporate reverse osmosis (RO) filtration in my water system is the need to conserve water and track water consumption. Reverse osmosis filters periodically rinse their membranes, flushing about 3-4 gallons of water down a drain for every one gallon of filtered water produced. This problem can be mitigated by returning RO waste water back to raw water storage, but this would seriously throw off water consumption tracking if an RO filter is located after the water meter. Returning waste brine back to clean water storage is definitely a bad idea since RO filters tend to become breeding grounds of biofilms. Nevertheless, RO filters are primarily point-of-use filters more suitable for under-sink use rather than in continuous circulation loops or as whole house filters. If your rainwater system is working properly then clean water delivered to your house is near distilled water quality and RO would not be needed anyway. Removing residual chlorine and chlorine by-products for drinking and cooking purposes is more easily and cheaply done with a 0.5 micron activated charcoal filter than with RO. And this does not require a return waste line back to storage tanks. The one situation where RO could be useful is with a single-tank emergency-only rainwater system. In this case neither high flow nor water usage tracking are a concern. Brine can be returned to the storage tank to conserve water. Emergency and off-grid applications are discussed briefly in Chapter 7.

Since there are strong differences of opinion among rainwater harvesting professionals over single-tank versus dual-tank designs, calming inlets versus turbulent inlets, and top versus bottom suction, I want to address these issues in a little more detail, particularly since they directly impact issues of filtration and disinfection. Sedimentation is used extensively by large centralized public water suppliers as a preliminary step in water purification, so it seems logical to incorporate it in rainwater harvesting systems. In such systems a raw water storage tank becomes part of the purification process, reducing dirt loading on final filters and eliminating a need for continuous circulation. This of course has the advantage of eliminating circulation pump electrical loads. The disadvantage of this approach is that, just as with sediment tanks at municipal water treatment plants, the sludge and sediment in a storage tank must be removed periodically. Additionally, very fine silt particulates can remain in suspension for several days only to be removed by filters anyway. Municipal water treatment plants add flocculant chemicals to treated water to encourage more rapid sedimentation of fine particulates. Treated water is separated from this sediment before further filtration and chlorination. However, a rainwater system that incorporates both sedimentation and chlorination will most likely have a severe THM problem. Organic sediment and biofilm growth increase chlorine demand and increase THM production.

Since cleaning or changing a filter cartridge is much easier than brushing and vacuuming storage tanks, I believe that filtration is far superior to sedimentation. This book strongly advocates continuous turbulent flow and filtration in all tanks, sucking from tank bottoms and discharging to top of tanks, so that filters remove suspended particulates and minimize sedimentation in storage tanks. A storage tank's purpose is only to store clean water, not aid in purifying rainwater through sedimentation. In addition to providing an audible indicator of sufficient flow through circulation loops, a turbulent top discharge also helps to dissipate chlorine-induced volatile organic compounds (THMs) to the atmosphere. The one exception where sedimentation may be justified is an emergency-only or off-grid application where electricity is scarce.

Those that promote sedimentation in rainwater tanks as part of a water purification process need to remember their freshman high school biology class. In my class students mixed a little dirt, grass, and cow manure in large glass jugs of water and then let it all settle out to the bottom of the jars. Over the following weeks, while water stagnated, students measured increasing populations of waterborne protozoa under a microscope each day. Even though the water looked clear, these jars were teeming with an ever increasing population of life due to an abundant food source in bottoms of jars. Swimming pool owners know full well that they cannot turn off filter pumps, let dirt and leaves settle to the bottom, and just add extra chlorine to their pool. Chlorine is consumed by organic matter. Leaving an abundant food source for bacteria and algae in a pool bottom and trying to counteract it with excess chlorine only ends up producing a stinking slimy mess that takes weeks to clean up! For this reason, many pool owners leave circulation pumps running continuously, even during winter months when their pool is unused.

I realize that Appendix K of the 2015 Uniform Plumbing Code (UPC) requires calming inlets on potable rainwater catchment systems (K 104.4.5) to prevent disturbing sediment. But regulatory codes are consensus decisions and by definition innovation never comes from the consensus! Furthermore, code writers tend to be clueless "armchair" professionals that have never built and maintained real systems. Let's use a little common sense here. Would you swim in a pool with sediment in the bottom and biofilm slime all over its walls? No? Then why in the world would you even think of drinking it! Like a swimming pool, if you find sediment and slime in your tanks then it's time to get out the pool brush and start scrubbing your tanks, stirring up sediment so that filters can remove it. And if sediment is heavy then get out your pool vacuum and clean your tanks. Just as you would maintain on your swimming pool or spa, you want crystal clean storage tanks with zero sedimentation.

We want continuous circulation, sucking from tank bottoms and discharging to tank tops in order to discourage sedimentation. The one valid reason for a circulation loop to suck from top surface water and discharge to bottom might be to increase contact time of ozone bubbles with water and to place ozone discharge closer to a distribution system entrance (assuming ozone on clean water tank). The house pressure pump still sucks from tank bottom to minimize risk of losing pressure pump prime, but freshly ozonated water from the purification loop is discharged through the same bottom port on the tank. This keeps the cleanest water close to house suction and helps reduce biofilms in distribution plumbing. If top suction is used for a circulation loop on clean water storage, a four-way valve should be incorporated to easily reverse circulation flow and restore pump prime if air gets in the upper suction line. Chapter 7 shows how this can be done.

For the sake of my government engineer readers let me make a more positive comment about regulatory codes such as the UPC. Most government code writers are taught to never mix performance specifications with hardware specifications (at least I was taught this decades ago). Whenever you see a hardware requirement, such as calming inlets, in an otherwise performance specification, you can be sure that this is the usually the work of some lobbyist. Codes should be written in such a way to guarantee a certain minimum performance *without* restricting innovation and competition. For example, instead of specifying "calming inlets," this section of the UPC could have been written, "System design shall not allow storage tank sediment to be deposited in household distribution plumbing."

Some rainwater harvesting professionals advocate encouraging biofilm growth in storage tanks as a means of helping to purify rainwater (e.g. Virginia Rainwater Harvesting Manual[21]). The concept is to use harmless bacterial growths to digest organic matter and absorb heavy metals. In order to accomplish this, chlorine or other disinfectants must not be added to the storage tank. Using this approach for potable rainwater harvesting is highly risky. Do a little research on biofilms and you will find that biofilms are implicated in over 80% of all infectious diseases, including in public water supplies. Study the EPA document Health Risks from Microbial Growth and Biofilms in Drinking Water Distribution Systems.[22] Widely divergent bacterial forms, including Cryptosporidium and Legionnaires disease, can thrive symbiotically in a biofilm and you cannot control which forms live in your storage tank if you promote biofilm growth. Allowing biofilms to grow uncontrolled in a storage tank can produce biologically unsafe stagnant pond water.

Other researchers promote confining biofilm to insides of filters (e.g. granulated activated charcoal) that can be backwashed or replaced when a biofilm gets too large. This application is sometimes used for pre-treating water to remove assimilable organic carbon to reduce THMs when chlorine is added further downstream in the process. This is certainly better than encouraging biofilm growth in a storage tank, but I remain skeptical even with this application. All living organisms take simple hydrocarbons and assemble them into more complex molecules. This is counterproductive to our goal of oxidation, which is to break down hydrocarbons into the simplest molecules such as water and carbon dioxide. Since biofilms slough off bacteria (which is how they propagate), a biologically active filter should be followed

---

21 Virginia Rainwater Harvesting Manual, Second Edition 2009, The Cabell Brand Center, Salem, VA.

22 Health Risks from Microbial Growth and Biofilms in Drinking Water Distribution Systems, U.S. EPA Office of Water (4601M), June 17, 2002.

by a UV lamp. I believe a better approach is to discourage all biofilm growth anywhere in a potable water system. It is hard enough keeping biofilms out of plumbing without turning storage tanks and filters into giant biofilm incubators! Biofilms are certainly beneficial in the natural environment, but they can be very dangerous in potable water systems. Using filtration and oxidation to remove metals and organic pollutants is far less risky than enlisting living biofilms in your potable water system. A well-designed system will have virtually no metals or minerals and little dissolved organic matter in raw rainwater to begin with. Keeping stored water clean is much easier than trying to purify biologically unsafe stagnant water.

As potable rainwater harvesters, our goal is not just to produce clear water, but to produce biologically and chemically safe water. Decades of experience have proven that the best approach is to first remove bacterial food sources (mold, algae, dirt) and then add sufficient free (unused) chlorine or other residual disinfectant to prevent a population of bacteria from getting started. For small scale potable water production, oxidation and filtration are the only practical ways to remove bacterial food sources from your water tanks. Sedimentation simply does not remove the food source for biological growth in your tanks. After municipal water has undergone sedimentation at water treatment plants it is no longer in contact with that sediment and is subsequently subjected to filtration and disinfection, whereas in a rainwater harvesting system utilizing sedimentation, water remains in contact with sediment until it is used by the household. If a rainwater harvesting system uses on-demand chlorine injection as its primary disinfectant, exposure times to disinfectant are nowhere close to those of typical public water supplies simply due to much shorter piping distances. Therefore, it is inappropriate to try to adapt water utility sedimentation/filtration/disinfection processes to small-scale rainwater harvesting. Rainwater harvesting professionals should consider lessons learned by private swimming pool owners rather than large-scale water treatment facilities.

If you insist on incorporating sedimentation, calming inlets, etc. in your storage tank, then you should treat your water as little better than stagnant pond water teeming with all sorts of living microorganisms and should incorporate a water purification process that can handle such biologically unsafe water sources. Commercial water filters do exist that can make stagnant pond water biologically safe to drink. One such filter is the Doulton silver-impregnated ceramic filter. Doulton manufactures a multiple candle filter that has sufficient flow for on-demand use. The candles will need to be cleaned often to remove biofilm buildup on candle outsides. Practically all filters that can handle low-quality water sources, such as stagnant pond water, have low flow rates and are generally only used in recreational applications such as backpacking and boating where high-quality water sources are unavailable. Filter maintenance is always higher with these types of filters. There may be situations, such as an off-grid application, where running continuous circulation pumps is not practical. In such cases a rainwater harvester has no other alternative but to deal with stagnant water and higher filter maintenance. But if you have a choice, you are much better off with continuous turbulent circulation in your water storage tanks. Since you are starting out with high quality rainwater from the sky, why downgrade it to stagnant pond water in a tank full of sediment and biofilms?

Chapter 3.
System Feasibility and Sizing

Before embarking on a detailed rainwater system design, you must determine feasibility of rainwater harvesting for your geographical area and circumstances. This involves determining roof area (or collection surface) and storage tank capacity required for average rainfall in your area to meet daily water consumption needs. As previously mentioned in Chapter 1, one's water "needs" under public water supply are always much greater than under rainwater harvesting. Many people look at their past water utility bills and conclude that rainwater harvesting is impossible for them, but they have not yet learned how to truly conserve water. Past water usage under public water supply is a reasonable place to start, but inevitably much improved conservation practices must be considered when determining true water needs. Water is like any other limited resource and most people learn to live within their available resources. If taking a long soak in a hot shower is a high priority, under rainwater harvesting you can instead install an outdoor hot tub rather than wasting water down the drain. Or you can wait till rain is plentiful and your storage tanks are overflowing before taking long hot showers. Rainwater harvesting does not mean you must go without normal comforts of life, it simply means you make adjustments in how and when you do things so as to conserve water, a limited resource. The first section below shows how to make a rough "back-of-the-envelope" calculation of system size and feasibility. That is not the whole story, as the next section will show, but it gives you a rough idea of whether average rainfall and roof area are sufficient to meet your potable water needs.

Potable rainwater harvesting does not necessarily imply a huge roof area nor a large water storage capacity. It all depends on your local weather conditions and your water needs. My own system is fairly modest with a 3200-gallon storage capacity occupying approximately 105 square feet of real estate (including tanks and pump house) and 1400 square feet of roof area. In all the years of operation we have never run out of water, even during severe drought years. Of course we keep track of weather, along with remaining water storage, and adjust water usage accordingly. This chapter shows how to properly size your system according to your specific conditions and needs. You want a system that adequately meets your needs, with reserve for droughts, but does not unnecessarily increase installation and operating costs.

3.1 Quick Calculation

The first step in system sizing is to estimate your daily (or annual) water usage under rainwater harvesting. This is easier said than done. You can start with water usage data on your past water bills as a worst case demand rate, but rainwater harvesting will force you to become conservation-minded in ways you never expected.

Get your family to pretend they are on rainwater harvesting or water rationing for several weeks and read your water meter directly. Find ways to conserve water as if your life depended on it and see how low you can reduce water consumption.

I began collecting water consumption data when we were renting a house and building our current house. I also installed a water meter on our cistern to track of water usage, even when we were on cave water. Not surprisingly, water consumption dropped dramatically over time as we learned to conserve water in normal daily activities, dropping from an initial average of 100 gallons per day to 50 gallons per day (for two people), which included watering plants and showering every day. Simple things like using a watering bucket rather than a hose to water plants, not letting water run continuously while brushing teeth or washing dishes, and not taking long showers hugely reduce water consumption. Low flow faucets and shower heads also help conserve water. Faucets with a single lever handle that can be controlled with one's elbow or back of the hand when hands are full of soapy dishes are helpful with conserving water. The biggest consumer of water in our house is the clothes washing machine, not the shower. We have an old-style top-loading machine that consumes over 40 gallons per load, but more modern water-conserving washing machines do exist.

The Texas Rainwater Harvesting manuals[23,24] provide guidelines to determine water usage. These are excellent publications on rainwater harvesting and should be studied. Table 3-1 shows various consumption rates measured on my own system.

Table 3-1. Measured Consumption Rates for a Rainwater Household

Backwash raw tank filter	25 gal.
Rinse prefilter	2 gal.
Dishwasher	6 gal.
Manual dishwashing	2 to 4 gal.
Clothes washer (large load)	41 gal.
Clothes washer (small load)	29 gal.
Normal shower & shave	10 gal.
Sea shower (warm)	2 gal.
Sea shower (cold)	1 gal.
Toilet flush	2 gal.

Second, determine average daily, monthly, or annual rainfall for your local area. This information can usually be found on the Internet by doing a search on "rainfall

23 The Texas Manual on Rainwater Harvesting, Texas Water Development Board, third edition, 2005, Austin, Texas

24 Harvesting, Storing, and Treating Rainwater for Domestic Indoor Use, Texas Commission on Environmental Quality, 2007, Austin, Texas

data" for your county or nearest large city. In northern Alabama this data is available from CHARM, (http://weather.msfc.nasa.gov/charm/) the Cooperative Huntsville Area Rainfall Measurements. Since rainfall is rarely evenly distributed over time, daily rainfall data provides better system sizing than monthly data. Monthly or annual rainfall data can provide initial feasibility of rainwater harvesting.

Third, determine your roof or collection surface area for collecting rainfall. This is the projected horizontal area as seen by a bird, not the actual sloped area of your roof. Essentially it is the horizontal length times the horizontal width of your roof. If your roof has a 16-inch overhang be sure to take that into account; don't just measure length and width from your walls. Gutter width should also be included in your calculation. Consult The Texas Manual on Rainwater Harvesting, where roof projected area is referred to as "roof footprint," if you are uncertain about how to determine this. Once you have determined your daily water usage, your local rainfall data, and collection area, you then calculate whether or not rainwater harvesting can meet your potable water needs.

Every inch of rainfall collected on one square foot of horizontal area produces 0.6234 gallons of water. This constant is easily derived by recognizing that 1 square foot captures 144 cubic inches of water with one inch of rainfall (by definition of rainfall amount). One US gallon equals 231 cubic inches. Converting 144 cubic inches into gallons (144 divided by 231) gives 0.6234 gallons per inch of rainfall per square foot of collection surface. (Those working in other units can derive a similar constant.) For example, if your household uses an average of 100 gallons of water per day then you need to collect at least 36,500 gallons of rainfall every year. If your average annual rainfall is 57 inches and your collection area is 1300 square feet, then you will collect:

$$(57 \, inches)(1300 \, square \, feet)(0.6234 \, gal/inch. \, sq. \, ft.) = 46194 \, gal \qquad (3\text{-}1)$$

Forty-six thousand gallons of rainwater more than meets your annual water needs and indicates rainwater harvesting is indeed feasible. Recalculate this for an unusually dry year in your area to determine if your collection area is sufficient to handle drought conditions (assuming you conserve water). Insufficient collection area during drought conditions does not necessarily mean that rainwater harvesting is not feasible. As shown in section 6.3, monitoring storage levels, usage rate, and rainfall allows you to begin implementing rigorous conservation measures in your household during drought conditions long before storage levels get dangerously low. During dry periods we easily cut our water usage in half with very little impact on daily life.

If your roof area is smaller than that required then you must either reduce water consumption or build a storage shed or carport to increase collection area. A word of caution: You should only use properly coated metal roofs for potable water systems. Avoid galvanized or bare metal roofs because rainwater is slightly acidic (carbonic acid). Do not use a typical asphalt shingle roof for anything other than plant watering due to petroleum pollutants in asphalt shingles that leach into rainwater. Other roofing alternatives to metal exist that are also suitable for potable water systems (e.g. slate, tile), but they tend to be much more expensive and heavier. Gutter inspections and maintenance are more important with rainwater harvesting, so keeping gutters within 10 feet of the ground is preferable. You do not want to be getting up on a 30-foot ladder in your old age! Also, a single story house will have a larger roof area than a two-story house of similar floor area. If you are designing a house and plan to incorporate rainwater harvesting, keep your roof design very simple. Complex roof lines and valleys drawn by typical architects are simply not suitable for rainwater harvesting. Roof valleys collect leaves which add their decay by-products to rainwater, which increases biofilm and THM problems. They are also not conducive to a long-lasting maintenance-free roof.

Once you have determined that your roof area and average rainfall are sufficient to meet water requirements, then storage capacity must be determined. Storage tank sizing accounts for the fact that you consume water every day but rainfall usually does not occur every day, nor every month. Your tank must be large enough to handle long dry periods and sufficiently minimize overflow during heavy rainfall. Determine typical longest number of days without rain for your area and multiply this by your daily water usage. Then add to that a reserve amount, which is the lowest level you ever want to see your water storage get to. Your total storage capacity must be at least this large. For example, if your local rainfall data shows rain always occurs every month then you can assume the longest period without rain would be 31 days. If rainfall amount every month is sufficient to meet your water usage needs and your daily water usage is 100 gallons per day then you would need at least 3100 gallons of stored water. If you then assume a reserve amount of 400 gallons, then your total storage tank capacity should be 3500 gallons. On the other hand, if there can be periods when little to no rain occurs for up to two months then your storage capacity would need to be closer to 6500 gallons.

If public water supply is available, but you want to do potable rainwater harvesting for health and conservation reasons, you can ignore the rare drought conditions and design your system with smaller storage capacity. You would use public water supply as an emergency water source to refill raw water storage should the need arise. You could connect public water supply to your distribution system through a county-approved check valve and 3-way valve, as shown later in Chapter 7, but that is neither necessary nor desirable if rainwater is your primary water source and public water is just your emergency source. If you have an operational potable rainwater collection system then you have a superior water purification system and you want to run public emergency water through a more rigorous cleaning process before you drink it. So you treat public water as raw surface water and dump it into your raw water tank. This also simplifies isolating your system from public water supply, as required by most counties and regulatory codes.

3.2 Sizing with WaterStorage1 Spreadsheet

The quick calculations above are only rough estimates of required roof area and storage tank size needed to determine if you are in the realm of feasibility for a rainwater harvesting system. These calculations do not take into account rainfall variability from month to month nor variability of water demand. Some months will produce above average rainfall which could overflow your tanks causing rainwater to be lost. Other months could have below average rainfall resulting in storage tanks running dry. Plants demand more watering when rainfall is scarce. We tend to adjust our water usage when rainfall is scarce and rainfall usually does not always occur on just one day of the month. A more accurate determination of tank sizing is obtained using our WaterStorage1 spreadsheet and actual daily or rainfall data for your area.

Table 3-2. WaterStorage1 Spreadsheet Setup

	A	B	C	D	E	F	G
1					Storage Level	Max Storage	Roof Area
2	Month	Rainfall	First Flush	Usage, GPD			
3	January						
4	February						

Setting up WaterStorage1 spreadsheet is quite easy. Table 3-2 shows the first four lines and labels in our spreadsheet for monthly rainfall data. If you use daily rainfall data change the labels in column A. Enter rainfall data in inches in column B. Enter the first-flush amount rejected to ground for roof washing in column C. Enter assumed daily consumption in gallons per day (GPD) in column D. Enter projected horizontal roof area (in square feet) in cell G2 and total maximum storage amount (in gallons) in cell F2. Enter initial water level (in gallons) at beginning of year in cell E2. Assume a level of about 50% maximum storage level so that you can observe level swings above and below this initial level over the year. For monthly rainfall data enter the following equation into cell E3 and propagate it downward to end of year.

$$=MAX(MIN(MAX(B3-C3,0)*\$G\$2*0.6234+E2-D3*30.42,\$F\$2),0)$$

For daily rainfall data remove the 30.42 constant (average days per month) and enter the following equation into cell E3. Then propagate it downward for the next 364 cells to end of year.

$$=MAX(MIN(MAX(B3-C3,0)*\$G\$2*0.6234+E2-D3,\$F\$2),0)$$

This equation first subtracts first flush amount from rainfall amount and then returns either that difference or zero if it is negative. If rainfall is less than first flush amount then no rainfall is captured. That is the purpose of the inner MAX function in this equation. This difference is then multiplied by roof area (cell $G2) and the constant 0.6234. This product is then added to previous storage level given by the cell above in column E. For daily rainfall data, the current daily consumption in column D is then subtracted from this sum. For monthly rainfall data, we multiply daily consumption by average number of days per month (30.42) and then subtract it from the sum. This is the new storage level displayed at the current cell in column E. However, if this new storage level exceeds maximum storage in cell $F$2, then maximum storage level is inserted instead, indicating additional rainfall is dumped to ground. The MIN function in above equations accomplish this. Lastly, if calculated storage level is less than zero (indicating tanks run dry) then zero is inserted into the current cell in column E, as provided by the outer MAX function in above equations.

Note that WaterStorage1 with monthly rainfall data assumes that rainfall all occurs on one day of each month. Total monthly first flush amount would actually be larger for rainfall spread out over several days within a month. You can easily adjust first flush amounts to account for this. Daily rainfall data more closely models reality, but it may not be available for your locality.

After setting up your spreadsheet with appropriate rainfall data (either monthly or daily), you then adjust numbers in cells E2, F2, and G2 and in column D to size your rainwater system. Cell E2, the assumed water storage level at the new year, should be less than or equal to the final level in this column. If the final value in column E is less than E2 then your tanks are slowly being drained, which is obviously not desirable. Generally, we want maximum storage amount in cell F2 to be as small as possible to minimize hardware costs and maintenance costs. However, if it is too small then too little rainfall is captured to carry us through during dry months.

As an example, 2007 was a particularly dry year with gauge #53 in the CHARM area recording a total of 36.6 inches for the year. Annual rainfall for our area is normally about 57.4 inches. This dry year produced an average monthly rainfall of 3.05 inches. Daily rainfall data shows rain occurred every month and the longest period without rain was 31 days from mid September to mid October. Section 3.1 quick calculations indicate that this rainfall is barely sufficient to meet a daily water use of 100 gal/day with a 1600 square foot roof and a 3500 gallon storage capacity, with zero first flushing. However, the WaterStorage1 spreadsheet gives a very different picture.

Using actual daily rainfall data from gauge #53 in the CHARM area, we enter rainfall amounts into column B and set first flush to zero.[25] Enter a daily usage rate of 100 gal/day, a roof area of 1600 square feet, a storage tank size of 3500 gallons, and a starting level of 2000 gallons. WaterStorage1 spreadsheet will then produce a plot of daily water storage levels as shown in Fig. 3-1. This figure shows a 3500-gallon tank runs dry in June and nearly dry in November, which is obviously not good. The long 31-day dry spell from mid September to mid October did not empty this tank, but the

---

25 Normally a minimum first flush amount of 0.02 inches would be used in dry periods. But the first-flush valve can be permanently closed for extreme drought conditions.

following month did empty it after a shorter dry spell. Also note the clipping of peaks at 3500 gallons in July and September. This indicates overflowing tanks and wasting rainwater to ground during this period. Also note that end-of-year storage level is lower than start-of-year storage level, which means we are gradually draining our storage tanks. This is obviously not good if the following year also happens to be a dry year.

Spring rains, although more frequent, were not sufficient to fill this tank, which allowed it to run dry in June. This would occur even with a larger tank size since peak water levels never rose to full tank capacity before June. The only way to alleviate this problem in June is to increase initial storage level on first of new year to 3000 gallons. Clipping of the peaks in July and September at 3500 gallons, indicating wasting precious rainfall, can be alleviated by increasing storage capacity. This will also bring up the storage level on December 31 to be more nearly equal to that on January 1.

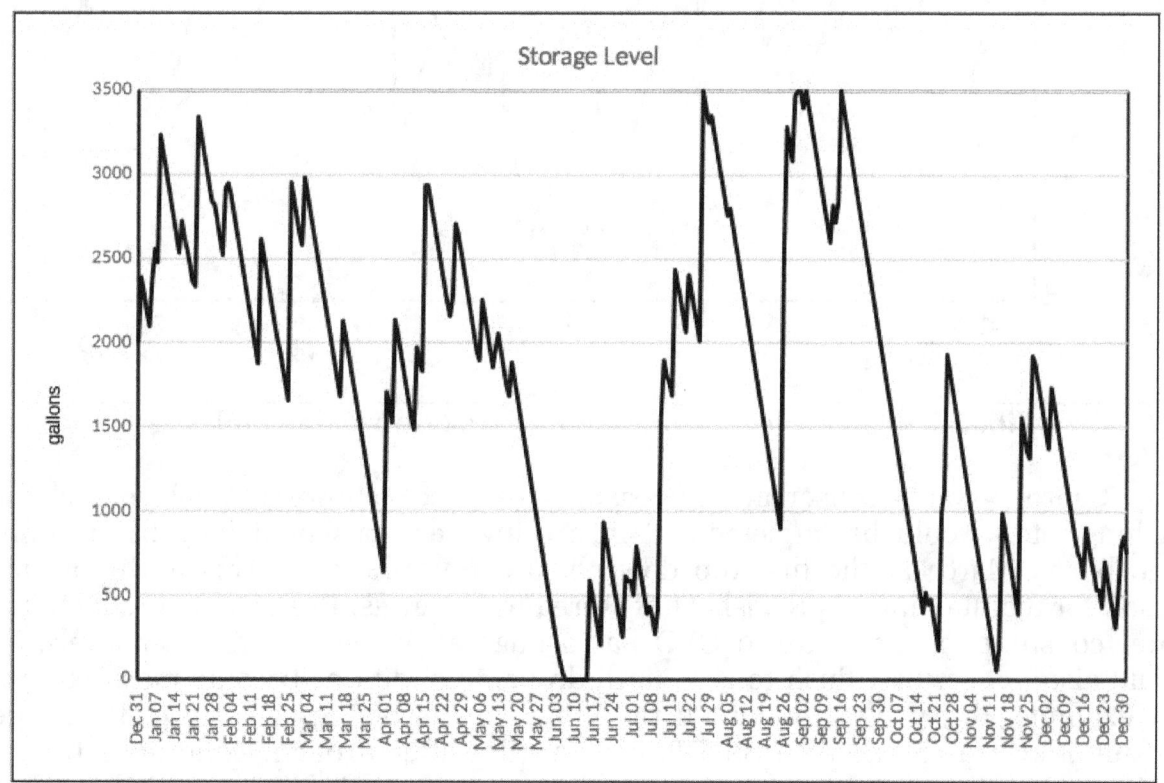

Fig. 3-1. Water level in storage tank with a 3500 gallon capacity.

All the problems in Fig. 3-1 are eliminated if we now increase storage capacity to 6000 gallons and initial water level on January 1 to 3000 gallons, as shown in Fig. 3-2. Water storage level gets uncomfortably low in June, but the tank does not run dry. This larger tank is nearly filled to capacity in September, but never overflows, indicating that no captured rainwater is lost. Ending level on December 31 is slightly higher than starting level on January 1, indicating this tank is gradually filling rather than draining. This assures survival in a subsequent dry year with similar rainfall conditions.

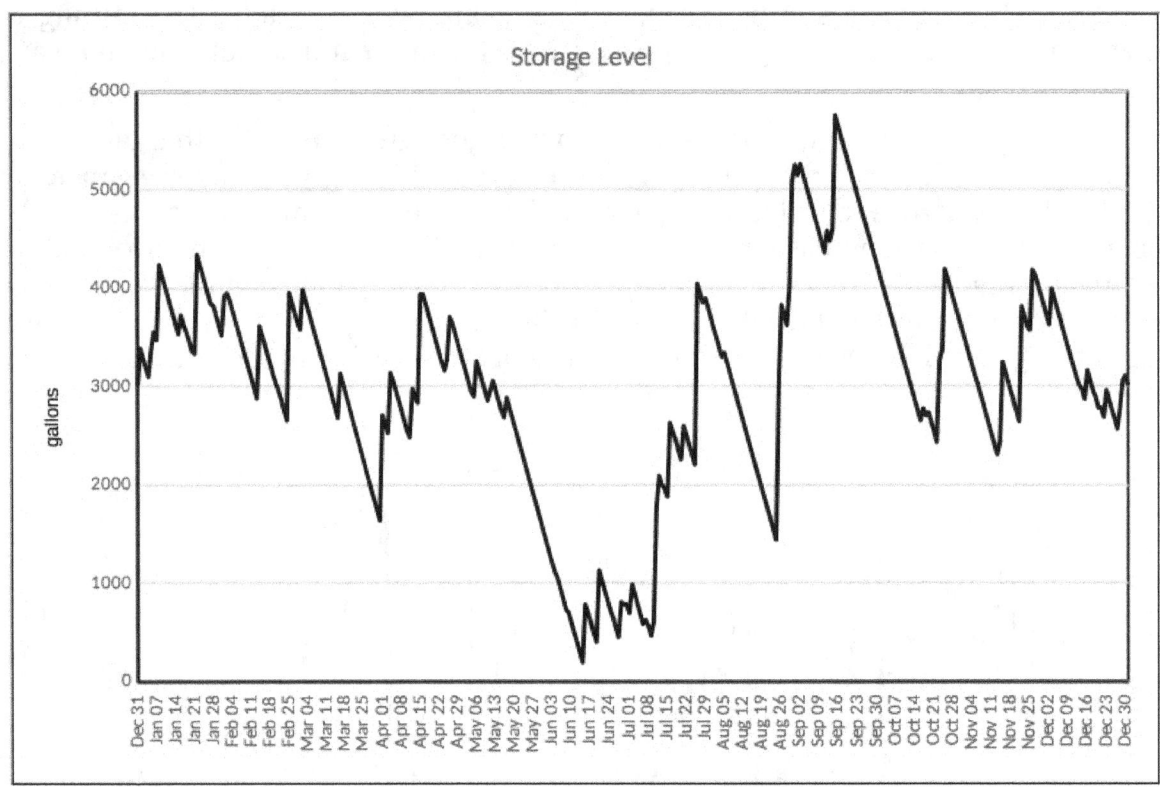

Fig. 3-2. Water level in storage tank for a 6000 gallon tank size.

Rigorous water conservation measures (e.g. sea showers,[26] delaying clothes washing, etc.) would be implemented in a rainwater household long before June, probably in March at the first dip down below the 50% level. This is the primary reason for maintaining a spreadsheet of stored water levels. Even a small reduction of water consumption down to 80 GPD has a huge effect on stored water levels. We would also not set first flush to zero until drought conditions became most extreme. For example, Fig. 3-3 maintains all the initial inputs of Fig. 3-1 except that water consumption was reduced to 80 GPM on April 1 after drought conditions became obvious. Note that water levels never became uncomfortably low and this household still dumped water to ground in July and September.

This example shows the value of using a simple spreadsheet and plot of stored water levels to size storage tanks. Actual daily rainfall data, preferably from an unusually dry year in your local area, can be very helpful in tank sizing. The

---

26 Sea showers, (practiced by Navy and Coast Guard when at sea) means quickly wetting down, turning off the water to soap down, then quickly rinsing off the soap. The water runs for only a few seconds.

WaterStorage1 spreadsheet allows you to quickly assess different collection areas, storage tank sizes, daily use rates, and first flush amounts.

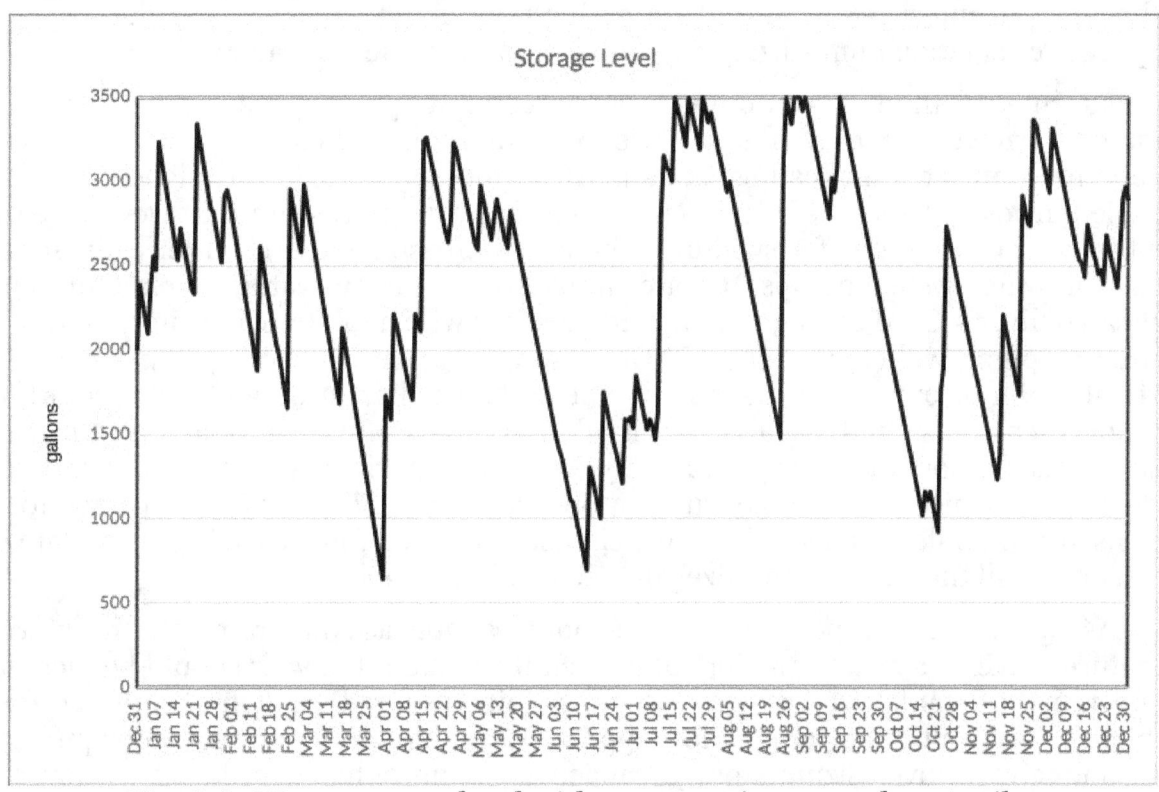

Fig. 3-3. Water storage level with conservation started on April 1.

During extreme drought conditions a rainwater harvesting household always reduces water consumption rates, even down to backpacker consumption levels if necessary. There is no reason for a properly designed and maintained rainwater storage tank to ever run dry, unless the operator is simply not paying attention. System sizing is really more of a question of how much water you want to waste each day in luxurious living, rather than how much water you actually need for survival. When our storage tanks are filled to capacity and no laundry is left to wash, and more rain is still forecast, I tell my wife that it is time to splurge a little and take extra-long hot showers! Rainwater harvesting does not mean doing without comforts of life; it simply means knowing when and how to indulge in those comforts.

Why not just oversize storage tank capacity to far greater levels than will ever be needed and be done with it? Unfortunately this unnecessarily increases both initial installation costs and annual operational costs. Like a swimming pool, storage tank circulation pumps must be properly sized. Larger tanks require larger circulation pumps in order to provide adequate turnover rates, and larger pumps consume more electricity. Higher flow rate pumps require larger filters, larger ozone generators, and larger UV lamps. If you oversize your tank and undersize its circulation pump, stored water will essentially become stagnant and could become a breeding ground for pathogens. An internet search on "swimming pool pump sizing" shows the necessity for proper circulation pump sizing on stored exposed bodies of water.

As rule of thumb, swimming pool circulation pumps are sized to provide complete turnover in 8 to 10 hours. Little to no information exists on sizing circulation pumps for rainwater harvesting tanks because most systems are designed for non-potable purposes. This is definitely a topic that needs more research. However, we can use turnover calculations for swimming pools as a reasonable first-order estimate for sizing rainwater system pumps. If, for example, you use a low wattage circulator pump that can deliver 5 GPM through filters and purifiers with a full tank of water, then in 10 hours this pump will circulate (5 GPM)(60 min/hr)(10 hr) = 3000 gallons. With this particular pump your storage tank should be no larger than 3000 gallons. If your tank is larger then you should use a larger circulation pump. However, a larger pump may require installation of a larger capacity UV lamp purifier (if you are using UV purification), which will also consume more electricity. Filter size might have to be increased to handle increased flow, which also increases filter cartridge replacement costs. So install the storage tank size you need and no larger.

As you can see, proper tank sizing is not as simple as some rainwater harvesting literature makes it seem. Variability of rainfall throughout the year should be included in deciding on tank size, as well as your ability to reduce consumption during droughts. Our continuous circulation approach to maintaining stored water purity is not conducive to over sizing storage capacity. We must have sufficient turnover of stored water to prevent biological growth in storage tanks. And the water must remain turbulent to prevent sedimentation.

One concept for handling rare drought events is to install a second raw water tank that can be taken off line when not needed. We usually know well ahead of time when drought conditions are approaching. During such times an extra tank could be added into our system without increasing circulation flow rate and simply deal with a lower turnover rate. Presumably, water usage would be significantly reduced during such times and a lower turnover rate could be tolerated. Take time to think carefully about your storage capacity needs for your particular situation, including historical weather patterns in your analysis.

Assuming that rainwater harvesting can only be implemented in high rainfall areas is simply not true. Although rainwater harvesting with man-made catchment surfaces is certainly harder in desert regions, it is still feasible. For example, consider Las Vegas, Nevada, that only receives 4.2 inches of rain per year on only a few days out of the entire year. If a household requires 50 gallons per day, its storage tank would need to be at least 18,250 gallons, about the size of a small to medium sized in-ground pool. Collection area would need to be about 7000 square feet. Roof area could supply part of this. A concrete or tile patio slightly sloped toward a central drain could provide additional collection area. The UPC does not allow a concrete driveway to be used for rainwater catchment (K 104.1) since cars sometimes drip oil that contains lead and other toxic metals. Prefiltering the rare heavy rainfall might be challenging, but mold would probably be non-existant in such a sun-baked environment. All this would likely still be cheaper than drilling a well several hundred feet deep, hoping to hit clean water.

Such desert areas might also be good applications for solar stills to recycle gray water from showers and washing machines. If a storage tank is set mostly in ground to keep stored water cool, water circulation through filters can also serve as cooling water in a distiller condenser. One benefit of rainwater harvesting in dryer areas is that organic matter in gutters is much less of a problem. Because of this, water storage might easily tolerate reduced turnover rates and smaller circulation pumps, particularly if stored water is well-protected with little exposure. There is no reason for Las Vegas residents to be totally dependent on government and Lake Mead to supply their potable water. Frankly, I see the hysteria over dropping water levels in Lake Mead as a great business opportunity for some innovative pool company in Las Vegas.

People have lived in desert regions for thousands of years without being dependent on centralized public water suppliers. They drank water, usually from a well or spring, that ultimately came from rain. Like any other rainwater harvester they also learned how to conserve water and how to keep stored water safe to drink. Every drop of water we drink today still ultimately comes from rainfall. Today centralized public utilities do the rainwater harvesting for us and we think they have an infinite supply of water, which is clearly false. This produces very little motivation in human society to learn how to conserve water. Practically all water shortage problems today could be solved by returning to private potable rainwater harvesting utilizing current advanced technology that was unavailable to our rainwater harvesting ancestors living in deserts!

3.3 Example System Sizing for Boulder Colorado

Since we would love to see Colorado bureaucrats end their ridiculous restrictions on rainwater harvesting, let's do an example system sizing design for a typical home in Boulder Colorado. We assume the home does not disconnect itself from municipal water, but uses municipal water only in emergency situations. So we can use an average rainfall year and ignore unusually dry years. Assume a 2000 square foot roof area and daily consumption rate of 75 gal/day with ability to reduce that to 50 gal/day during dry periods. According to the NOAA.gov website, average annual rainfall in Boulder for the past decade (2007-2016) was 21.6 inches. Equation (3-1) shows this roof area collects 26,931 gallons of rainfall per year. At 75 GPD this house would need to collect 27,375 gallons, which exceeds roof collection amount. At 50 GPD this house would only need to collect 18,250 gallons, which is well below roof collection amount. So we conclude that potable rainwater harvesting is potentially feasible for this house and we need to look more closely at details to size storage tank capacity.

Table 3-3. Stored water level in Boulder home with constant 75 GPD usage rate.

				Storage Level	Max Storage	Roof Area
Month	Rainfall (in)	First Flush	Usage, GPD	3000	6000	2000
January	0.71	0.05	75	1541		
February	1.30	0.05	75	818		
March	1.63	0.05	75	507		
April	3.05	0.05	75	1966		
May	3.57	0.05	75	4073		
June	1.53	0.05	75	3637		
July	2.17	0.05	75	3998		
August	1.20	0.05	75	3151		
September	3.09	0.05	75	4660		
October	1.57	0.05	75	4273		
November	0.69	0.05	75	2790		
December	1.16	0.05	75	1892		

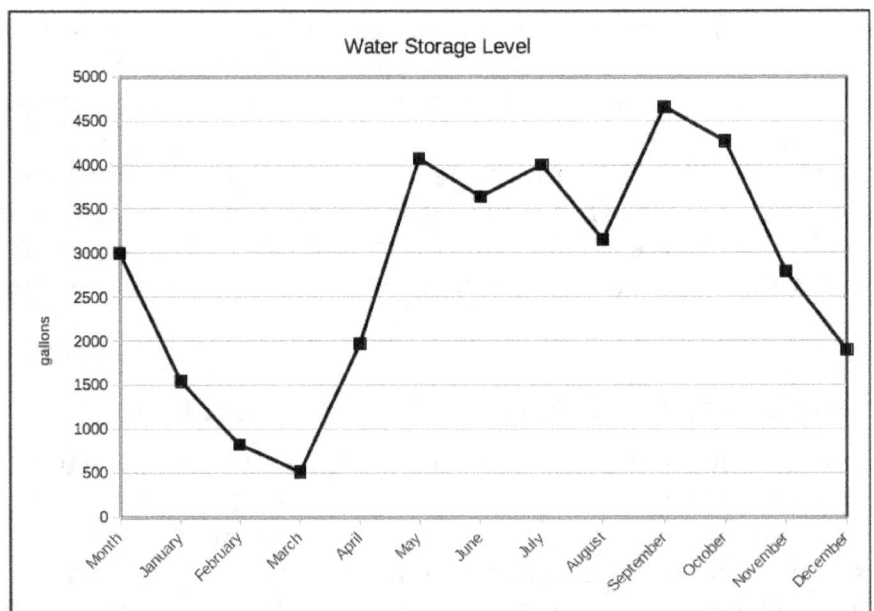

Fig. 3-4. Plot of water storage levels for Table 3-3.

We set up WaterStorage1 spreadsheet with twelve monthly average rainfall amounts entered in column B, with averages determined for the past ten years.[27] We enter a monthly first-flush amount of 0.05 inches, a daily consumption rate of 75 GPD,

---

27 Ten years is arbitrary. One can choose either a longer or shorter time frame for average rainfall.

and initial water level of 3000 gallons, a storage capacity of 6000 gallons, and a roof area of 2000 square feet. Table 3-3 and Fig. 3-4 show the results. As expected this home is slightly consuming more water than collected, as indicated by a lower level at end of year than at beginning of year.

Table 3-4. Stored water level in Boulder home with reduced usage on two months.

Month	Rainfall (in)	First Flush	Usage, GPD	Storage Level 3000	Max Storage 6000	Roof Area 2000
January	0.71	0.05	50	2302		
February	1.30	0.05	75	1579		
March	1.63	0.05	75	1267		
April	3.05	0.05	75	2726		
May	3.57	0.05	75	4833		
June	1.53	0.05	75	4397		
July	2.17	0.05	75	4759		
August	1.20	0.05	75	3911		
September	3.09	0.05	75	5420		
October	1.57	0.05	75	5034		
November	0.69	0.05	50	4311		
December	1.16	0.05	75	3413		

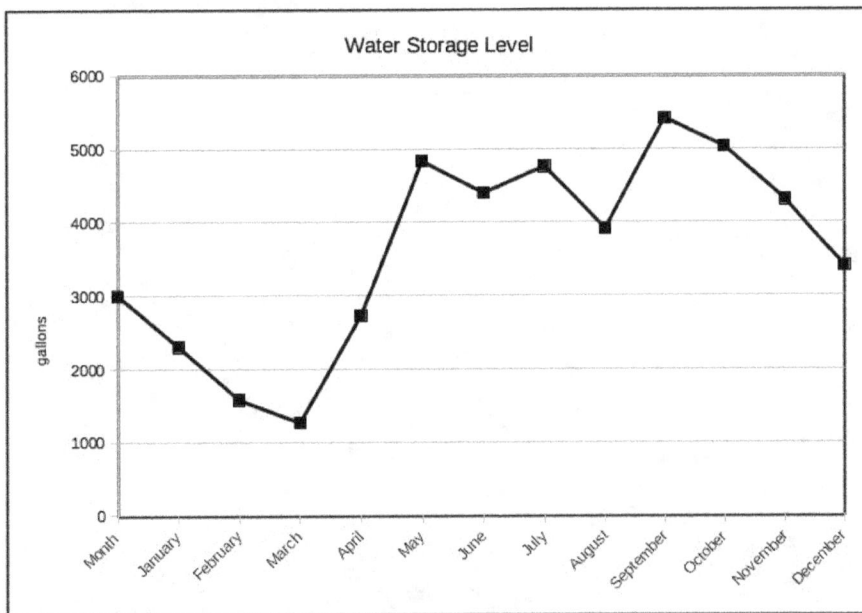

Fig. 3-5. Plot of water storage levels of Table 3-4.

Noticing that months January and November are dry months, this household reduces its consumption to 50 GPD for these two months. Table 3-4 and Fig. 3-5 now show a much more comfortable March storage level of over twice the level in Fig. 3-4 and a year end level that is higher than start of year. This was achieved by simply watching rainfall amounts, seasonal weather prediction, and implementing extra conservation when needed.

Other alternatives exist for this hypothetical rainwater harvesting home to make water storage level plots show feasibility. This home could reduce first-flush amount to zero during winter months of November through March and only have to reduce water consumption in November. If it remains connected to municipal water it could purchase some extra water during dry months. We conclude that potable rainwater harvesting is indeed feasible for this Boulder Colorado home with the assumed roof area, storage capacity, and consumption rate. As with all rainwater systems, it must be actively managed, adjusting first-flush amounts and consumption rates when needed. Our simple WaterStorage1 spreadsheet helps evaluate feasibility, storage capacity requirements, and adjustments needed to make rainwater harvesting work. In Chapter 6 we show a more detailed spreadsheet for actively managing an operational system in real time.

Chapter 4.
Mechanical Design and Construction

We now begin our discussion of mechanical system design and construction. This chapter is generally organized in the order that rainwater encounters various hardware components in a potable rainwater system. We start with roof gutters then proceed to drainage system, first-flush diverter valve, prefilter, storage tanks, circulation loops and disinfection hardware, and finally to the household pressure pump and pressure tank.

4.1 Gutters, Gutter Screens, and Downspouts

Gutters, gutter screens, and downspouts are probably among the most under-appreciated components of a potable rainwater system. I certainly did not appreciate how important these hardware components were to water quality when I first installed my system. When I finally solved my gutter problems much of the disinfection byproduct problems disappeared and my prefilter stayed much cleaner. Gutters and gutter screens suitable for rainwater harvesting is an area that needs some serious research and development. Most homeowners simply want to get the roof water away from the foundation and dump runoff to ground. Rainwater harvesters have very different goals. We want to dump leaves and dirt to ground but collect as much of the clean rainwater as possible.

My gutters and leaf screens were installed by a professional gutter installer, but these leaf screens proved to be ineffective for rainwater harvesting. Tree flowers and leaf matter got past gutter screens and often clogged my prefilter during rainfalls, causing much collected water to be spilled to ground through the pressure relief vent. I purchased Leaf Eater Ultras to install on all downspouts to reduce prefilter clogging, but then discovered the real source of my problem was improperly installed gutters and gutter screens for rainwater harvesting. Professionally installed gutters tend to be installed too close to the drip edge and with very little slope on them. This is not a problem if their only purpose is to move water away from the foundation and look nice on the roof. However, this is completely unsuitable for potable rainwater harvesting.

We do not want rotting leaves sitting on top of our roofs and gutter screens helping promote biofilm growth and mold in gutters, drain lines, and storage tanks! For a long time I had concluded this was unavoidable since I could not find a single gutter screen that worked satisfactorily, until I learned that my problems were due to a gutter installation that was not designed for rainwater harvesting. Most professional gutter installers do not understand these issues of rainwater harvesting. They install gutters for appearances sake rather than maximizing collected rainwater quality. Professionally installed gutters are usually installed nearly level, or with only a very slight slope, allowing water to pool in gutters after rain has ended. This stagnant gutter water and leaf debris promote biofilms, algae, and mold growth in gutters and drain lines, which then clogs the prefilter at the next rainfall event.

Make sure your gutters are installed with sufficient slope along their entire length to completely drain gutters after a rainfall. If you put a 4-foot level along gutter bottoms there should be no place along its entire length where it is level. For long runs put the downspout either in the middle or on both ends to reduce height change needed to obtain proper slope. The advantage of professionally installed aluminum gutters is that these gutters are seamless with little or no connecting joints to trap organic matter and encourage mold and algae growth. You will need to make sure your installer understands these gutters are a vital part of your potable water system and that you have certain requirements that are different from his standard installation procedures. If he is not willing to depart from standard installation procedure and follow your requirements, then find a different gutter installer.

Secondly, professionally installed gutters are usually installed too close to the roof drip edge with insufficient slope on gutter screens, allowing leaves to collect on top of screens and begin rotting. Leaves often get snagged in typical screens preventing the wind from blowing them off the roof. Again, this is not a serious problem if gutters are installed only to move water away from a house foundation, but this is totally unacceptable for rainwater harvesting. Rotting leaves on top of gutter screens add decay by-products to our rainwater. These decay by-products, also referred to as AOC, become food sources for biofilm buildup in gutters, tanks, and house plumbing, which increases chlorine demand and formation of THMs. Practically all gutter screens on the market today are designed to minimize clogging of gutters with leaves, not to immediately separate leaves from rainwater and spill those leaves to ground. Although typical gutter screens are better than nothing at all, these screens are really not suitable for potable rainwater harvesting.

In spite of having gutter screens on my gutters from the very beginning, I still had to get up on a ladder twice each year, usually after trees stop shedding flowers and pollen during springtime and falling leaves in autumn, and pressure wash my gutter screens. Some smooth top gutter guards may separate large leaves from rainwater under certain rainfall conditions, but they also provide attractive nest-building areas for wasps, lizards, and other small creatures. And they still do not do an adequate job of separating organic matter from rainwater for potable rainwater harvesting. The ideal gutter screen for rainwater harvesting is one that flushes all organic matter immediately to ground with minimal water loss, leaves no organic matter sitting on top of gutters after a rainfall, and does not allow insects to get inside gutters.

After an extensive internet search, I finally discovered a gutter screen that solves all these problems and more nearly approaches the ideal gutter screen. I came across Tim Carter's "Ask The Builder" site (www.askthebuilder.com) where he discussed various types of gutter screens. He too was disappointed by the current technology of gutter screens. After testing various types of screens over a 10-year period, Tim found a screen he liked: the MasterShield gutter guard (www.mastershield.com) invented by Alex Higginbotham. MasterShield uses a stainless steel micro-mesh screen that will not snag leaves and tree flowers like other coarser screens. This screen is supported underneath with a stiff perforated plate that helps wick water through its micro-mesh screen.

Fig. 4-1. Typical professionally installed gutters and gutter screens become leaf traps.

Figure 4-1 shows my original professionally installed gutters and gutter screens. As can be seen in the figure these gutters were installed too close to the drip edge requiring gutter screens to lay flat and making them become effective traps for leaves and debris. Gutters should be installed in such a way that if a straight edge is placed on the roof the gutter's outer edge should just barely touch the straight edge or sit slightly below it. When gutter screens are sloped, rather than flat, they help immediately dump leaves to ground. Furthermore, my gutters were installed relatively level so that they did not completely drain after a rainfall, which promoted algae growth and provided attractive sites for mosquitoes. I removed all my gutters and re-installed them with greater slope and hung them lower on the facia. I removed the old screens and installed micro-mesh gutter guards (www.micromeshgutterguards.com), which is a do-it-yourself product made by Alex Higginbotham of MasterShield. Figure 4-2 shows reinstalled gutters and micro-mesh gutter guards after their first autumn. All leaves had fallen from trees but none were caught on my roof or gutters, unlike previous years. Raw water quality significantly increased after installing these micro-mesh screens. Micro-mesh screens still need periodic cleaning, like any other screen, but cleaning is much easier and less frequent. Gutter cleaning is discussed in more detail in Chapter 6.

Fig. 4-2. Properly installed gutters with screens on same pitch as roof.

When I removed the original gutter screens to fix my gutters, I found gutter bottoms coated with thick greenish-black sludge, mold, and algae due to decaying leaves sitting on top of gutter screens and insufficient gutter slope. This slimy mess contaminated captured rainwater and clogged my bag prefilter practically every rainfall. After lowering gutters on the facia and increasing their slope, I pressure-washed the gutters, sanded off stubborn stains, and painted gutter insides with an NSF-approved epoxy coating suitable for painting insides of potable water tanks. I then installed the new micro-mesh screens at approximately the same slope as my roof. Most commercial aluminum gutters have a thin clear coating on their inside surface that will last only a few years and keep aluminum out of rainwater. But eventually this thin coating wears out, exposing bare aluminum to naturally acidic rain. This coating must be thin to prevent it from cracking during on-site aluminum bending and gutter fabrication. But once gutters are formed their insides can be painted with a more durable thick coating that will last far longer than its original standard thin coating. The best time to paint gutter insides is immediately after they are formed or hung on your roof, before mold starts to form. But if your gutters are old you will need to thoroughly clean them and sand off stains before painting them. Spraying down the inside first with a 50/50 mixture of bleach and water helps loosen mold. Follow this with pressure washing and then manual sanding. Unfortunately, I have not found a way to apply a similar coating to insides of aluminum downspouts.

Several companies produce epoxy coatings meeting NSF Standard 61 for potable water tanks. Since a Sherwin-Williams store was nearby I chose Sherwin-Williams Macropoxy 646-PW, standard white. It is a polyamide epoxy coating used for potable water immersion and tank linings. Make sure you specify to your vendor that your use is for potable water, not just general water immersion. There are multiple Macropoxy 646's, but only one is suitable for potable water. The Macropoxy 646-PW is fairly easy to use, but you do need to take certain safety precautions with it. Wear rubber or nitrile gloves and eye protection when handling this stuff. It is not water-soluble and you do not want any of it splashing into your eyes! It comes with a Part A and Part B that are mixed in equal proportions. After mixing you let it set for 30 minutes, called the "sweat-in time," then mix one more time before applying with a brush. You have about 4 hours after sweat-in before the product gets too hard to brush. Acetone can be used to clean up tools before it hardens. If you paint your gutters after they are hung, you should have an acetone-soaked rag handy in case you accidentally brush up against the drip edge of your nice metal roof. After 8 hours of drying you can apply a second coat.

If you have your gutters professionally installed, which I do recommend for high quality seamless gutters, explain to your installer your unique requirements and how you intend to modify these gutters after he installs them. Most likely an installer will not want to spend the time doing these modifications; they like to get on and off a job as quickly as possible. But understanding your requirements for micro-mesh screens, tight sealing, and proper slope will help him install your gutters properly. Gutters must have sufficient slope toward their downspouts at every point along their length so that they completely drain. Test them after installation by dumping some water in at their high ends. Gutters should be installed so that screens can be installed with a slope similar to that of the roof slope. Use a food-grade silicone adhesive sealant to seal up all cracks between gutters and fascia to prevent insects and lizards from getting into your gutters from behind. I used EMI5005 silicone sealant purchased on-line from EMI Supply in Monroe, NC (www.emisupply.com). After sealing and painting gutter insides, install micro-mesh screens, securing screens to outer edges of gutters with #6 half-inch stainless steel screws. Caulk the upper edge, lower edge, and butt joints with silicone to keep out insects. Also seal up ends with a bent-over portion of screen or with aluminum flashing, and then apply silicone caulk to all seams. Do not glue screens down permanently to gutters because you may need to remove them at some point for inspection or to clean gutters. A thin bead of silicone along outer edges after screwing down screens, that can easily be removed if necessary, should suffice. Taking time to do these simple modifications to your professionally-installed gutters will give you a high-quality gutter system that minimizes organic matter in your rainwater, minimizes THM production, and minimizes dissolved aluminum in your water.

Fig. 4-3. Downspout coupling and cleanout with molded rubber seal.

Some amount of mold growth inside gutters, downspouts, and drain lines is unavoidable. But you can discourage large clumps of mold from taking hold by keeping all inside surfaces smooth. For this reason you should avoid using typical flexible corrugated pipe for connecting downspouts to drain lines. PVC long sweep elbows provide smoother inside surfaces and discourage trapping organic matter and large clumps of mold growth. If your drain lines are buried or rather long, be sure to install PVC cleanouts at downspout couplings to drain lines to facilitate cleaning when necessary. There are no suitable couplings on the market for attaching rectangular gutter downspouts to circular PVC drainpipes. Flexible thin plastic couplings found in home building supplies for this purpose are too flimsy and their corrugations tend to trap and promote algae growth. Smooth inside walls all along drain lines help prevent algae and mold from attaching. Also, black rubber couplings for this purpose may not be NSF-approved for drinking water. There's not much demand for NSF-approved downspout-to-drainpipe couplings! The easiest way to solve this is simply fill gaps between coupling and downspout with food-grade silicone caulk.

Molded couplings can also be easily made with food-grade silicone rubber from Smooth-On (www.smooth-on.com). Smooth-On's Smooth-Sil 940 and Sorta-Clear 18 or 37 are suitable for this purpose. If your downspouts are the standard 3"×4" rectangular aluminum type, they will fit inside a 4-inch PVC coupling. You can easily make a mold with a PVC coupling or 3"×4" reducer and a short length of aluminum downspout. Temporarily plug leaks at the bottom of the mold with modeling clay or hot glue (see Smooth-On web site for mold building). Apply a thin coating of vegetable oil or other suitable mold release and then mix liquid silicone rubber and pour into the mold. After the silicone is cured disassemble the mold and pull out your custom-made rubber coupling. Make enough rubber couplings for all your downspouts. Figure 4-3 shows one of my downspouts coupled to a 3-inch cleanout using a custom-molded silicone rubber coupling. Finally, seal up all connecting joints in downspouts with food-grade silicone caulk to keep out small insects and to prevent sprayed garden insecticides from entering your water system.

Gutters, gutter screens, and downspouts are essential components of potable rainwater harvesting. After the roof, these are the first components that captured rainwater comes in contact with. Poorly installed gutters, gutter screens, downspouts, and drain lines increase organic matter buildup in raw water storage and increase disinfectant demand and THM production. Gutters and gutter screens are the ideal place to separate rainwater from natural organic matter, where water flow per unit area is lower, not the prefilter where water flow rates are much higher. Pay close attention to design and installation of these often overlooked hardware components. If you are building a house to be used with rainwater harvesting, make sure you can easily service your gutters when necessary. Keep roof lines simple to avoid trapping leaves in roof valleys and adding their decay by-products to your rainwater. When I got my gutters and gutter screens installed properly raw water quality dramatically improved.

4.2 Drainage System Layout

We depend on gravity to move water from roof gutters to raw water storage. Therefore tank height must be considered when selecting and locating storage tanks. The top of raw water storage must obviously be lower than roof gutters. More specifically, the raw water tank top must be lower than three higher critical points in your gutter drain lines as shown in Fig. 4-4. And we want to maximize the height difference between downspout couplings (highest point) and the third highest point immediately after the prefilter in order to push rainwater through the prefilter. We use 3-inch or 4-inch PVC foam core DWV pipe for drain lines, unless the pipe is buried under a driveway, in which case we use schedule 40 PVC pipe for its greater compressive strength. As Fig 4-4 shows, the highest points in your PVC drain line system are the couplings between aluminum downspouts and PVC drain lines. Typical aluminum downspouts will fit inside a 4-inch PVC coupling. So we use 4"×3" PVC couplings to connect downspouts to a 3-inch PVC drainage system. Gaps between PVC couplings and aluminum downspouts are filled with food-grade silicone caulk.

However, we do not rely on these silicone seals to hold any pressure. So we assume hydrostatic gauge pressure at this "highest point" is zero.

The lowest point in our drain line system must be the first-flush diverter valve because we want all drain lines to completely drain when the first-flush valve is opened after rainfall has ended. We do not want any stagnant water anywhere in our drain line system. The prefilter, consisting of a bag filter and custom fabricated housing immediately following the first-flush valve, is turned upside down in order to provide some automatic backwashing of the prefilter. Rainwater rises in elevation through the prefilter to the "3rd highest point" in the drainage system. This 3rd highest point provides water pressure for prefilter backwashing when the first-flush valve is opened after a rainfall event. This 3rd highest point must be lower than all downspout couplings and higher than raw water tank tops. Hydrostatic pressure pushing water through the prefilter is proportional to height difference between downspout couplings (highest point) and the 3rd highest point immediately following the prefilter. It is not the height difference between downspout couplings and raw water tank top because this line entering the tank is usually not flooded; it has air in it.

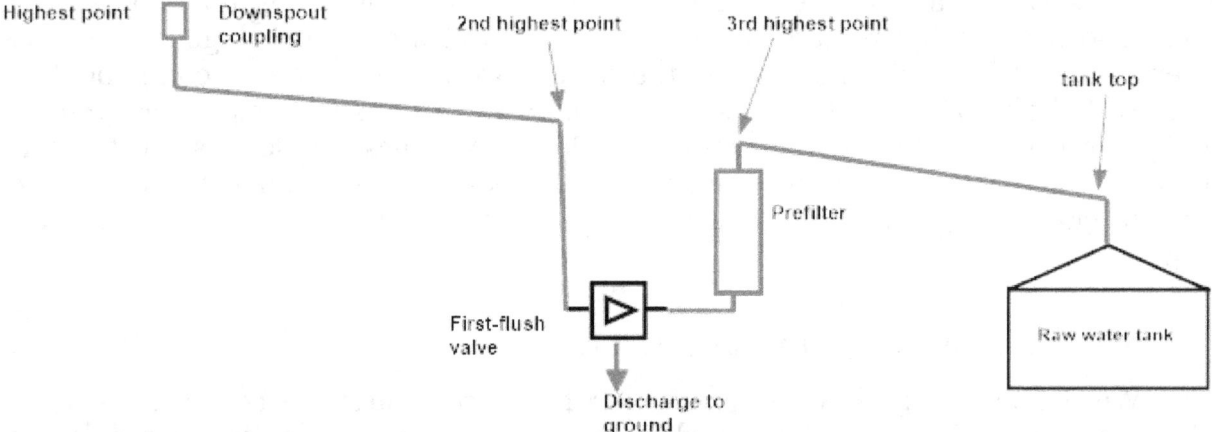

Fig. 4-4. Critical points for locating raw storage tank.

A larger distance between downspout couplings and this 3rd highest point provides greater head pressure to push rainwater through a partially clogged prefilter. Flow rate through this bag prefilter will progressively slow as it captures mold, pollen, and other particles not removed by gutter screens or first flushing. Like a strainer on a swimming pool, this bag prefilter must be cleaned more often than other system filters, depending on time of year. So locate your prefilter in a convenient spot for servicing it. The self-backwashing capability of this prefilter helps reduce frequency of manual cleaning, but does not completely eliminate it.

Checking prefilter status during rainfall to determine whether or not it is clogged can be inconvenient, especially if your only recourse is to remove the manhole cover and look inside the tank. One way to solve this problem is to install a short 6-inch section of clear PVC in the horizontal line between the prefilter and tank. A visual check of relative flow through this clear section during rainfall can immediately tell you whether or not the prefilter is clogged.

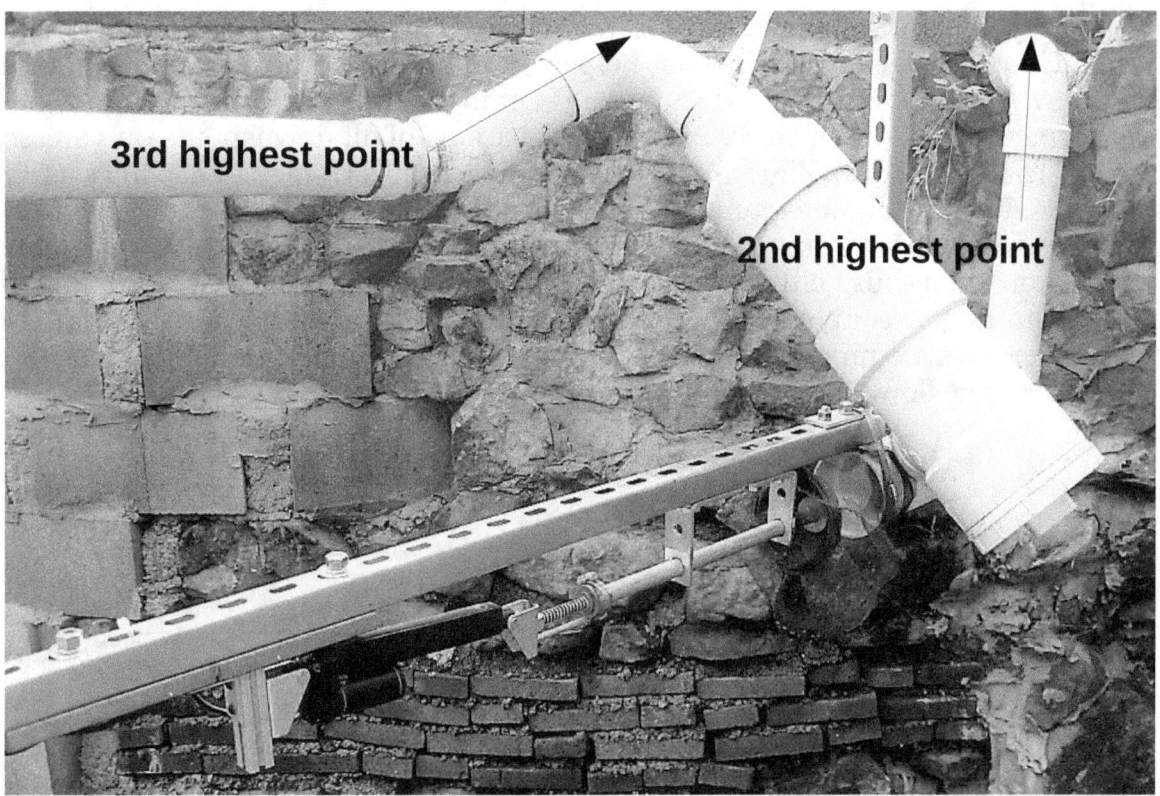

Fig. 4-5. First-flush valve and prefilter showing two critical points.

All horizontal drain lines must obviously slope toward the raw water storage tank. Sometimes this is difficult to accomplish on long runs, but this is an absolute requirement. Otherwise you will have stagnant water in your drain lines growing mold and biofilm. Use a 4-foot level to make sure there are no dips anywhere in your drainage system that can trap water. Furthermore, and this is very important, the 2nd highest point in Fig. 4-4 immediately ahead of the first-flush valve must be higher than the 3rd highest point immediately following the prefilter. If it is not then a huge amount of water will back up in the long horizontal sections from downspouts to prefilter that will not get into your tank.[28] It will dump to ground when the first-flush valve opens, but we want these long sections to completely drain when the valve is closed, not just when it opens. The only water wasted to ground upon opening of the first-flush valve should be in the two vertical sections immediately before and after the first-flush valve. Twice I made the mistake of installing the 3rd highest point slightly higher than the 2nd highest point when building a drainage system, and had to go

28 The number of gallons per foot for 3", 4", and 6" pipe are: 0.367, 0.653, and 1.47 gal/ft respectively.

back and lengthen the vertical leg immediately ahead of the prefilter. I point this out so you don't make the same mistake. Figure 4-5 shows these two critical points on upstream and downstream sides of my first-flush valve and prefilter.

After your storage tanks are located, make a scale drawing of your drainage system. This drainage layout drawing helps you minimize both cost and resistance to water flow from roof to raw water storage. Drainage system layout is best done on graph paper or in CAD software. First draw locations of all downspouts on your scale drawing. Then route drain lines from downspouts to tanks trying to keep lines as short as possible. Keeping all lines running either horizontally, vertically, or at 45 degrees to other lines will help save on number of required fittings. When two lines are combined into one you should use either a Wye fitting (45-degree coupling) or a Combination Tee Wye (90-degree coupling). Do not use Sanitary Tees for combining two drain lines because it creates too much back pressure during high flow conditions. Use long sweep elbows to change flow direction by 90 degrees.

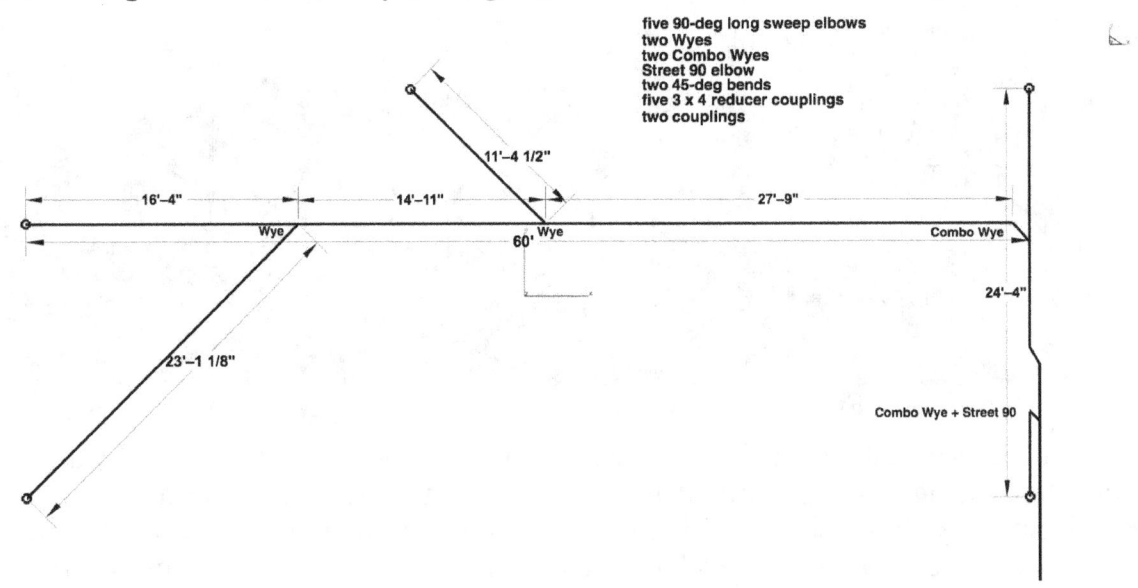

Fig. 4-6. Example drainage system layout.

Figure 4-6 shows an example drainage layout I did for a customer (viewed from top). The foundation of this house allowed running drain lines under the house to storage tanks located downhill. The house has five downspouts, shown as small circles in the drawing. Drainage is toward the lower right corner where the first-flush valve and prefilter are located. Storage tanks (not shown) are also located near the lower right corner in the drawing. Drain lines were combined in four places; using two Wye fittings, one Combo Wye fitting, and one Combo Wye plus Street 90 fitting. Since we planned to purchase 3-inch foam core DWV in 20-foot lengths, the drawing showed that at least two couplings were required since two line lengths were over 20 feet. Two 45-degree elbows were also needed to offset the lower right vertical line to accommodate first-flush valve location.

Table 4-1. Plumbing costs for example layout of Fig. 4-6.

3-in foam core DWV PVC pipe, 20-ft length	Park Supply	$19.22	7	$134.54
3-in PVC Wye	Lowe's #23381	$5.28	2	$10.56
3-in PVC Combo Tee-Wye	Lowe's #23427	$9.31	2	$18.62
3-in PVC Coupling	Lowe's #23283	$1.54	3	$4.62
3-in PVC Street 90° Elbow	Lowe's #23358	$3.79	2	$7.58
3-in PVC 90° Long Sweep Elbow	Lowe's #23361	$6.92	5	$34.60
4"×3" PVC Coupling	Lowe's #23320	$6.34	5	$31.70
3-in PVC 45° Elbow	Lowe's #23339	$3.14	2	$6.28
3-in PVC 22.5° Elbow	Lowe's #23333	$5.49	2	$10.98
4"×3" PVC Bushing	Lowe's #23313	$5.17	1	$5.17
Plastic pipe hanger	Lowe's #302463	$5.83	1	$5.83
Oatey Purple Primer, 8 fl. oz.	Lowe's #23781	$5.98	1	$5.98
Oatey Medium Clear PVC Cement, 16 fl. oz.	Lowe's #23468	$7.98	1	$7.98

Table 4-1 shows the pipe, fittings, and costs required to install this drain line system. Some of additional fittings not indicated by the layout drawing were required to route around foundation posts and connect the prefilter. A street 90 was used at the discharge into tank top to keep this drain line as low as possible. Lowe's part numbers are shown in this table because Lowe's happened to be the closest plumbing supply to this house. But these fittings can obviously be found at many other plumbing supply retailers, including Home Depot and small hardware stores (although part numbers will be different). Prices listed in the table do not include tax and were current prices at the time this list was made. Total hardware costs for this drainage layout was $284.44.

When purchasing PVC fittings and drainage pipe, try to select clean fittings and drainage pipe. DWV fittings and pipe is most often used for sewage and so plumbing stores usually do not try to protect them from getting dirty during storage. But we do not want dirt inside our rainwater drainage system. If insides of newly purchased PVC fittings or pipe are dirty, clean them first with a water-soaked rag before installing on your drainage system. You can run a rag tied in the middle of rope down the inside of dirty pipe until the rag comes out clean. This is most easily done with two people working the rope on each end. Use a shop vacuum to initially suck the rope through the long pipe. After assembling your drainage system, before connecting to your storage tanks, you can attach a rubber stopper at the lower end and fill your drainage system with water and bleach at high shock levels of chlorine to oxidize any remaining dirt. Let it sit in the pipe for a few hours and then dump it to ground. Obviously, you can avoid all this extra work if you get clean PVC DWV pipe and fittings.

Cutting DWV pipe is easily done with a common hand saw. Wrap a flexible straight edge around the pipe, such as a strip of vinyl plastic used for flashing, and mark the pipe all the way around. Cut pipe on this mark while rotating the pipe to achieve a nice perpendicular cut. Sand off burrs on inner and outer edges with sandpaper. Bevel outer edge with sandpaper so that sharp edges do not scrape glue off insides of socket when you assemble it. When assembling apply purple primer to both surfaces (inside of socket and outside of pipe) and then apply PVC cement to both surfaces. Rotate pipe slightly about ¼ turn as you push it into socket to help distribute glue along joined surfaces and hold in place for about a minute until plastic sufficiently fuses. (Rotating may not be possible if pipe is already joined at other end.) If you release too quickly the pipe may move in its socket and fuse in a less-than-ideal position. Let entire assembly dry overnight before using it to collect rainwater.

To decide whether to use 3-inch PVC or 4-inch PVC for your drainage system, calculate maximum flow rate through your drainage system and available head between highest point and third highest point in Fig. 4-4. First determine maximum rainfall rate, in inches per minute, for your geographic area. Maximum rainfall rate tables can be found for various cities in storm drainage sizing sections of building codes or on the Internet. Multiply maximum rainfall rate by 0.623 and footprint area of your roof to determine maximum flow rate in GPM through drainage system.

$$(\text{max GPM}) = (\text{max rainfall rate, in/min})(0.623 \text{ gal/in/ft}^2)(\text{roof area, ft}^2) \qquad (4\text{-}1)$$

For example, maximum rainfall rate for Huntsville, Alabama from U.S. Weather Bureau published data is 3.3 inches per hour, or 0.055 inches per minute. Roof area footprint for the house of Fig. 4-6 drainage layout is 1468 square feet. This produces a maximum drainage flow rate of 0.055 × 0.623 × 1468 = 50 GPM. Use this calculation to also select an appropriate filter bag for your prefilter.

Next, determine maximum flow capacity of drainage system based on selected pipe inside diameter and average slope between highest point and third highest point as defined in Fig. 4-4. From Fig. 4-6 layout drawing we determine that the longest run from downspout to the third highest point just after the prefilter is from the lower left downspout in the figure. From this layout drawing we estimate this distance to be 79 feet. If we then set our downspout couplings to be 3 feet above the third highest point then average slope in this longest run is 3 ft divided by 79 ft which gives a slope of 0.038 (rise/run). Of course this is a conservative estimate since the entire 50 GPM does not flow this entire length. But this calculation is quick and easy.

Several different empirical equations exist for calculating water flow rates in drain lines. Some of these include the Darcy-Weisbach formula, the Hazen-Williams formula, and the Manning equation. The Manning equation is a more commonly accepted formula for gravity-induced flow in a circular pipe. According to the Manning equation flow rate, Q, in GPM, is given by:

$$Q = \frac{0.275}{n} \cdot d^{2.667} \cdot S^{0.5} \qquad (4\text{-}2)$$

where d is inside pipe diameter in inches, S is slope of pipe in vertical feet drop divided by horizontal feet of run, and n is the Manning flow coefficient. Flow coefficient is a frictional coefficient specific to pipe material; a larger coefficient implies greater frictional head losses. Head loss is the vertical distance that water backs up in a pipe to maintain a specified flow. Not surprisingly, plastic pipe has the lowest Manning flow coefficient of all common pipe materials due to its inherent smoothness. A Manning coefficient of 0.009 is typically used for plastic pipe.

Calculating flow capacity for a 3-inch (ID) PVC pipe on a slope of 0.038 using Eq. (4-2) shows this drain line has a flow capacity of 112 GPM, which significantly exceeds our requirement of 50 GPM. Thus 3-inch PVC is more than adequate for our drainage system shown in Fig. 4-6. We conclude that primary flow resistance will not be the drain lines, but the prefilter. The 3-ft head in our design is mainly to push water through a partially clogged prefilter.

Equation (4-2) can also be used to calculate head loss (in feet) by solving for S and then multiplying by horizontal pipe length. For example, with 50 GPM flowing through 3-inch pipe the Manning equation gives $S=0.008$, which when multiplied by the above 79-foot length gives a vertical head of 0.63 feet. This is the height water will back up in the drain pipe to maintain a 50 GPM flow, assuming no blockage at the prefilter.

The Hazen-Williams formula, which can also be used for pressurized flow, provides another calculation for head loss. The formula is also material specific, using a Hazen-Williams flow factor, similar to the Manning flow coefficient. Head loss (HL) in feet per 100 feet of PVC pipe is given by:

$$HL = \frac{0.0983 \cdot Q^{1.852}}{d^{4.8655}} \qquad (4\text{-}3)$$

where Q is flow rate in GPM and d is inside pipe diameter in inches. Using 50 GPM for Q and 3 inches for d this equation gives a head loss of 0.66 feet per 100 feet, or 0.52 feet for a 79-ft run, which is reasonably close to the Manning calculation. These head loss calculations confirm 3-inch drain pipe is plenty adequate and that water will not back up above the highest point and blow out downspout coupling seals for this particular drainage system design.

In summary, drainage system design and layout is a very important part of rainwater system design. The drainage system must efficiently transfer roof water to storage, with minimal wasted water to ground. Keep in mind the four critical points: downspout couplings (highest point), the second highest point, the third highest point, and elbow discharge into tank top. Select tank heights and locations to make sure these four critical points are successively lower in the sequence. Try to maximize height difference between highest point and third highest point to reduce risk of a partially clogged prefilter causing captured water to overflow at downspout couplings and be dumped to ground. Use Eq. (4-1) to determine maximum flow rate through drain lines. Then use Eq. (4-2) and (4-3) to make sure slope is sufficient for your installation.

4.3 Automatic First-flush Diverter

The automatic rainwater diverter serves as both a first-flush device for initial roof rinsing before collecting rainwater and a diverter to prevent overflowing raw water storage and unnecessarily introducing excess organic matter into storage. The next chapter shows how this diverter is controlled with an Arduino microcontroller, allowing easy adjustment of first-flush amount to compensate for seasonal variation of tree litter. Lower rainwater quality during spring and autumn due to heavy pollen and tree flowers requires greater first-flush amounts. This first-flush valve is driven by a 12 VDC linear actuator. The diverter valve is located up-stream of the prefilter to reduce prefilter clogging and required cleaning. Window screen is attached around diverter valve opening to prevent rodents and insects from entering drain lines.

This electrically-driven first-flush valve is a major change from that shown in my first edition of <u>Modern Potable Rainwater Harvesting</u>. This first edition showed a mechanical diverter and mechanical AND gate operated by an imbalance of forces between a bucket of water, a counterweight, and a float in the raw water tank. Drain line pressures had to be controlled for it to work properly, which meant locating the diverter downstream of prefilter. This new electrically-driven valve does not have that problem and it is now located upstream of the prefilter where it belongs.

Figure 4-5 shows my first-flush valve and Fig. 4-7 shows a customer's first-flush valve (before installing screen around valve opening). Both figures show the valve open, allowing rainwater to be dumped to ground. It is constructed from a 4-inch Wye fitting, a rubber toilet plunger, a galvanized ½-inch pipe nipple, Superstrut channel and angle brackets[29], and a linear actuator with 4 inches of travel. All this is off-the-shelf hardware either locally obtained or through mail order. The only custom fabricated components are two molded plastic spacers for clamping the Wye fitting to the Superstrut channel. (Wooden spacers deteriorate under moisture and insects.) Table 4-2 shows a list of all the hardware and their costs at the time of fabricating this. The Smooth Cast from Reynolds Advanced Materials included silicone for making the

---

29 Superstrut hardware is usually found in the electrical department of Home Depot or other industrial suppliers.

rubber molds for the spacers. The rubber toilet plunger was later replaced with a molded food-grade silicone rubber plunger.

Fig. 4-7. Diverter valve and inverted prefilter immediately downstream.

The operation of the valve is actually quite simple. Rainwater from roof enters the 4-inch Wye from the right, through the port opposite the rubber plunger in both Fig. 4-5 and Fig. 4-7. The side branch is angled upward so that gravity forces water out the open end to ground when the plunger is not pressed up against the Wye. When the controller wants to close the valve and send water to storage, it drives the rubber plunger up against the Wye and closes the open end. Rainwater is then forced through the side branch and up into the 6-inch section of pipe, which is the prefilter (discussed in next section). This prefilter consists of a 6-inch sanitary Tee with a 4-inch side port. The cleanout plug on the lower right is for accessing and cleaning the 4-inch filter bag. The filter bag sits inverted in the 12-inch long section (6-inch pipe) sloped upward toward the left. The pipe diameter is then reduced back down to 3" or 4" after the filter and goes to water storage. The few gallons of water in this 6-inch pipe section serves to backwash the filter bag when the valve opens back up after rainfall has ended.

Table 4-2 shows the materials needed to fabricate this first-flush valve. Select a galvanized 1/2" pipe nipple about 12" long. The pipe must be a few inches longer than your linear actuator travel. A longer pipe will put less bending moment on the actuator-to-pipe coupling if the plunger is not perfectly lined up with the Wye fitting. Screw a 3/4" washer down onto threads of one end of 1/2" galvanized pipe nipple.

When it reaches the end of the threads tap it down further on the pipe until the washer is about 1.5 inches from end of pipe. Tighten a hose clamp behind the washer to prevent it from moving any further down the pipe. Drill through the center of a rubber plunger where the handle normally screws into it and screw it onto the pipe threads down onto the washer. Secure rubber plunger in place by screwing a galvanized cap onto the pipe. Paint the galvanized cap with water tank epoxy (the same as used for gutters) to minimize metals leaching into your rainwater.

Table 4-2. First-flush diverter hardware costs.

4" PVC Wye	Lowe's 23384	1	$10.29
Rubber plunger	Lowe's 795241	1	$5.16
18" long 1/2" galvanized pipe nipple & cap	Lowe's 24004, 22460	1	$9.32
3/4" flat washer & 1" hose clamp	Lowe's 25522, 80887	1	$3.43
2 1/2" × ¼-20 bolt, washers, and lock nut	Lowe's 63314, 63403, 63306	1	$0.59
6" hose clamp	Lowe's 121290	2	$4.32
AEI linear actuator 6104TP, 4" travel, 12 VDC	DCActuators.com	1	$122.86
Superstrut angle brackets, 4-hole	Home Depot	5	$13.30
Superstrut hex head bolts 1/2"	Home Depot	10	$10.77
Superstrut spring nut 1/2"	Home Depot	10	$10.55
Superstrut channel 12 ga	Home Depot	1	$21.73
Schedule 80 PVC 1/2" pipe coupling	Lowe's or Home Depot	1	$2.19
Smooth Cast 321	Reynolds Advanced Materials	1	$71.34

Drill out half the threads of a PVC schedule 80 1/2" NPT coupling so that it fits snugly over the linear actuator shaft. Drill a 1/4" hole through the side to secure it to the linear actuator shaft with a 1/4" bolt and lock nut. The linear actuator and plunger assembly will be supported on a Superstrut channel with two 4.125"×3.5" 4-hole angle brackets, one at rear end of the linear actuator and another on the pipe nipple just behind the plunger. Drill out the outer hole on long legs of both brackets. This will provide maximum distance of actuator-plunger assembly from the channel. Outside diameter of 1/2" galvanized pipe is slightly smaller than 7/8", so a standard 7/8" drill will work just fine. Similarly, diameter of actuator rear mount next to the gear box is slightly less than 7/8" (actuator listed in Table 4-2). Slip one of the drilled out angle brackets onto the pipe nipple and screw the nipple and plunger into the PVC coupling attached to linear actuator shaft. Slide the other angle bracket onto rear of actuator and secure in place with a 2" long bolt, two washers, and lock nut. Washers will keep the actuator from rotating in the bracket.

Cut a length of Superstrut channel long enough to hold this actuator-plunger assembly and PVC Wye fitting. Using a 12 VDC wall adapter, run the actuator shaft all the way in until its limit switch stops motion. Make sure the coupling does not touch the actuator housing. If it is too close unscrew the shaft a few turns to back it away from the housing. You want the limit switch to stop motion, not the coupling! Mount the two angle brackets to the Superstrut channel using 1/2" bolts and spring nuts. Now run the actuator out to its full extended length to set the position of the Wye fitting. The Wye fitting must be secured to the channel about 1/2" below the channel. This is the purpose of the molded plastic spacers. Six-inch hose clamps hold the Wye in place on spacers and channel. Push Wye fitting up against the extended plunger until it is clearly sealed and then tighten hose clamps. Alternatively, loosen the rear angle bracket and push the actuator against the Wye.

Smooth-On's website (www.smooth-on.com) contains several instruction videos for mold making. I first made a spacer model out of wood, making sure that it held the Wye fitting at the correct height centered on rubber plunger, and coated it with several layers of acrylic sealer. I then used Smooth-On OOMOO silicone molding rubber to make a mold of the wooden spacer. After curing the rubber mold I used Smooth Cast 321 to make two plastic spacers. No doubt other methods of fabricating durable mounts for the Wye to sit on exist. But this was quick and easy to replicate. The main objective is to securely hold the Wye in place on the same Superstrut channel that the linear actuator is mounted to because the latter will exert considerable force on the Wye when the valve closes.

Easily adjustable first-flush diverter valves are essential components on potable rainwater harvesting systems. They divert the first few hundredths of an inch of rainfall to ground to rinse roof and drainage system of dirt and debris before sending rainwater to storage. This diverted amount should be easily adjustable by a system operator to accommodate seasonal changes in rainwater quality. This diverter also dumps all remaining rainfall directly to ground when storage tanks are filled to capacity, to avoid unnecessarily flushing storage tanks with new rainfall. Most rainwater harvesting manuals show an overflow pipe on water storage, but this unnecessarily dumps to ground relatively clean rainwater that has been continuously circulating through filters. If chlorine is used as a primary disinfectant/oxidant in raw water storage then overflowing wastes chlorine to ground. This automatically controlled diverter valve does not let that happen.

If there is no circulation/filtration loop on raw water storage, such as with an emergency-only system briefly discussed in Chapter 7, then overflowing a raw water tank and flushing it out with fresh rainwater may be one's only means of helping to reduce stagnant water problems and control biological growth. But with the system shown in this book we don't actually overflow our tanks. I may sometimes refer to dumping excess rainwater as "overflowing tanks," but excess rainfall never actually flows *into* storage tanks. This valve simply dumps excess water to ground when tanks are full.

4.4 Rainwater Prefilter

The rainwater prefilter downstream of the first-flush diverter valve is the last line of defense against pollen, mold, and algae before rainwater enters raw water storage. Rainwater prefiltering is an area that needs much more research and development. There are no commercially available low-cost bag filter housings suitable for rainwater harvesting. Rainwater prefilters must handle high flow rates of 50 GPM or more and be easily cleaned without wasting a lot of water. Look in any industrial catalog and you will quickly realize that typical bag filter housings cost thousands of dollars, especially those with 3" or 4" inlets and outlets. And none are easily backwashed like the one we show here.

Pollen sizes can range from tens of microns up to about 200 microns, which gives high flow rate bag filters only limited success with pollen removal. If prefilter pore size is reduced then filter area must be increased to handle required high flow rates, which this design easily accommodates. Pollen densities can either be greater than or less than that of water, which makes vortex type separators also have limited success against pollen. Most vortex separators used for rainwater harvesting today remove waste from the center of vortices. This assumes waste of lower density than water. But higher density pollen will get through these vortex separators and end up settling to tank bottoms and promoting biofilm growth. Vortex separators also require a portion of rainwater to be continuously diverted as waste. We prefer a prefilter that wastes no water to ground.

Realizing the limitations of vortex separators and wanting to minimize wasted water, I chose to use a bag filter as my prefilter. The first edition of this book showed a low-cost two-stage prefilter. We show here an even simpler version that works better than that two-stage filter. This bag filter housing is easily fabricated from PVC fittings and a 4"×14" size 4 filter bag that can be purchased from most industrial hardware suppliers. You can also use longer filter bags if you extend the 6" PVC pipe section length in this housing. Our PVC filter housing is far cheaper to build than typical commercial bag filter housings and lightweight enough to be suspended upside down above the first-flush valve. This filter bag should be carefully chosen for your particular system. A smaller mesh size will reduce particulates passing through the bag but will also reduce flow rate and increase potential for backup and overflowing at gutters or blowing out coupling seals, unless you also increase its length.

Use maximum flow rate calculations determined in section 4.2 for your system, based on roof area and maximum rainfall rate. A clean filter bag must at least handle this maximum flow rate. The example shown in section 4.2 had maximum flow rate of 50 GPM. Therefore, a filter bag capacity of 50 GPM or higher was required for that system. If you are unable to find a filter bag with your required capacity then you can install two prefilters in parallel to handle the larger flow rate. Split discharge from first-flush valve with a Wye fitting and then recombine with another Wye after the prefilters. The filter bag I chose for my system is a 100-micron polyester mesh bag 14 inches long by 4 inches diameter capable of handling 50 GPM (Grainger item number 1EUE9). This filter bag fits perfectly inside the 3.5" socket of a 3"x4" reducer bushing. Unfortunately, Grainger has discontinued this item. The closest replacement I have found so far is MSC item no. 04560694, but it will not fit in 3"x4" reducer bushing without machining a 3.75" diameter hole, 3/8" deep, into the socket. MSC has other size 4 filter bags that will work. I personally prefer the polyester mesh bags over the polyester felt bags since they are easier to clean.

Table 4-3. Prefilter hardware costs.

6" × 6" × 4" PVC Sanitary Tee	Park Supply	1	$43.03
6" × 4" PVC Reducer (female)	Park Supply	1	$19.86
6" × 4" PVC Reducer Bushing	Lowe's 23411	1	$16.99
4" × 3" PVC Reducer Bushing	Lowe's 23313	1	$5.35
6" PVC Cleanout Adapter	Lowe's 51873	1	$12.06
6" PVC Cleanout Plug	Lowe's 53292	1	$7.01
11" long x 6" diameter PVC pipe	Park Supply	1	$15.27
4" × 14" bag filter, 100 micron, 50 GPM, 1EUE9	Grainger.com	1	$5.34
¼-20 nylon screws & wing nuts	Lowe's 139010, 138998	2	$3.23
3/4" PVC plug	Lowe's 22697	2	$2.14
MAX CLR food safe epoxy resin, 24 oz	Polymer Composites, Inc	24 oz	$46.22[30]
Oatey PVC purple primer, 8 oz	Lowe's 23781	8 oz	$5.98
Oatey Medium Clear PVC cement, 16 oz	Lowe's 23468	16 oz	$7.98

This bag prefilter housing is made from 6-inch PVC DWV pipe and fittings shown in Table 4-3. Some machining is required, but all of this can be done with tools found in most home-based shops. Water enters the prefilter through the Sanitary Tee's 4-inch side port, then flows upward through the filter bag and into raw water storage. When the diverter valve later opens, water in this vertical pipe section above the filter bag reverses direction and backwashes the filter bag out through the diverter valve to ground. I often find my filter bag turned inside out due to this backwashing action after a rainfall. Two wing nuts hold the filter bag's plastic flange in its seat, so that new rainfall will re-invert the filter bag. This automatic backwashing significantly reduces a need for manual cleaning of the filter bag. However, when necessary, the filter bag is easily removed by loosening two wing nuts. The reason we use a Sanitary Tee rather than a Wye is that the shallower Tee provides easier access to these wing nuts.

The two reducer bushings are used to fabricate a socket to seat the filter bag flange against. These reducer bushings have molded cavities that must be filled to provide a smooth seat for the bag flange. Filling them also reduces mold and dirt

---

30 Includes hazardous materials shipping

buildup in these cavities. We use food-safe MAX CLR casting epoxy to fill these bushing cavities. This cast resin is then drilled for two nylon screws which hold the filter bag in place.

Fabrication begins by assembling and machining the bag filter seat. Use Oatey Medium Clear PVC cement to glue a 4"×3" reducer bushing into a 6"×4" reducer bushing. Use proper PVC gluing techniques including beveling sharp edges with sand paper (so as not to scrape away glue when assembling) and priming surfaces with purple primer before applying PVC cement. Since this fitting must be glued into the 3-inch deep socket of a 6-inch Sanitary Tee along with a length of 6-inch pipe, we must machine this fitting down to 2 inches, leaving 1 inch of glue surface for the pipe.

Fig. 4-8. Machining the filter bag seat.

Measure the depth of the cavity in the rim of the 6"×4" reducer bushing and mark its location on the outside surface of the bushing. You can see this mark near the fitting's bottom edge in Fig. 4-8. This mark provides a reference for maximum depth of machining on the smooth conical side of the bushing. Machine or cut this side of the bushing down to within about 1/8" of the mark so as not to puncture through into its hollow cavity.

Flip the fitting over and then machine the cavity side outer wall until it is 2 inches high or less, as shown in Fig. 4-8. Do not machine the inner bushing wall or cavity. When finished the top of the inner 4"×3" reducer should extend slightly more than 1 inch above the top of the 6"×4" bushing outer wall. Machining can be done on a table saw with properly set fence and blade height or a milling machine equipped with a turntable as shown in Fig. 4-8.

Fig. 4-9. Gluing filter bag seat into Sanitary Tee.

Set the 6"×6"×4" Sanitary Tee on a table so that its 4" side inlet is curved downward as it enters the 6-inch section. The machined fitting above will go into the lower socket, along with an eleven-inch length of 6-inch diameter PVC pipe. The cleanout adapter (providing access to the filter bag) will be cemented into the upper socket after cutting it down about 1.5 inches (to provide easier access to filter bag). The 6-inch diameter socket depth should be about 3 inches; 1.5 inches is plenty sufficient for our purposes. Place a mark 1.5" down on the side from the top of the flange and cut the flange off at this point. This is easily done on a table saw with the fence set at 1.5" from the blade, or with a hand saw after circumscribing a mark all around the socket. Similarly, cut the threaded cleanout adapter down 1.5 inches. Bevel sharp edges of surfaces to be glued with sandpaper. However, do not glue the cleanout adapter into the Tee at this point, until all other fabrication steps below are completed.

Set the modified Tee on a table with its machined socket down and 4" side inlet curving upward into Tee as shown in Fig. 4-9. The machined reducer bushing will now be cemented into upper full depth socket with its hollow cavities facing down (smooth side up). Apply purple primer to both socket and bushing. Then apply PVC cement to both socket and bushing and push bushing all the way down to the bottom of the socket, leaving about 1 inch of socket surface for the pipe to be glued into, as shown in Fig. 4-9. Wipe off excess glue. Cut a section of 6-inch PVC pipe 11 inches long[31] and glue this into the Sanitary Tee flange above the bushing fitting. Allow this assembly to completely dry for 24 hours before doing any more work on it.

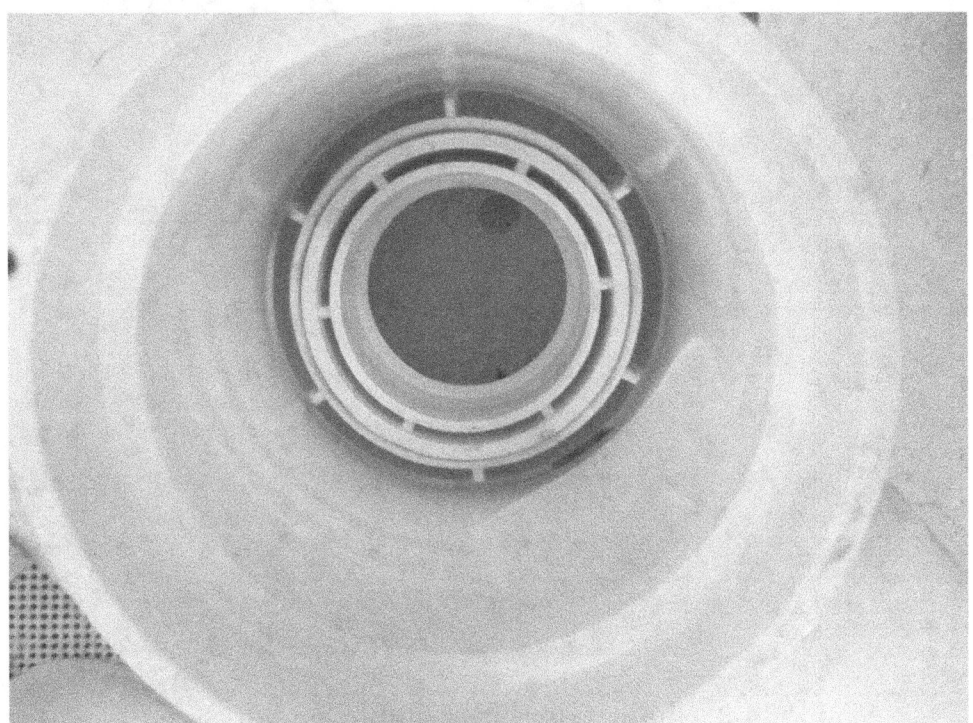

Fig. 4-10. Inside sanitary tee before filling cavities with food-grade epoxy.

Now invert Sanitary Tee so that the bushing hollow cavities are facing up, as shown in Fig. 4-10. We will fill these bushing cavities with food safe MAX CLR epoxy resin (manufactured by Polymer Composites, Inc.). A simple way to determine amount of epoxy to mix is to fill all cavities with water and measure amount of water required to fill them. Make sure cavities are completely dry before pouring epoxy into them (blow dry or let dry overnight if necessary). Mix up desired amount of epoxy, following manufacturer's instructions, and pour into bushing cavities, forming a smooth flat surface across top of cavities from Tee wall to 3-inch socket of inner bushing. Epoxy will partially enter side inlet when this is done. Let this assembly dry for another 24 before proceeding further.

Insert your selected filter bag into its seat and place two marks on dried epoxy surface just outside the filter bag flange. Remove the filter bag and using a long 1/4" bit, drill a vertical hole at each of these marks about 1/2" deep so that these holes clear the filter bag flange by about 1/8", but are otherwise as far away from the Sanitary Tee wall as possible to allow room for turning cogs and wing nuts. These holes should be

31 This assumes using a 14" long filter bag. Increase this length if a longer bag is used.

drilled as vertical as possible. If the machined Sanitary Tee will fit under your drill press without the 6-inch pipe then hold off on gluing the 6-inch pipe into the Tee until these holes are drilled. Otherwise use a slow-speed hand drill and carefully drill these two holes. Cut the heads off two 2-inch long ¼-20 nylon screws and epoxy these into the drilled holes.

Fig. 4-11. Filter retainer cogs to hold filter bag in seat.

Now fabricate two cogs, as shown in Fig. 4-11, that will hold the filter bag in its seat. Fill the hollow cavities of two 3/4" PVC plugs with epoxy and drill a 1/4" offset hole through them. Then grind to an oval shape with a bench grinder so that when turned sideways on the nylon screws the filter bag can be pulled out of its socket. Grind or file a step on one end of these cogs of depth equal to the filter bag flange thickness, as shown in Fig. 4-11. This step will prevent the nylon screws from bending backwards when tightening the nylon wing nuts. Install the two cogs onto the glued nylon screws and hold in place with two nylon wing nuts.

With all interior machining of Sanitary Tee complete, finish up assembly by gluing the cleanout adapter onto Tee. If not already assembled, glue the 6-inch pipe onto Tee on top of bushing. Glue a 6"×4" reducer on end of 6" pipe section. Your prefilter is now complete and ready to be installed onto the 4" side port of the diverter valve Wye, as shown in Fig. 4-7. The cleanout plug should be angled downward with body of prefilter housing pointing upward above diverter valve. This provides head pressure for automatic backwashing of filter bag. Install with cleanout easily accessible so that filter bag can be inspected and cleaned when required.

The first flush diverter valve uses a small amount (user selectable) of initial rainfall to rinse roof and drain lines of dirt before sending rainwater through the prefilter and into storage. After this initial rinse the diverter valve closes and sends additional rainwater to raw water storage. Once raw water storage is full, as indicated by a small float switch in the top of raw water storage, the diverter opens back up and dumps remaining rainwater to ground, avoiding raw water tank overflow. Even if you have properly installed micro-mesh gutter screens a prefilter should be included in your system to capture mold and algae growing in gutters and drain lines. It also captures pollen that passes through micro-mesh gutter screens. The prefilter is the last line of defense against these pollutants getting in your tank, and it is an immediate indicator of the condition of your gutters and drain lines. If your prefilter is constantly getting clogged with a slimy film after every rainfall, then you have a biofilm problem in your gutters and drain lines that needs cleaning out.

4.5 Storage Tanks

Our system uses two separate storage tanks; a raw water tank that immediately receives rainwater from roof or other catchment surfaces and another for purified potable water. Splitting total storage into two tanks with separate continuous circulation loops allows purification to occur at slower, more practical rates, rather than at high on-demand rates of household use. To reduce size and cost of purifier equipment we purify rainwater at slower rates and store it in clean water storage large enough to meet maximum household demand rates. UV systems, filters, and pumps are all cheaper at lower flow rates. Low flow pumps also use less electricity and run more efficiently. Most importantly, this dual tank design puts an extra layer of protection between you and contaminants from roof and gutters. It allows different disinfectants or oxidants to be applied sequentially, such as AOP or ozone in the raw water purification loop followed by low level chlorination in the clean water circulation loop. Rainwater is filtered and aerated many times before you drink it. Chlorine serves only as a residual disinfectant, not a primary oxidant, to protect the clean water.

This dual-tank system allows a more stable, consistent supply of high quality water to your house, independent of water quality from roof and gutters. On my system raw water quality tends to be low during spring and fall due to heavy pollen and leaf matter. Raw water quality can fluctuate greatly during these times, from very low immediately after a heavy rainfall to crystal clear several days later after sufficient filtration and oxidation. Isolating purified water from raw water in a separate tank, and only transferring small amounts of water from raw water storage as make-up feed to top off clean water storage, provides a buffering effect that maintains water quality stability in the clean water tank. As previously discussed in Chapter 2, this dual-tank design also provides flexibility in applying different disinfection processes to solve various disinfection problems.

Total water storage capacity equals the sum of capacities of both tanks, and should equal or exceed minimum storage capacity calculated for your specific situation according to methods discussed in Chapter 3. Although splitting storage into two equal capacities is not essential, cost per gallon of storage generally goes down with increased tank capacity and minimum cost usually occurs when both tanks are approximately equal in size. You need two circulator pumps, one for clean water storage and another for raw water storage. If pumps are appropriately sized for storage capacities, minimum electrical consumption usually occurs when both pumps are identical. If different sized tanks are installed or you use three or more tanks to meet storage requirements, use a smaller tank or fewer multiple tanks for clean water storage. This provides higher turnover rates through purification equipment for clean water, assuming identical circulation pumps on both loops.

Shorter tanks require less head pressure (psi) from circulation pumps when water levels become low (assuming bottom suction and top discharge). Head pressure due to tank height can be held constant using top suction and bottom discharge, but this has other problems as discussed in Chapter 7. The increase in hydrostatic pressure for every foot height of water is approximately 0.43 psi. We want to minimize circulator pump electrical power consumption and shorter tank heights help accomplish this. The tradeoff is that shorter tanks are necessarily larger in diameter for the same storage capacity and may make shipping more difficult. Nevertheless, the top of raw water storage must remain sufficiently below roof gutters.

If multiple tanks are needed to meet maximum storage requirements the easiest solution is to use tanks of identical heights with all raw water tanks located at the same elevation. Water from the roof is discharged into one of these raw tanks and then flows to others through bottom connections until water levels equalize. If one of the multiple raw water tanks were lower than others it would overflow before the others were full and thus limit total storage capacity. As discussed later in section 4.7, the clean water tank should be located at a slightly higher elevation to prevent raw water from flowing into it when the transfer pump is off.

Figure 4-12 illustrates a 3-tank storage system with two raw water tanks connected together. The two tanks are connected through their bottom ports via valves C and D in the figure. Every storage tank should have a ball valve on its bottom port so that hardware or plumbing repairs can be made without draining tanks. These are 1.5" or 2" ball valves, depending on pipe size for bottom connection. For most mid-sized installations 1.5" ID PVC pipe is sufficient for this bottom connection. Normally these lower valves remain open at all times so that gravity equalizes water levels in both tanks. The circulation pump draws from both tanks simultaneously and returns purified water to both tanks via valves A and B in the figure. These top valves are adjusted to equalize flow rates into both tanks. Adjusting is simple. Start with both valves A and B fully open. If discharges into both tanks are not equal then partially close the valve with greatest discharge rate until rates are equal.

Multiple raw water tanks at identical elevation also simplifies tank cleaning without having to dump stored water. For example, if you want to clean Tank 2 in Fig. 4-12 and lower its water level to facilitate cleaning, you can use the circulation pump to transfer its water to Tank 1, assuming raw water storage is not completely filled. You would completely close valves B and C to move water from Tank 2 to Tank 1. When water level is low enough for cleaning close valve D and reopen valve C to maintain circulation in Tank 1.

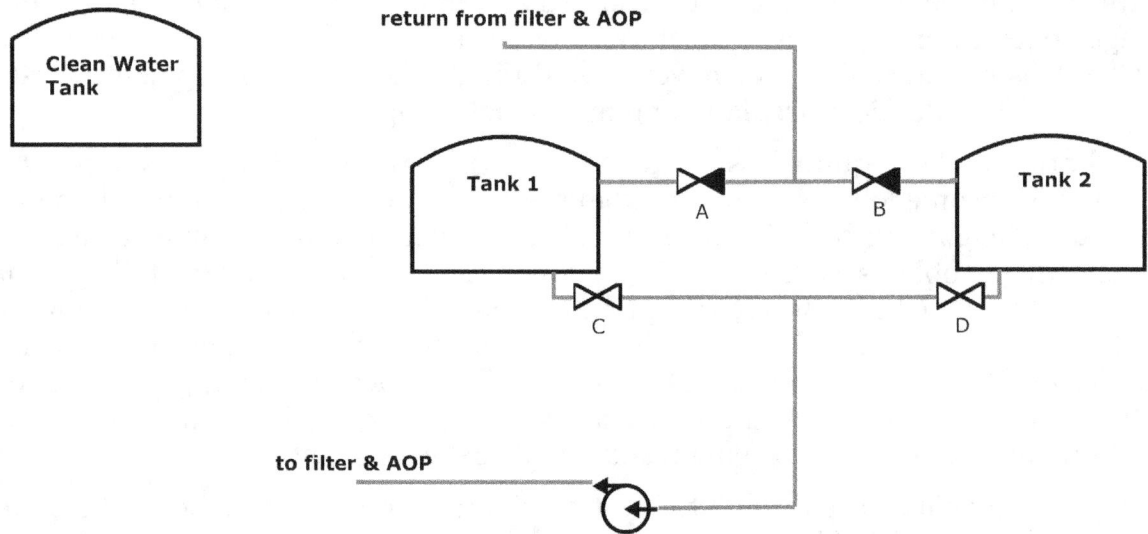

Fig. 4-12. Two raw water tanks at identical elevations.

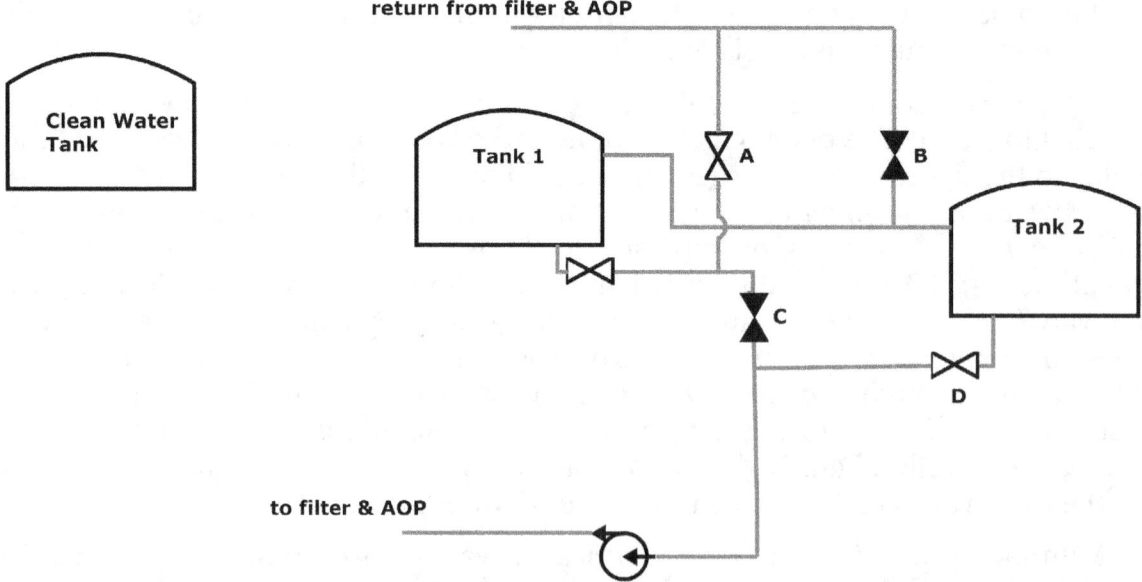

Fig. 4-13. Two raw water tanks at different elevations.

If land topography does not allow locating raw water tanks at identical elevations, raw water tanks can be located at different elevations and still maintain full storage capacity and continuous circulation through all tanks, if plumbing connections are modified slightly. Figure 4-13 shows how this can be done. We add normally-closed valve C and move discharge from valve A to the bottom port of higher tank (Tank 1 in the figure). Circulation is now in series, rather than parallel as in Fig. 4-12. Roof water from prefilter is discharged into lower Tank 2. Circulation water enters bottom of Tank 1 and spills out of its top port into Tank 2, and is then sucked out of bottom port of Tank 2 and passed through purification equipment.

Figure 4-13 shows the discharge line from Tank 1 coming out horizontal and then dropping to the level of Tank 2 intake. The top ports on most tanks are usually well below the manhole covers and this causes a significant amount of storage space in Tank 1 to be wasted. You can minimize this wasted space in Tank 1 by letting the discharge line rise a few inches before dropping down to the Tank 2 level, but keep the rise lower than the manhole lid to prevent overflowing Tank 1. Alternatively, you can install an L-fitting and short piece of vertical pipe inside the tank on the upper port to raise water level to just under the manhole cover.

Under normal conditions valves B and C in Fig. 4-13 remain closed. Tank 1 generally remains filled to capacity with Tank 2 fluctuating in level with rainfall amount. The bottom float switch (discussed in Chapter 5) is located in Tank 2 since this is the lowest point in raw water storage. When raw water levels drop to the point of tripping this float switch in Tank 2, you open valves B and C and close valve A to move some of the water in Tank 1 to Tank 2 to restore operation of transfer pump. Unfortunately, this does stop circulation in Tank 1. If water levels continue to drop below 50% total raw water storage capacity then you can transfer all remaining water in Tank 1 to Tank 2 by temporarily closing valve D until Tank 1 is empty. Then close valve C to isolate Tank 1 and reopen valve D to restore circulation in Tank 2. When rainfall returns move some water back into Tank 1 to make room to capture rainfall. Temporarily close valve B and open valve A until you have moved a sufficient amount of water. Then isolate Tank 1 and restore circulation in Tank 2. When raw water storage rises back above 50% then you can return to the normal sequential circulation in both tanks.

Due to the sloped topography at my house I was able to locate the raw water tank top below my driveway, allowing gutter drain lines to be buried under my driveway. My storage tanks are low cost polyethylene tanks that are NSF-approved for potable water storage. They are not meant to be buried,[32] but can be if you build a concrete block retaining wall around the tank, which is what I did with my tanks in order to accommodate sloped topography and place the tanks at a specific elevation.

---

32 The winters in northern Alabama are usually not cold enough to freeze water in above-ground tanks, especially if water is continuously circulated. But burial or partial burial may be necessary in more northern climates so that water can absorb ground heat to prevent freezing.

Alternatively, there are more expensive reinforced tanks suitable for burial, but if you use these make sure they have smooth inside surfaces to facilitate periodic tank cleaning. Keep manhole covers and tank vents high enough above ground to prevent infiltration by ground surface runoff during heavy rain.

Above ground tanks should be opaque (black or green) to inhibit algae growth. Of course, proper filtration, adequate circulation, and disinfection are primary means of preventing algae growth in tanks. Like a swimming pool, tanks still need periodic brushing or vacuuming to remove biofilms and dirt. So locate tanks to facilitate this periodic maintenance task. Smooth-walled tanks make brushing and vacuuming much easier. Make sure that you can see all tank bottom areas from its manhole access and that you are able to reach all areas with a pool brush or vacuum on an extension pole. Practically all rotationally molded plastic tanks allow this. There are some tanks being sold for rainwater harvesting that I would not recommend for potable applications because cleaning them is virtually impossible. Do not install a tank that you cannot periodically access and clean with a pole brush or pool vacuum.

As previously discussed, we want to filter dirt out as quickly as possible and keep tanks as clean as possible. Layers of dirt in tank bottoms encourage biofilm formation that can serve as breeding grounds of pathogens. You should not have to scrub down insides of tanks as often as swimming pool owners must do, but periodic tank cleaning will be a part of your regular maintenance program to ensure biofilms do not proliferate in your tanks. So make sure you have convenient, safe, and easy access to manhole ports on top of your tanks, and that you can reach all areas inside tanks with an extension pole. A sturdy permanent platform to stand on for tank maintenance is highly recommended.

Generally, we try to avoid underground tanks because it makes tank servicing more difficult and increases cost of tank installation. Inspecting bottom ports and tank connections for leaks is nearly impossible with buried tanks, unless some provision is made for this. However, there may be circumstances that require at least partially buried tanks, such as locating tanks below gutters or protecting them from freezing weather or hot sun, particularly if tanks are low-cost rotationally-molded plastic tanks. Partial burial can allow stored water to absorb ground heat and prevent freezing of stored water. Burying circulation lines below the frost line also helps with ground heat absorption. Seal bottom tank fittings with silicone before connecting plumbing.

Smooth-walled plastic tanks are ideal for controlling biofilm growth, but most are not made for burial. Wet ground can collapse a partially empty plastic tank if it is not strengthened by some other means. An inexpensive and easy way to do this is to build a concrete block retaining wall around the tank before backfilling with dirt around it. Dig a hole for the tank with a backhoe. Carefully level a dirt or sand pad that the tank will sit on and remove all rocks that could puncture the tank. Sand is easy to level and provides a nice cushion for tanks to sit on. Set the plastic tank on the pad and add some water to it to hold it in place while you build the retaining wall. Pour a concrete footer around the tank reinforced with #3 or #4 rebar bent in a circle. While the footer is still wet, place a row of 4"×8"×16" hollow cinder blocks on the footer around the tank leaving about a 1-inch gap between the tank and cinder blocks. The inside edges of the blocks should touch, leaving a wedge-shaped gap between adjacent blocks that will be later filled in with concrete. Push vertical lengths of rebar down into the footer through every other cavity in the block row and begin filling all the cavities with concrete. You will have to lift remaining blocks over these lengths of rebar, so do not make them longer than necessary.

Once concrete in block cavities has set up sufficiently to prevent this row of blocks from moving, then add 8 inches of mortar mix between block wall and tank. This step should be done incrementally, as each row is added, to ensure mortar completely fills the gap between tank wall and block wall. The tank should be filled with water while this block wall is built so that the tank is expanded to its maximum diameter. This will also help provide back-pressure against mortar to keep mortar from deforming the tank. Stagger each row half a block, as normally done, as you build the concrete block wall around the tank. Use mortar mix between each row of blocks and between tank and block wall. Use concrete mix to fill all block cavities. Pack concrete mix into wedge-shaped gaps between blocks with a gloved hand. The final result will be a steel-reinforced concrete wall around tank about 5 inches thick that can be backfilled with dirt. Total cost will still be less than buying plastic tanks suitable for burial, especially if you do the masonry work yourself.

To protect tank tops you can add a steel-reinforced stucco cap tied into the walls. If you do this, do not add stucco all at once because this could collapse the top of your tank. Add an annular ring of stucco a few inches wide from the wall toward the center of the tank and let it dry before adding more stucco. Alternatively, you could add a shed roof over the tank and use this roof as part of your collection surface. Whatever means you use to protect tank tops, make sure you have sufficient room to get a swimming pool brush and vacuum inside its manhole access. Tank tops should not need as much insulation from cold as the tank sides since a layer of air usually exists between tank top and water surface which reduces heat conduction. Tank bottoms should be located below the frost line so that ground heat is transferred into stored water.

An alternative means of insulating tanks from freezing and provide a structural retaining wall to prevent an empty tank from collapsing is to insulate tank sides and top with steel-reinforced lightweight insulating concrete or stucco. You can find a number of different recipes for lightweight concrete on the Internet. They are basically regular cement mortar mixed with an insulating aggregate such as perlite, vermiculite, or expanded polystyrene beads. Water-reducing additives and chopped fiber are usually included in the mix. Wrap a reinforcing wire mesh (e.g. chicken wire or heavier) around tanks, held in place off tank surfaces with plastic rebar chairs, and spray or apply several inches of insulating stucco onto tank walls and top. Note, wire mesh does no good if it is in contact with tank walls; it must be imbedded inside the concrete layer. Installing an insulating horizontal skirt around tank perimeter just below ground surface will help raise the frost line under tank bottoms. Heat transfer from ground to cold air above is lower when the ground is dry and a perimeter skirt helps keep the ground dry. Tank insulation needs to be thick enough so that heat lost through sides and top is less than heat gained through the bottom in contact with warmer ground.

You can provide some additional heat to circulating water by locating filters, circulation pumps, and purifiers in an insulated pump house or basement along with all other mechanical hardware, such as hot water heater, pressure pump, and pressure tank. Heat generated by electrical hardware will help keep the space warm and transfer some of that heat into circulating water. If water freezing on a few extremely cold days of the year is still a concern, you can wrap thermostatically controlled heat tape (found in most hardware or plumbing stores) around metallic sections of water circulation lines. These sections can be short lengths of stainless steel tubes. Copper can be used but be prepared to change it out when pinhole leaks develop from corrosion! Do not wrap heat tape around PVC or other plastic pipe unless it is specifically designed for use with plastic pipe.

Numerous suppliers of FDA-approved or NSF-approved potable water tanks exist. Rotationally molded plastic (polyethylene) tanks are the least expensive of all the various types of storage tanks. They are easier to seal up against insect penetration, unlike steel tanks with pool liners and corrugated steel roofs. Use black or dark green polyethylene tanks to minimize algae growth and make sure they meet NSF standards for drinking water storage. You can find drop shippers and manufacturers of these tanks with an internet search on "plastic water storage tanks" or "rotationally molded water tanks." A few suppliers include: www.plastic-mart.com, www.norwesco.com, www.ntotank.com, www.tank-depot.com, www.tanksforless.com, and many others. Shipping these bulky items can be a significant portion of tank cost, especially if they are shipped over long distances. So shop around and don't forget shipping costs. Drop shippers usually have access to a number of different fabricators and can help you save shipping costs. You can also save on shipping by picking up tanks yourself at a nearby fabricator, if you have a large enough trailer.

4.6 Storage Tank Lid Vent and Gasket

Preventing insects from bringing dirt and organic matter into your tank and other water system components is particularly challenging. Insects sense the cool moist air around tank lids and vents and try to build nests around these areas, particularly during dry conditions. They know how to find water when they need it! Several times, when performing routine tank inspections, I opened my clean water tank lid and found ants building a dirt nest under the lid rim! (My raw water tank lid is much higher off the ground and has less of a problem with ants.) Typical lid seals and vents on most plastic tanks are insufficient to keep out insects. Two relatively simple modifications to your tank lid are required to keep insects out of your tanks. The first is to replace the stock lid vent baffle with a custom micro-mesh screen vent. The second is to add a soft rubber gasket to the lid to close up all cracks between lid and manhole rim on tank.

Depending on particular brand, the manhole screw-off lid will likely have a 1 ½-inch to 4-inch vent hole in its center with a pressed-in baffle to keep out larger rodents. This vent is necessary to allow air displacement when adding or removing water from tank. But it is also a prime entry point for insects. Window screen can keep larger insects out, but not the tiny ants that build nests under tank lids. This baffled lid vent must be replaced with a micro-mesh air filter that can survive outdoor weather. Our recommended gutter guards use 100-micron stainless steel screen, which is perfect for these tank vent air filters.

Fig. 4-14. Simple insect-proof tank vent.

A simple insect-proof air filter vent can be fabricated with PVC pipe and fittings and left-over micro-mesh screen from your gutter guards. Drill ½-inch holes through the side of a short length of PVC pipe of appropriate size for the lid hole. Saw a coupling in half and use the two halves to attach the perforated pipe to the tank lid (one half glued to pipe on top of lid and other half glued to pipe on bottom of lid). Glue a cap on top. Wrap the pipe with stainless steel micro-mesh screen and secure it with plastic wire ties, as shown in Fig. 4-14.

The simplest solution for a tank lid seal is to cut a gasket from a sheet of soft rubber after drawing cut lines on the rubber. Use FDA or NSF approved silicone or neoprene rubber one sixteenth inch thick (which can be purchased from an industrial supplier such as MSC). Glue one side of the gasket down to the tank manhole using food-grade silicone caulk to prevent it from sliding when screwing the lid down and also caulk edges where lid seat is screwed down to tank, wherever cracks exist for tiny insects to penetrate. Be careful that you do not apply too much caulk so that it squeezes out onto the lid's threads and glues it to the tank! Apply a thin coating of vegetable oil to lid threads to help prevent this. EMI5005 is a good food-grade adhesive silicone that can be purchased from EMI Supply Inc in Monroe, NC (www.emisupply.com). The disadvantage of cutting gaskets is that you need large rubber sheets (at least 18"x18" for my tanks) and there will be considerable waste, which can be somewhat costly. These gaskets will eventually wear out from ozone, chlorine, and weather. So you will have to replace them every few years.

Fig. 4-15. Mold for molded tank lid gaskets.

Another method for fabricating a tank lid gasket is to cast silicone rubber gaskets in a mold. In order to do this you can either make a mold to cast multiple spares at once, or cast the gasket directly on the lid using the lid as a mold. If you do the latter you will need to cover the tank opening with something else for about a day. Do not try to mold a gasket with silicone caulk directly on the tank by screwing the lid part way down. Even with mold release on the tank or lid surfaces you may find your lid permanently glued down! Smooth-On, Inc in Easton, PA (www.smooth-on.com) makes various castable rubbers and plastics. Smooth-On's Sorta-Clear 37 and Smooth Sil 940 are food-grade two-part castable silicones suitable for this purpose. This can be less costly than buying large sheets of rubber to cut out annular gaskets, especially if you use the silicone rubber to cast custom downspout couplings (discussed in the next section below).

Figure 4-15 shows a picture of my gasket mold made from two particle board circles, white board, and vinyl flashing. A circular annulus for the bottom surface of the mold can be easily cut from smooth white marker board on a table saw. Cut two of these annuli, the second to press down on the silicone. Drill two holes in your saw tabletop in line with the motor axis at the desired radial distances from the saw blade. Drill a hole in the center of a square piece of marker board and insert a nail through the holes in marker board and saw tabletop, allowing the marker board to rotate about nail on the table saw. Start with the blade lowered all the way down. With the saw turned on, rotate the hardboard about the nail while gradually raising the blade. Cut the outer radius of the annulus first, and then cut the inner radius. Use the same technique to cut two circles out of particle board equal to the inner and outer diameters of the annulus. Stack the two circles with the smaller diameter on top and put a nail through the pivot holes to hold them centered. This forms the flat base of the mold. Place the smooth marker board annulus on top of the base. Use plastic flashing for the vertical walls of the mold and hot-glue it in place to the particle board.

Spray the inside with mold release or wipe a very thin coating of vegetable oil on the mold surfaces. Do not use anything with silicone in it as a mold release as it may serve as a primer and bond the silicone rubber to the mold. Make sure the mold is perfectly level and simply pour the mixed liquid into it to the desired gasket thickness. Calculate the volume of your gasket and mix up only the amount needed. For example, my tank gasket has an inner diameter of 16 inches and outer diameter of 18.25 inches and is 1/8 inch thick. The volume is the difference in the areas of the circles times the thickness, which in this case is 7.57 cubic inches. This equals 4.2 US fluid ounces. Apply mold release to the second annulus and gently press it down on top of the liquid silicone, sandwiching the silicone between two smooth annuli. Follow Smooth-On's instructions for curing the gasket. Gaskets will eventually wear out from chlorine or ozone gas and repeated unscrewing of the lid. You can make several spares at once and replace the lid gasket when necessary. As with the cut rubber gasket, glue the bottom surface of the gasket to the tank with silicone caulk to prevent the gasket from slipping out of its seat when screwing down the lid.

4.7 Circulation Plumbing Layout

Figure 4-16 shows the basic plumbing layout of our dual-tank circulation and purification system with separate circulation loops. Valves (indicated by the hour-glass symbols, solid black are normally-closed, hollow white are normally-open) are placed in strategic locations for water testing and system maintenance. Be sure to install cutoff valves on all tank bottom outlets so that hardware components connected to outlets can be serviced without draining tanks. Two 60-watt circulator pumps (e.g. Laing Thermotech E10) continuously circulate raw water and clean water through their respective purification loops.

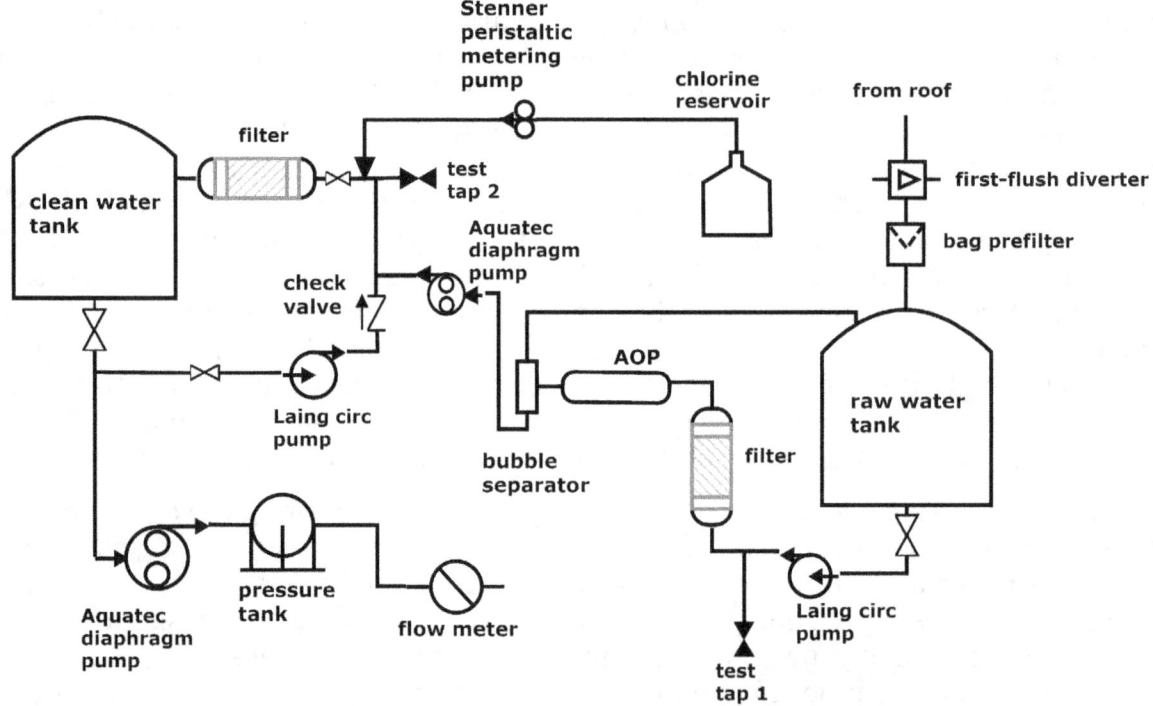

Fig. 4-16. Basic dual-tank rainwater system with disinfection flexibility.

Clean water is sucked from bottom of tank and passed through a 1 micron (or less) filter and discharged to top of tank. A check valve just above this circulator pump forces transferred raw water to pass through the clean water filter. The raw water circulator pump passes circulated water first through a 20-micron filter and then through a 34-watt Prozone Eco Master advanced oxidation process (AOP) purifier[33], then back to top of tank. When required, a diaphragm pump transfers some of this filtered raw water to the clean water tank. A bubble separator on suction side of transfer pump separates most of the ozone bubbles from transferred water to reduce

---

33 Eco Master is an AOP unit manufactured by Prozone Water Products in Huntsville, AL. It combines ozone and UV into a single unit.

cavitation wear in transfer pump. However, this only works if flow rate of circulation pump is greater than flow rate of transfer pump.

Since circulator pumps run continuously, we want these to be as low wattage as possible while still meeting pressure and flow requirements. They must deliver sufficient pressure to lift water the height of the tank and push water through filters. They must also meet minimum flow requirements for proper operation of an AOP purifier. I found the Laing Thermotech pumps to be among the most efficient on today's market. The E10 model delivers a maximum head of 18 feet and maximum flow rate of 11 GPM, which are usually sufficient for our purposes. This model only consumes a maximum of 60 watts. If you have a steeply sloped topography with circulation pumps significantly lower than tanks, you may need a higher pressure pump. For complete water turnover in 10 hours or less, this E10 model will work on tanks up to 6600 gallons (11 GPM × 60 min/hr × 10 hr).

Flow rates of circulator pumps change with pressure head, which must be taken into account to make sure they still meet minimum flow requirements for AOP when the tank is near empty. This data is usually provided in manufacturers specifications. For example, the E10 flow rate drops to 8 GPM with a pressure head of 10 feet. This is still above the minimum flow rate required by the Eco Master AOP unit, so we judge this pump to be adequate. If you build an AOP system from separate UV and ozone purifiers, make sure your UV lamp is adequately sized for your selected circulation pump. Several companies manufacture UV water purifiers of various sizes and capacities. Sterilight UV systems (http://viqua.com/) are available for maximum flow rates of 2 GPM up to 15 GPM, with power consumptions of 19 W to 48 W respectively. These maximum flow rates (2 and 15 GPM) correspond to maximum tank capacities of 1200 and 9000 gallons respectively, assuming a 10-hour turnover rate.

As an aside, we do not use UV or ozone as on-demand purifiers. It should be obvious that UV lamp purifiers and ozone generators must be located in continuous circulation loops with proper flow in order to be effective, yet I still see plumbing drawings by rainwater harvesting professionals showing a UV lamp purifier located in the main water supply pipe leading up to a house! UV purifiers do little purifying when water is not flowing through them, and do inadequate purifying when water is flowing too rapidly through them. For this reason UV lamp purifiers must not be located where flow is driven by household demand or a high flow rate pressure pump. UV purifiers and ozone generators must be located only where flow is controlled by a properly-sized continuous circulation pump. (This does not apply to direct injection of ozone into a storage tank using a small air pump.) Size the circulation pump to provide an adequate turnover rate for a full tank of water but keep its flow rate within maximum rate specified for the UV lamp.

The transfer pump is controlled by two float switches, one in the clean water tank and one in the raw water tank (discussed in more detail in Chapter 5). Clean water storage is kept topped off at all times, as long as water is available for transfer in raw storage. When the clean water level drops by about 100 gallons, the transfer pump turns on and moves raw water to clean water storage, as long as the raw water tank is not near empty. If raw water storage is near empty (water level just above bottom outlets) then the transfer pump will not turn on. When the transfer pump turns on a relay activates a Stenner peristaltic metering pump to inject liquid chlorine into the transferred water. Keeping clean water storage topped off maximizes available raw tank space to receive additional rainfall. Transferring only a small amount (~100 gallons) of water at a time also provides a buffering effect that keeps clean water quality relatively stable, regardless of the quality of raw water. Transferred raw water travels through all filters and purifiers before being discharged into clean water storage. This assures maximum cleanliness of transferred water from raw water storage to clean water storage.

We use a low-wattage diaphragm pump for water transfer (e.g. Aquatec, Shurflo, Flojet) because its built-in check valves prevent backflow of water when the pump is off. However, it is possible for water to flow in the forward direction through these pumps if upstream pressure minus downstream pressure is greater than back pressure provided by these check valves. For this reason we locate the transfer pump suction side on the low pressure side of the raw water loop. Furthermore, elevating the clean water tank above the raw water tank provides some additional downstream pressure to prevent raw water from flowing forward through an off state transfer pump. Alternatively, you can elevate the transfer line above raw water tank discharge to provide some back pressure. A running clean water circulator pump will also provide additional back pressure to keep this from happening. You can easily check that your clean water loop return is sufficiently elevated above the raw water loop by running only the raw water circulation pump. If you get no discharge into the clean tank with the clean water circulator pump turned off, then water is not leaking through the transfer pump in the forward direction. And you can be confident that it will not happen when the clean water circulator is running.

If the clean water tank is close to the same height as raw water tanks, or even lower, then a 3-way solenoid valve can replace the diaphragm transfer pump. Since 3-way valves are somewhat uncommon, two 2-way solenoid valves can also be used. One of these two is a normally closed valve that opens when voltage is applied and the other is a normally open valve that closes when voltage is applied. They would be inserted in the transfer section plumbing as shown in Fig. 4-17. The electrical transfer controller is identical to that with a transfer pump. When one or the other tank float switches are open, no current flows to the two solenoid valves and the transfer line remains closed. Circulation is maintained in both loops. When both float switches are closed current flows to both solenoid valves, opens the transfer line and closes the raw water circulation loop. The raw water circulator pump then pushes raw water through the transfer line and into clean water storage.

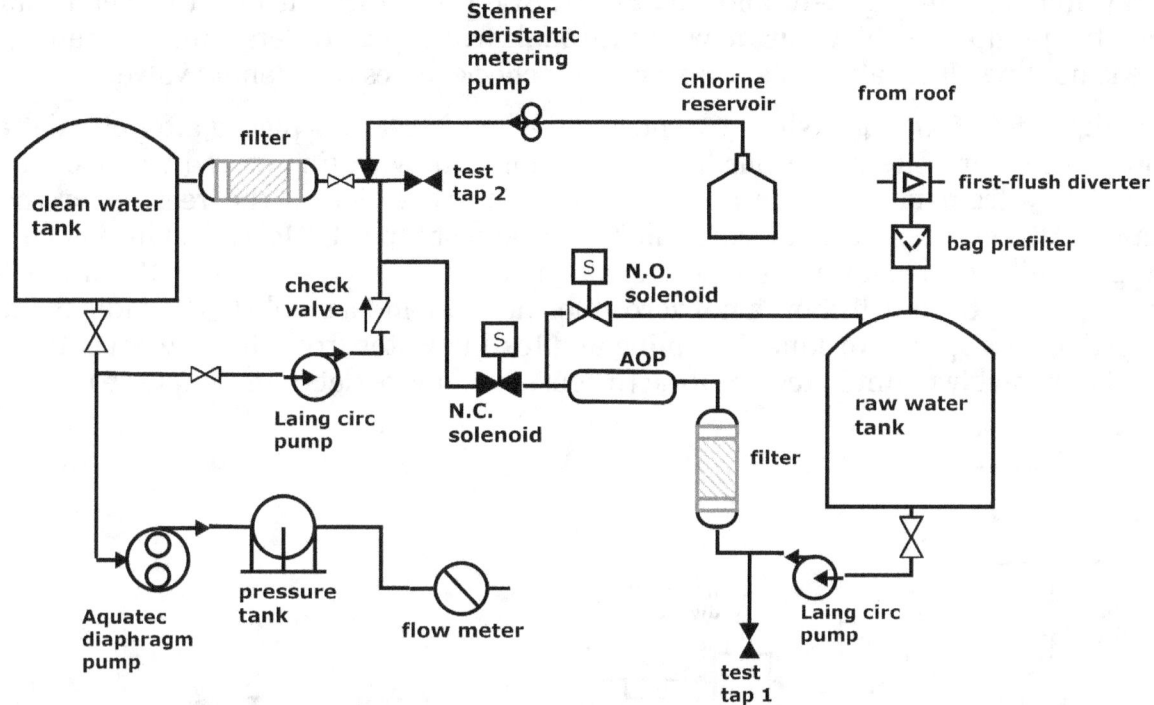

Fig. 4-17. Solenoid valves for water transfer instead of pump.

Most low-cost plastic body electric solenoid valves are used in irrigation systems, which do not require NSF 61 certification. NSF certification usually drives up hardware cost and manufacturers avoid it unless driven by market demand. You can ask a manufacturer about internal valve materials in contact with water that might prevent their valve from obtaining NSF 61 certification if the manufacturer pursued it. The Alcon plastic body solenoid valves sold by FreshWaterSystems.com are NSF 61 certified. Total cost of two solenoid valves will probably be about the same as a diaphragm transfer pump. Advantages of using solenoid valves instead of a pump include lower power and possibly longer life. I was not able to use solenoid valves on my system due to the significant height difference between clean tank and raw tank.

The check valve following the clean water circulation pump forces transferred water to pass through the clean water filter before entering clean water storage. Mechanical check valves do not last forever. The PVC check valve on my system stopped working after a few years of operation. During a major plumbing modification I found the valve was not closing properly and replaced it with a new valve. Cutting open the PVC check valve showed the problem was due to hard iron bacteria deposits on the rubber seal. The rubber itself was also showing signs of decay. So this valve should be tested periodically to make sure it is still closing properly. This check valve can be easily tested through test tap 2 in Fig. 4-17 if you provide an addition ball valve downstream of this test tap. This additional valve is shown on the right side of the clean water filter in Fig. 4-16 and Fig. 4-17. Close this valve, turn off the clean water circulator pump, and blow air or water through test tap 2 to determine if the check valve is holding. (This also tests transfer pump check valves or solenoid valve.)

Figures 4-16 and 4-17 show two normally-closed test taps placed after circulation pumps but before filters and purifiers. These taps allow testing stored water coming immediately from tanks, not water exiting purifiers, which is the reason for their location. Although residual chlorine will be zero at test tap 1, ORP is often initially high and gradually decreases in a test cup as hydroxyl radicals from AOP dissipate. Contrarily, test tap 2 will show a nonzero chlorine level and a stable ORP reading when the system is properly working. Sampling and testing water from these two test taps is part of our weekly maintenance routine (discussed in more detail in Chapter 6).

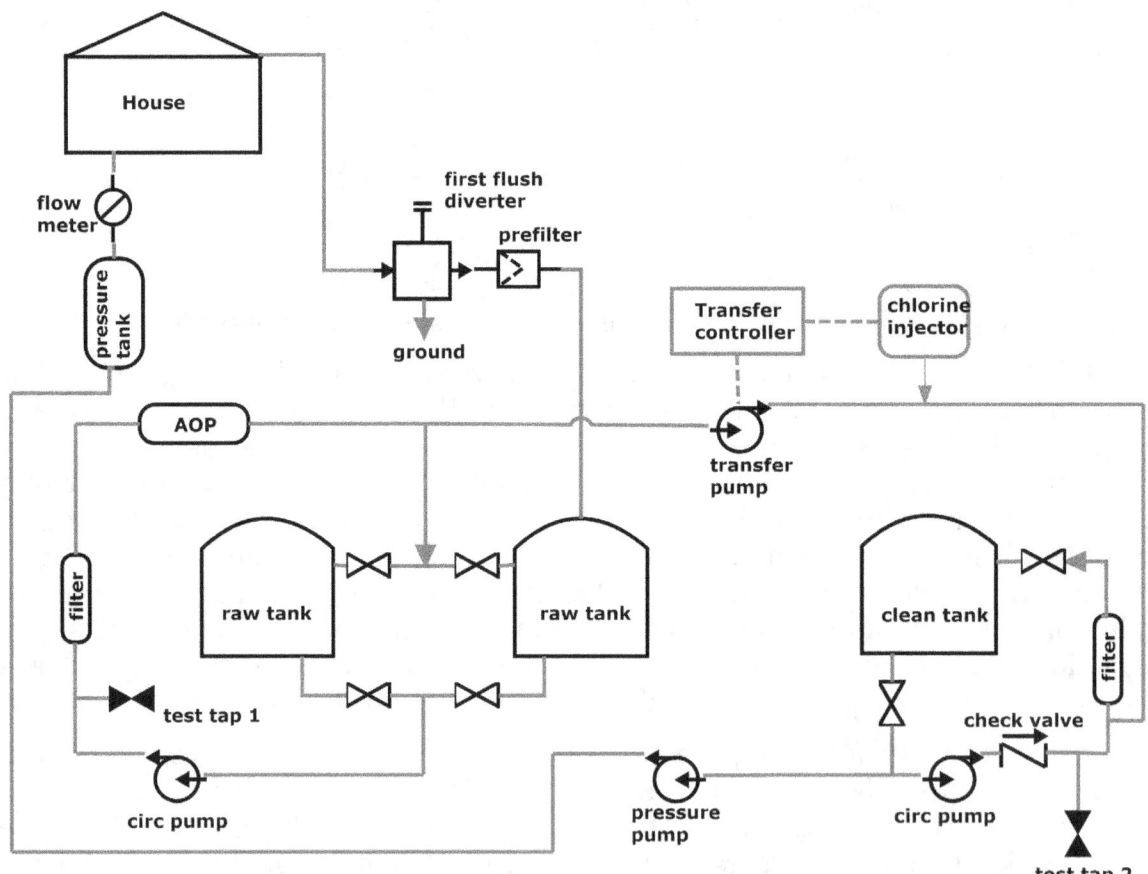

Fig. 4-18. Circulation plumbing layout with two raw water tanks.

Figure 4-18 shows the circulation plumbing layout for a system with two raw water tanks located at the same level. Flow through the two raw water tanks is in parallel. All four interconnecting valves between these two tanks are normally open if both tanks are exactly at the same level. If one of these tanks is slightly higher than the other (>0.5") you may have to partially close one of the two top valves to equalize flow into both tanks. If one of these two tanks is significantly higher than the other by several inches or more, you must use the sequential flow method shown in Fig. 4-13.

The transfer controller is a simple electrical controller discussed in the next chapter that controls both the transfer pump (or solenoid valves) and the chlorine injector pump. Input signal to this controller is provided by two float switches, one near the top of clean water storage and another near the bottom of one of the raw water tanks. We want to keep circulation going continuously in both raw and clean storage. So if raw water drops to a level just above the bottom outlet, the float switch in this tank opens and prevents the transfer pump from turning on. Setting correct height on this bottom float switch is easier to do if water level is low in raw storage. So this should be done before rainfall fills up raw water storage. Lower raw water level to just above the bottom port by either moving water to another raw water tank or to clean water storage. Then set the float switch so that it just turns off at this water level. A simple clamping mechanism for holding float switches at correct heights can be fabricated with PVC plugs. Fill the plug with epoxy and then drill a hole through it for the float switch cable. Then saw through one side of the plug all the way to the drilled hole. Thread float switch cable through hole in top of tank and then through this PVC plug. Attach a stainless steel hose clamp around plug to clamp down on cable and hold in position.

As shown in Fig. 4-16 through Fig. 4-18, roof water first enters the first flush diverter valve which either dumps water to ground or sends it through the prefilter and into raw water storage. Raw water is gradually purified by continuous circulation through a 20-micron filter and through an AOP purifier. This AOP purifier, consisting of an ozone generator and UV lamp, generates hydroxyl radicals that oxidize dissolved organic matter. Using AOP to oxidize organic matter rather than chlorine reduces THM formation. No doubt this raw water is clean enough to drink after circulating through AOP. However, hydroxyl radicals are short-lived and cannot prevent reinfection of water in distribution plumbing where water stagnates. Maintaining continuous circulation through household distribution plumbing is usually not practical. So we inject a small amount of chlorine during water transfer from raw tank to clean tank to serve as a residual disinfectant in house distribution plumbing. We use a 1-gallon glass jug for a chlorine reservoir so that remaining chlorine levels can be easily checked. One gallon glass jugs are hard to find nowadays. But you can still find them in the cheap wine section of Walmart.

Storage capacity can always be increased later by adding additional tanks on the raw water loop of the system. These additional tanks can be connected in parallel as shown in Fig. 4-18 if they are at the same elevation or in series as shown in Fig. 4-13 if they are at different elevations. The goal is to maintain continuous circulation in all tanks and prevent stagnation by drawing water from one end of a tank and returning it to the opposite end. This concept of continuous circulation through all tanks is much like that of a swimming pool, but on a smaller scale. If you increase storage capacity at a later date, check the new turnover rate and increase circulation pump capacity if necessary. But do not exceed maximum flow capacity of your UV lamp or AOP system.

Determining location of pumps, filters, AOP unit, and controllers can be a challenge. Ideally you want all this equipment together in a place that is easily serviced and free of mold, bugs, and rodents. This location should easily handle water spillage when servicing filters, not be susceptible to freezing, and have good fresh air ventilation for ozone generators. These hardware devices should be relatively close to storage tanks to minimize flow-induced back pressure. A garage or storage shed might be an acceptable location, but these tend to be used for storing volatile organic compounds like gasoline, paint, insecticides, and solvents. Ozone generators use ambient air, so do not place them anywhere near off-gassing chemicals, to prevent stored drinking water from absorbing chemical odors and tastes. Additionally, an outdoor storage shed can become a breeding ground for mold and musty smells. Insects can be problematic. Mud daubers often build nests in small holes, such as the ground plug on AC outlets. Wood-boring bees are not stopped by either pressure-treated lumber or painted wood surfaces. In our area anything made of wood eventually becomes hollowed out by these insects. I left my wooden handled broom outdoors a few days one time and one of these bees started to drill a hole in the broom handle! Take extra precautions to protect your system from insects, birds, and rodents.

The solution I chose was to build a small, well-ventilated concrete block pump/filter house and paint both inside and out with waterproof concrete paint. I then built a custom insulated door, with both upper and lower vents that can be easily closed during freezing temperatures. The door is weather-stripped around the perimeter to keep out most insects. When necessary I vacuum out spider webs and other insects and spray with a water/chlorine mixture if a musty odor develops.

Figure 4-19 shows the inside of a customer's pump house. The floor is concrete and walls and ceiling are insulated. His pressure tank and hot water heater are also located in this space. Its door is padlocked to prevent tampering. Note the easy access to filters, test taps, water meter, and pump controls. Pumps are connected with either union couplings or vinyl tubing so they can be easily replaced if they fail. Wiring is enclosed in gray PVC conduit and junction boxes. A 20-micron polypropylene filter and AOP system are installed on the raw water circulation loop (left) and a 1-micron polypropylene filter is installed on the clean water circulation loop (right). Filter housings are clear to assist with determining when to change filters. Both circulation pumps are placed on 120 VAC switches for convenience in changing filters. The receptacles on the left-hand switch (raw tank circulator) are also switched so that the AOP unit is shut off when the raw tank circulation pump is shut off. The 4-plug receptacle just below the transfer pump controller (Fig. 4-19) is not switched. The pressure pump runtime monitor (discussed in detail in Chapter 5) is located on the right hand wall just above the pressure tank.

Fig. 4-19. Inside of pump/filter house.

Although I have a back-washable filter on my system (left over from trying to purify cave water), I do not recommend it for rainwater systems because it consumes too much water during back-washing. My back-washable filter consumes 25 gallons during backwashing! So I only backwash it when flow becomes weak as indicated by insufficient ozone bubbles. Filter cartridges are fairly low cost and mine last for several months before they need changing.

Circulation pumps are centrifugal type pumps and must remain flooded at all times. Unlike diaphragm pumps, they cannot pump air. Therefore, they must be located near the bottom of tanks to prevent loosing their prime or becoming air-locked. Sloped topography may prevent locating circulation pumps in your pump house. For this case you can enclose a circulation pump in a PVC junction box located at the base of your tank as shown in Fig. 4-20. Cut some screened ventilation holes in the top and bottom sides of the box to remove heat generated by the pump. Figure 4-20 shows this pump housing mounted on two steel Superstrut posts concreted into the ground. A pressure-treated wood step attached to the top of these posts serves as a platform to stand on for tanks inspections and cleaning. All water lines and electric cables between tanks, pump/filter house, and main house are buried underground to protect them from freezing. Wires are enclosed in conduit to protect them from gnawing rodents and garden tools. Also note the concrete block retaining wall on the right side of the tank in the figure. This is a semi-circular block wall on the uphill side of the tank to permit installing a level tank pad on steeply sloped ground. It also redirects surface runoff around the tank.

Fig. 4-20. Raw water tank circulator pump housing near tank bottom.

If you must locate a circulator pump remote from your pump/filter house and run 120 VAC wiring to it, do not put AC power cable and low-voltage signal wire in the same conduit like I originally did in Fig. 4-20, especially for long runs. As a result of putting the AC power cable and signal cable in the same conduit I measured up to 36 volts of 60 Hz noise on my low-voltage signal wires with an oscilloscope! This induced 60 Hz noise creates problems for sensitive digital electronics. I later changed this and ran shielded signal cable in separate conduit. So trying to save a few dollars on conduit by combining 120 VAC and low voltage circuits in the same conduit run is just not worth it. Go ahead and run two buried conduits, one for shielded low-voltage signal cable and a separate conduit for 120 VAC power cable with at least a foot of separation between the two. Low-pass or notch filtering is possible for removing 60 Hz noise, but minimizing noise with proper installation from the start is easier. Both raw water and clean water tanks will require signal wires run to them for float switches and level sensors, which are discussed in more detail in Chapter 5.

Table 4-4 shows plumbing and electrical hardware costs to build the circulation loops shown in Fig. 4-19. Total cost is about $1750 not including tax. This circulation system was built with ¾-inch PVC, but PEX can also be used. Whether you use PVC or PEX use fittings that allow easy removal and replacement of pumps. Include valves around pumps and filters to allow servicing without shutting down the entire circulation system.

Table 4-4. Circulation Loop Plumbing and Electrical Costs.

Aquatec 5513-1E01-B606	FreshWaterSystems.com	1	$151.49
1/2" x 6" sch80 nipple, #56123	Lowe's	1	$0.71
1/2" PVC union fittings, #188211	Lowe's	2	$4.96
3/4" PVC coupling, #23850	Lowe's	2	$0.78
3/4" x 1/2" PVC bushing, #23923	Lowe's	2	$0.96
Pentek 20" clear filter housings, #150568	FreshWaterSystems.com	2	$105.98
Neo-Pure 20" polypropylene filter, 20 micron	FreshWaterSystems.com	1	$4.97
Neo-Pure 20" polypropylene filter, 1 micron	FreshWaterSystems.com	1	$5.07
3/4" sch80 NPT to PVC adapter, #52001	Lowe's	4	$9.76
Laing Thermotech E10 circulation pump	Amazon.com	2	$440.00
3/4" barb to NPT male, #22539	Lowe's	6	$2.28
3/4" NPT to PVC adapter, #23862	Lowe's	6	$3.78
3/4" CPVC to PVC adapter, #65322	Lowe's	2	$2.96
3/4" CPVC stop valve, #23772	Lowe's	2	$11.66
3/4" PVC ball valve, #21485	Lowe's	4	$12.68
1.5" PVC ball valve, #21488	Lowe's	3	$35.94
3/4" PVC check valve, #79589	Lowe's	1	$6.78
1.5" x 3/4" PVC Tee, #317743	Lowe's	1	$1.68
1.5" hose clamps, 10-pack, #57188	Lowe's	1	$8.46
3/4" PVC 45-deg elbow, #23891	Lowe's	5	$4.85
3/4" PVC elbow, 10-pack, #26055	Lowe's	3	$12.27
3/4" PVC Tee, #23874	Lowe's	7	$3.50
3/4" PVC sch 40 pipe, #23971	Lowe's	7	$20.79
3/4" x 1/2" PVC Tee, #23933	Lowe's	2	$1.96
Prozone Eco Master AOP 15,000 gal	Prozoneint.com	1	$574.00
Stenner Econ FP metering pump, 4.5 GPD, #F tube	FreshWaterSystems.com	1	$256.97
Clear 1 gal glass jug for sodium hypochlorite	Walmart	1	$15.00
1/2" PVC gray conduit, 10 ft, #72808	Lowe's	2	$4.42
1/2" PVC gray conduit L connector box, #75783	Lowe's	3	$7.44

1/2" PVC gray conduit T connector box, #115991	Lowe's	1	$2.48
1/2" PVC gray conduit 90-deg bend, #50916	Lowe's	7	$4.55
1/2" PVC gray conduit box adapter, #72855	Lowe's	8	$3.84
PVC electrical box, #145087	Lowe's	1	$5.58
PVC electrical box, #145115	Lowe's	1	$5.98
PVC electrical box, #115855	Lowe's	1	$4.58
Electrical switch, #246962	Lowe's	2	$1.38
Receptacle, #246862	Lowe's	3	$2.04
Electrical box wall plate, #158792	Lowe's	1	$1.38
Electrical box wall plate, #52250	Lowe's	1	$1.39
Toggle switch plate, #72507	Lowe's	1	$0.47

4.8 Pressure Pump and Pressure Tank

If you live on the side of a mountain you could potentially locate your clean water storage tank uphill from your house and let gravity provide water pressure. However, when you do the math this usually turns out to be impractical. Typical household water pressures run between 30 psi to 60 psi. When pressures drop below 30 psi users notice it. So if you wanted to supply your house with a minimum pressure of 30 psi you would have to locate your clean water tank 70 feet higher than your house (30 psi divided by 0.43 psi per foot). The actual plumbing run would be longer, depending on the slope of your land, and all pipes would need to be insulated or buried below the frost line. Water moving through long pipe runs looses head pressure, so you would need to have adequate increased pipe diameter to reduce frictional losses. You would also need two separate pump houses, one for clean water circulation equipment and another for raw water circulation equipment. When you add up all these additional costs, a simple pressure pump and pressure tank doesn't seem all that expensive!

The pressure pump's job is to pump water from clean water storage and pressurize it for household use. The pressure tank's job is to protect the pressure pump and keep it from rapidly cycling on and off and burning out the pump during water use. Pumps can usually deliver water faster than you can use it. Therefore, a pressure pump will continuously cycle on and off without a pressure tank when water is used. Distribution plumbing *holds* pressure but it is not a good *store* of pressure because it does not change volume. That's what the pressure tank does; store pressure by compressing a volume of air above several gallons of water in a tank. As water is used pressure in the tank gradually drops. A pressure switch on the pump or pressure tank is set to turn on the pump when pressure drops below a certain point and turn off the pump when pressure rises above a certain point. Typical set points are 30 psi for turn on and 55 psi for turn off, but these are often user adjustable.

Rainwater is naturally acidic and aggressively attacks metals, as anyone who has left metal tools outdoors knows. Atmospheric carbon dioxide is absorbed by rainwater to form carbonic acid. Nitrogen oxides formed by lightning during thunderstorms are absorbed by rainwater to form nitrous acid. So it is best to minimize use of metals in a rainwater system wherever possible. Also, some studies indicate that biofilm buildup is lower with PVC than on metal pipes. All plumbing in my system, including house distribution plumbing, is either PVC or CPVC. If copper, brass, or other metal fittings are required, ensure they are lead-free and NSF-approved for potable water use. Unfortunately most cast iron shallow well pumps (used for supply pressurization) do not have any NSF approval in spite of being widely used for potable water purposes throughout the U.S. Such pumps often become primary food sources for iron bacteria in rainwater harvesting systems. You can minimize this problem by using stainless steel or plastic rotor/housing pumps, preferably with an NSF approval, throughout your system. The primary NSF/ANSI standards for our purposes are Standard 61 for drinking water components and Standard 60 for drinking water chemicals (e.g. chlorine).

Unlike typical water well installations, we do not need a high power pump motor to pressurize our water. Well pumps first have to lift the water 50 feet or more from below ground and then pressurize it. We only need to pressurize it. So we can use high efficiency low wattage pumps, as long as they can achieve about 60 psi of pressure and deliver about 6 GPM under open flow. Rainwater harvesters rarely require high flow rates, other than to fill a bucket or fill a washing machine. And the pressure tank can usually provide those short-term high flow rates when necessary. A high efficiency pressure pump that I recommend for my customers is the Aquatec 5513-1E01-B606 sold by FreshWaterSystems.com and others. This is a 5-chamber diaphragm pump, delivering 60 psi, an open flow of 6 GPM, and consumes only about 230 watts. This pump costs much less than typical water well pumps.

The inlet and outlet of this Aquatec pump is 1/2" FPT. But we use 3/4" PVC for our circulation loops and suction line from the clean water tank. Pumps should be attached to your system with either union couplings or vinyl tubing and hose clamps to make them easy to replace without having to cut and re-glue PVC pipe. To attach this Aquatec pump I use 1/2" PVC union couplings. I cut a 6-inch long 1/2" schedule 80 pipe nipple in half and glue each half into a union coupling. The threaded end of the nipple is then screwed into the pump body. A 3/4"-to-1/2" adapter is glued onto the other end of each union to attach the pump to 3/4" PVC plumbing.

Do not oversize your pressure tank. Again, we rainwater harvesters consume much less water than typical well owners or municipal water users do. In order to minimize stagnation and biological growth in your pressure tank, you want to exchange the pressure tank's water at least once per day. So a 25-gallon pressure tank is adequate for most rainwater harvesting households. I recommend the Well-X-Trol tanks (WX-202 or WX-202XL) manufactured by Amtrol for my customers. These are quality American-made pressure tanks with a long history of fabrication and a solid warranty. The smaller size tanks have a 1-inch NPTF connection. Use a schedule 80 PVC adapter to attach 3/4" PVC to this tank. Install a Tee near this tank with a pressure gauge so that you can quickly determine system pressure and proper operation of the pressure pump.

A water flow meter that records total gallons used should be installed between your pressure tank and household distribution plumbing. This flow meter may seem like an unnecessary expense, but I assure you that it is essential for good maintenance and management of your rainwater system. You periodically enter water meter readings in the WaterStorage2 spreadsheet (discussed in section 6.3) to track water consumption and predict when your tanks will run dry if rain does not arrive. An abnormal spike in water consumption may indicate a potential water leak in the system that needs repairing. The water flow meter provides immediate and essential feedback on the effectiveness of your household's water conservation practices or a potential slow leak (e.g. leaky faucet or toilet valve). You use it to measure exactly the amount of water consumed by every activity in your household (e.g. dish washing, clothes washing, bathing, etc.). Chapter 6 shows how to use the water meter to determine average roof collection efficiency, which is needed to accurately track water storage. Chapter 5 shows how to use it to properly set the pump runtime counter. So the water meter is an essential part of your potable rainwater system.

If you have a professional plumber install your system or if a county building inspector is required to inspect it for compliance to regulatory codes, I recommend that you obtain a copy of the UPC used by your county and study it. Inspectors often do not know the regulatory codes as well as they should and sometimes misapply requirements, particularly on systems they are unfamiliar with. Sometimes I have pointed out a requirement to a county inspector only to hear him say, "Oh, I did not know that was there!" The supply line from storage tanks to house is an item that both plumbers and inspectors do not understand. Table 610.4 in the 2015 UPC specifies the building supply pipe size based on length and number of fixture units. This table is based on a municipal water main being able to supply pressure under maximum water demand from a house. Plumbers and inspectors assume this table equally applies to a rainwater system with storage tanks. It does not! When I first installed my system a plumber told me that I needed to install 1.25-inch pipe from my storage tanks to my house. I now regret that I listened to him because these oversized pipes are a major source of distribution system biofilms.

Maximum water flow rate through a supply line from storage tanks to house is *not* determined by number of fixture units in your house. Rather it is determined by the maximum flow rate of your pressure pump. Unlike a supply line attached to a municipal water main, water will never flow through that line from your tanks any faster than the pressure pump can pump it. Minimum pipe sizes are specified in plumbing codes to keep flow rates below a point where friction and water hammer become problematic. For example, section 610.12 of the 2015 UPC specifies that maximum velocity in copper pipe shall not exceed 8 feet per second. Flow rate equals velocity times inside cross sectional area of pipe. So if our pump delivers a maximum flow rate of 6 GPM, what is the maximum velocity through 1.25-inch pipe? Since velocity equals flow rate divided by area and that 231 cubic inches equals one gallon we can easily write the formula for velocity (in feet per second) as a function of flow rate (in GPM) and pipe ID (in inches):

$$velocity = 0.408 \frac{GPM}{ID^2} \qquad (4\text{-}4)$$

The above equation shows that a flow rate of 6 GPM and pipe ID of 1.25 inches gives a maximum velocity of only 1.57 feet per second, which is way below the maximum rate specified in the code. Obviously, we do not need a 1.25-inch pipe!

Equation (4-4) only calculates maximum velocity with zero back-pressure on pump outlet (i.e. a ruptured distribution system). Velocity with the pump working against back-pressure in a pressure tank is actually much lower. Furthermore, when we consider the amount of water stored in a long pipe compared to the amount of water we typically use in one day, we begin to see that this supply line water is starting to approach stagnant water velocities that can promote biofilm growth. For example, if we use 50 gallons per day then average daily flow rate through the pipe is 0.035 GPM and Eq. (4-4) gives an average daily velocity of only 0.0091 ft/s or 33 feet per hour.

We want to keep flow rates in pipes high enough to continually strip biofilms off inside surfaces, but not so high as to risk water hammer damage or significantly reduce pressure. Equation (4-4) shows that 6 GPM flow in ¾-inch pipe gives a flow velocity of 4.4 ft/s, which is still well below the 8 ft/s maximum specified in the UPC. Equation (4-3) gives the head loss in feet per 100 feet of pipe length due to flow friction. This equation shows that ¾-inch pipe under 6 GPM of flow would have a maximum head loss of 11 feet per every 100 feet of length, which is equivalent to a maximum pressure loss of 4.8 psi. As the pressure tank fills up and flow slows down, this pressure loss approaches zero. So it will never be noticed at any of the household fixtures. It is only a frictional loss that means the pump runs a few seconds longer to fill the pressure tank. Using 1-inch pipe reduces maximum pressure loss to 1.2 psi per every 100 ft of length and produces a maximum flow velocity of 2.4 ft/s.

So if a plumber or county building inspector tries to tell you that you must install 1.25-inch ID pipe or larger from your tanks to your house, stand your ground and explain to him why those UPC requirements do not apply to your rainwater system. Your pressure pump does not have the flow capacity of a municipal water main and you do not want biofilm buildup in your main supply line! Table 4-3 shows hardware costs for a recent pressurization system installation. Storage tanks were located a few feet from the pressure pump so the above main supply line issue did not apply.

Table 4-3. Pressurization Plumbing Costs.

Aquatec 5513-1E01-B606	FreshWaterSystems.com	1	$151.49
1/2" x 6" sch80 nipple, #56123	Lowe's	1	$0.82
1/2" PVC union fittings, #188211	Lowe's	2	$4.96
3/4" PVC coupling, #23850	Lowe's	2	$0.68
3/4" x 1/2" PVC bushing, #23923	Lowe's	2	$0.96
Amtrol Well-X-Trol 202 20-gal. pressure tank	Amazon.com	1	$209.00
Badger engineered polymer flow meter	Instumart.com	1	$104.00
3/4" sch 80 adapter, #51691	Lowe's	2	$5.56
Pressure gauge, #64124	Lowe's	1	$9.98
1/2" x 1/4" FIP bushing adapter, #748380	Lowe's	1	$4.59
3/4" slip x 1/2" threaded PVC bushing, #23932	Lowe's	1	$0.73
3/4" PVC ball valve, #21482	Lowe's	1	$3.17
1" sch 80 PVC adapter, #52563	Lowe's	1	$4.25
1" x 3/4" PVC sch 40 bushing, #23913	Lowe's	1	$0.80
3/4" PVC Tee, #23874	Lowe's	2	$1.00
3/4" PVC sch 40 pipe, #23971	Lowe's	1	$2.97

The two-stage purification process and dual-tank system shown in this chapter has worked very well for our home for several years. It is the system I recommend for my customers. Like any mechanical system it still requires proper maintenance, but it is far less problematic than our previously used cave water and my neighbors' well water. Rainwater is completely free of hazardous chemicals often found in groundwater today. More importantly, it allows me to have oversight, control, and confidence in supplying clean healthy drinking water to my home rather than blindly trusting someone else to supply it.

I close this chapter with one last very important plumbing issue. Many of my neighbors with much newer and more expensive homes than mine must drip their faucets during hard winter freezes to keep their plumbing from freezing, and sometimes even that does not work. The need to drip faucets indicates poorly insulated plumbing. Builders often cut corners and do not properly insulate house plumbing. We rainwater harvesters cannot afford to drip our faucets in winter! Dripping faucets to prevent freezing can quickly drain your storage tanks. I have never dripped my faucets and my plumbing has never frozen because all water lines are properly insulated, including cold water lines. Make sure your plumbing is properly insulated before installing a rainwater system. If you must, remove drywall and properly insulate distribution plumbing, especially on exterior walls.

Chapter 5.
Electronic Controls, Design and Assembly

   Electronic circuits are ubiquitous in our homes today, controlling and automating systems so that they work as designed, and watching over us to keep us secure from fire, burglary, or other disasters. So it should be no surprise that electronics can be effectively applied to potable rainwater harvesting to automate system processes and improve system reliability. Obvious uses for electronics include water quality monitoring, storage level monitoring, electronic rain gauges, controlling electronic first-flush valves, and automatic disinfectant injection with a peristaltic metering pump. In colder climates an electronic controller might monitor water temperature and sound an alarm or turn on an auxiliary heater before stored water freezes. Or it might trigger an alarm if collection efficiency drops below a certain value.

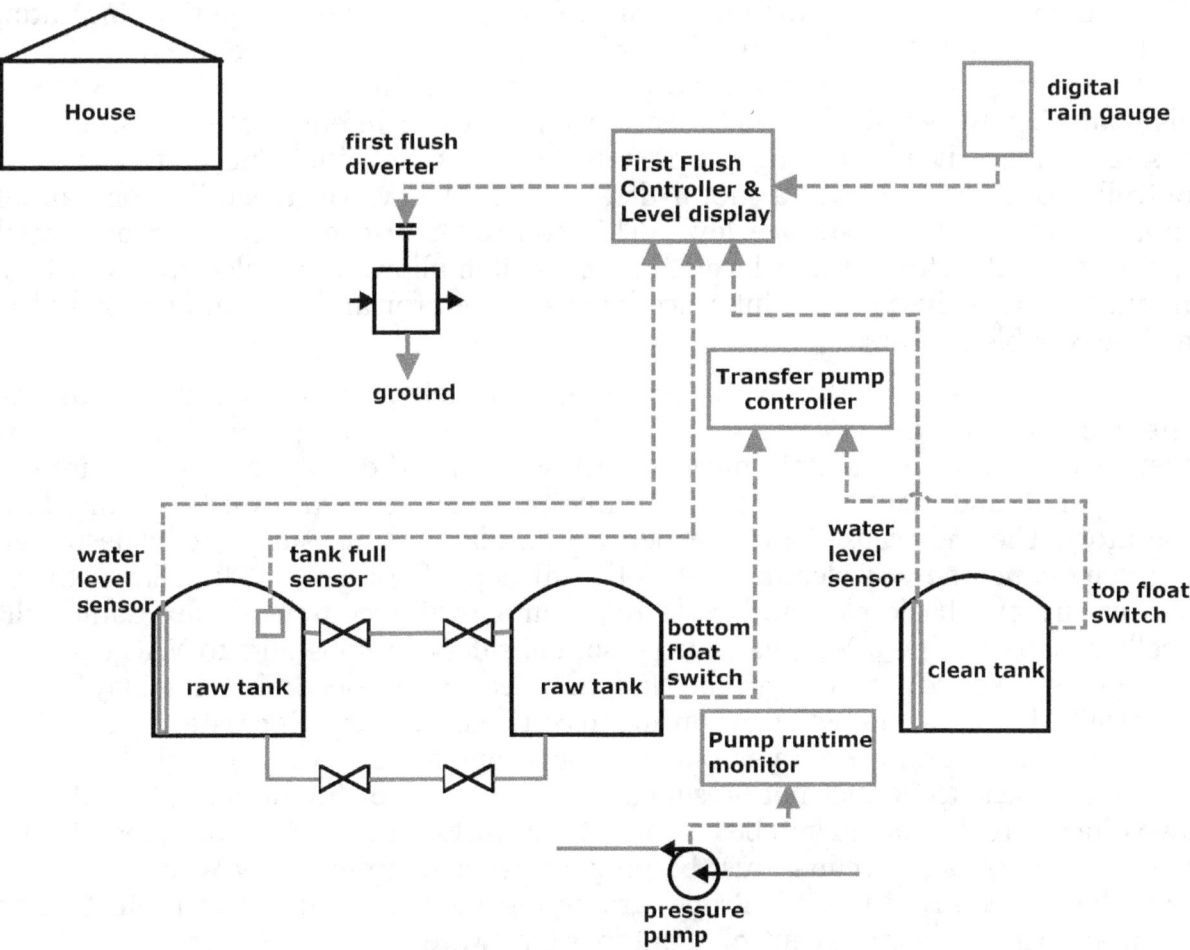

Fig. 5-1. Electronic controllers for rainwater system.

In this chapter we discuss three essential electronic control circuits, illustrated in Fig. 5-1; a transfer pump controller (section 5.1), a pressure pump runtime monitor (section 5.2), and a first-flush valve controller (section 5.3). These three controllers are ordered according to complexity, with the first-flush controller being the most complex. Two hardware components that provide inputs to the first-flush controller are discussed in following sections. These include capacitive tank level sensors (section 5.4) and a digital rain gauge (section 5.5). In section 5.5 we discuss two different types of digital rain gauges; a standard tipping bucket gauge and a capacitive rain gauge. The first has mechanical problems which are not easily solved and the second has electronic problems which can be solved with microcontrollers.

The transfer pump controller automates water transfer from raw water storage to clean water storage, and turns on the chlorine metering pump. This controller receives inputs from two float switches; one switch near the bottom of raw water storage and another near the top of clean water storage. The pressure pump runtime monitor shuts off pump power if it detects the pump runs too long in any 24-hour period, indicating a potential major leak in pressurized plumbing. This circuit has twice saved me from a pressure pump trying to completely drain my storage tanks! We provide a switch on this controller to remove its safety feature for times of intentional long running of pressure pump, such as during maintenance or tank cleaning. The first-flush valve controller opens and closes a motor-driven first-flush valve depending on rainfall amount and raw water storage level. This controller also displays current rainfall amount and total stored water level. A rotary switch allows user selectable first-flush amounts. This is the primary interface between operator and system and should be easily accessible to users.

Before doing any work with 120 VAC electricity, make sure you know the difference between the hot and neutral wires and how to properly wire a 120 VAC receptacle! I once had a PhD physics professor who did not know the difference between hot and neutral wires and wired a receptacle backwards in the laser laboratory. The error caused a very expensive CO_2 laser power supply to be destroyed. Fortunately, no one was electrocuted in the mishap. If you are not familiar with 120 VAC wiring and basic electrical code requirements, I recommend Rex Cauldwell's excellent book Wiring a House.[34] For personnel safety reasons all 120 VAC electrical wiring must be done according to the National Electrical Code. Something may "work" electrically, but if it is not wired according to code requirements it creates a potentially hazardous condition for both personnel and equipment. Low voltage digital wiring is more forgiving since it does not present a potential shock or fire hazard. Nevertheless, low-voltage circuits should still be enclosed to protect them from weather, insects, and rodents. Low voltage circuits must be properly isolated from 120 VAC circuits when using them to control 120 VAC. Proper grounding, shielding, and surge protection are important with low voltage control of outdoor hardware.

---

34 Rex Cauldwell, Wiring a House, The Taunton Press, 2002.

5.1 Transfer Pump Controller

Automating water transfer from raw water tank to clean water tank is a must. I cannot count how many times I accidentally overflowed my clean water tank while manually controlling water transfer because I forgot that the transfer pump was running! (I sometimes had to resort to manual control when the original circuit failed due to lightning.) This control circuit evolved over time and the one presented here is the simplest version that I currently install on customer's systems. Earlier versions were more complex logic circuits resulting from using low-voltage float switches that had little to no hysteresis (opening and closing at different water levels). If you use only a single float switch with no hysteresis then your transfer pump will quickly cycle on and off from water turbulence as water level approaches the switch point. I initially used two such miniature magnetic reed float switches separated vertically by a few inches to provide electronic hysteresis. Hysteresis was built into the electronic circuit rather than the switch. This made the circuit more complex and more susceptible to surge-induced failure.

I later replaced these separated reed switches with float switches containing built-in hysteresis, specifically designed for controlling pumps. These are mechanical float switches on a tether. Tether length is adjusted sto set desired amount of hysteresis. One manufacturer of these float switches is SJE Rhombus (www.sjerhombus.com) in Detroit Lakes, MN. The company makes both mechanically-activated and mercury-activated switches. Never use any mercury switch in your tanks; mercury switches are not NSF-approved for potable water applications! SJE Rhombus manufactures 120 VAC float switches with ANSI/NSF 61 certification. These switches significantly simplify transfer pump control.

Figure 5-2 shows a schematic of our very simple transfer pump controller. Note that all 120 VAC switches are on the hot side of the circuit. Never switch the neutral side of any device (e.g. motor, light, etc.) otherwise you will have a device that is electrically hot when it is turned off, which presents a shock hazard. A three-position toggle switch (on-off-on) is mounted on the controller that allows the operator to select either manual pump control or automatic pump control. This switch is normally in automatic control position so that tank float switches turn the pump off and on. However, there are times when manual control is required, such as system maintenance and adjusting chlorine metering.

The figure shows this 3-position switch in the manual pump-on position. A 120 VAC relay is wired in parallel across the pump motor, which activates the relay every time the pump motor turns on. This provides an electrically isolated switch (sometimes referred to as "dry contacts") to provide a turn-on signal for the chlorine metering pump. The Stenner peristaltic metering pump can be programmed to either run for a fixed time after receiving this turn-on signal or run at a specified speed for the entire time that the turn-on signal is present.

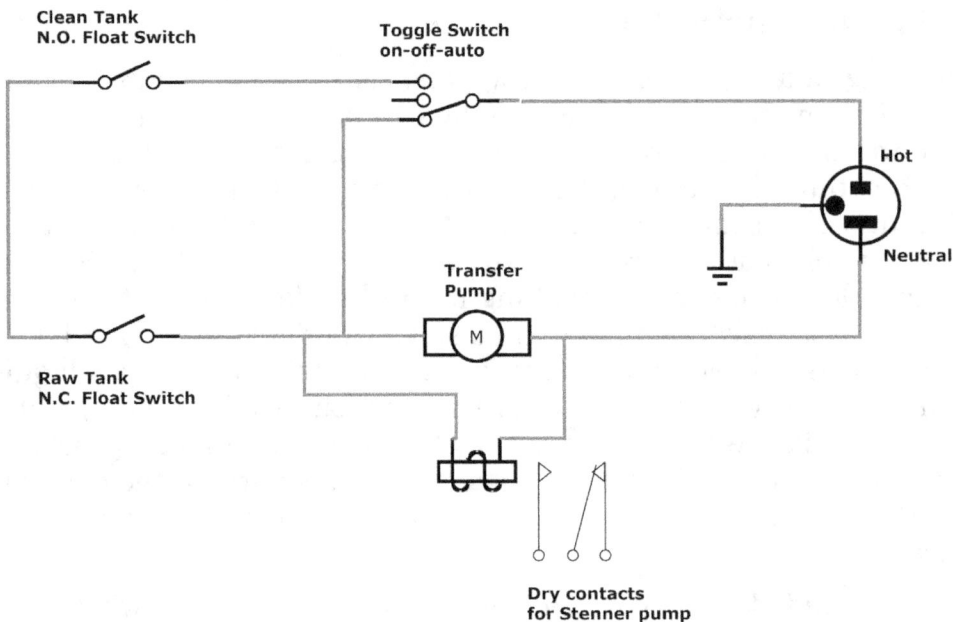

Fig. 5-2. Transfer pump and metering pump control circuit.

When switched to "auto" position, the transfer pump and metering pump are controlled by two float switches wired in series; a normally open (N.O.) switch near top of clean water tank and a normally closed (N.C.) switch near bottom of raw water tank. The clean water tank N.O. float switch closes when clean water level drops to a certain point. We set this turn-on point to be about 100 gallons less than a full tank. This switch opens (turning off pump) when clean water level is restored to completely full. This keeps clean water storage topped off at all times and maximizes available space in raw water storage to receive additional rainfall. The N.C. switch in the bottom of the raw water tank prevents the transfer pump from turning on when raw water level drops to just above the bottom tank outlet. This allows raw water circulation to continue when this tank becomes near empty and stops any further water transfer until new rainfall arrives. This switch is a normally closed (N.C.) switch that opens when water level drops to just above the bottom tank outlet.

The NSF 61 approved float switch we use is the SJE Rhombus Pumpmaster WPS float switch, which can be wired in either N.O. or N.C. configurations. This switch has three wires; red, black, and white. Red and white are used for the N.C. configuration and black and white are used for the N.O. configuration. We should note that SJE Rhombus has defined N.O. and N.C. completely opposite to how we have defined these terms in Fig. 5-2 and in the paragraph above. SJE Rhombus defines a N.C. switch as one that closes when water level drops below a certain point, and a N.O. switch as one that closes when water level rises above a certain point (for a different application than ours). If you use this switch in your system simply wire it in the N.C. configuration in the clean water tank and in the N.O. configuration in the raw water tank.

Figure 5-2 assumes separate circuits for the circulation and transfer pumps. You run two 14-2 cables with ground to the tank, one for the float switch/transfer pump and one for the circulation pump. If one of the tanks is located some distance from the pump house, transfer pump, and controller, you can save on copper by running outdoor 14-3 cable with ground to the remote tank with its primary hot wire (black) tied to both the circulation pump and one side of the float switch. Use the other 14-3 hot wire (red) as a return from the other side of the float switch to be tied in series with the second float switch on the other tank and toggle switch on controller. This keeps all switches on the hot side of transfer pump and eliminates having to run two 14-2 cables through conduit. Again, you never want to have equipment (e.g. pumps, lights, etc.) floating hot when they are turned off, so never put switches on the neutral side of any 120 VAC equipment. Unplugging the transfer pump control circuit for maintenance then de-energizes both transfer pump and circulation pump on the remote tank. Make sure all wiring is done according to electrical code. In addition to using UV-grade outdoor cable, enclose all 120 VAC cable in gray PVC conduit to protect it from gnawing rodents and gardening shovels.

The three-position auto-off-on toggle switch on the transfer pump controller allows turning off the transfer pump without de-energizing the circulation pump. This is useful for lowering clean water levels for tank cleaning or other maintenance operations. The manual "on" position is also useful for adjusting chlorine metering. If the peristaltic metering pump is set for time rather than speed, you can manually adjust chlorine levels up or down with this toggle switch. If clean water chlorine level gets too low you can easily raise it without removing the tank lid by toggling this switch on and off a few times. Then increase chlorine pump metering slightly. If chlorine gets too high, turn off the transfer pump a few days to let clean water drop more than 100 gallons. Then turn it back on. The metering pump will run for its set fixed time, but the transfer pump will run until the clean tank is refilled, thus effectively reducing chlorine level in the water. You then make a slight reduction in chlorine metering on the peristaltic pump.

As previously mentioned, you want to minimize induced 60 Hz electrical noise from 120 VAC circuits onto low-voltage signal circuits. High impedance digital circuits can act like antennas and easily pick up induced 60 Hz noise. For long runs put low-voltage digital circuit cable and 120 VAC cables in separate conduits. Use shielded cable for digital circuits. If you have a noisy AC ground you may wish to provide a separate quieter earth ground near your storage tanks for digital circuits, but avoid ground loops. Do not use two different earth grounds for digital circuits. If the run is short from tanks to transfer pump controller you can probably get away with putting both the 120 VAC and low-voltage shielded signal cable in the same conduit. The 120 VAC cables for float switches will only carry current when the transfer pump is turned on. Therefore, even if it does induce some 60 Hz noise onto your low-voltage cables it does so for only a short period of time when the transfer pump is running.

Table 5-1 shows typical components and costs for building the transfer pump controller. This was for a short run from tanks to controller and both 120 VAC float switch cables and low-voltage sensor cables were included in the same conduit. The transfer pump controller was located within only a few feet of storage tanks. The low-voltage signal cable separates from the 120 VAC cable at the junction box inside the pump house and then runs to the first-flush controller located in the main house. The 1" PVC plugs and hose clamps in the table are used for fabricating adjustable cable clamps for the float switches. These float switches are supported by their own cables. Fill the plug with epoxy and drill a hole through the center of the plug equal to the diameter of the float switch cable. Then saw a slit through the side of the plug to the drilled hole in the center. After threading the float switch cable through a hole in top of tank, thread it through this cable clamp. Then use a hose clamp to compress the cable clamp around the cable. Total hardware cost for this transfer pump controller is about $252.

Table 5-1. Transfer Pump Controller Costs.

4"x4"x4" gray PVC junction box, #10029	Lowe's	1	$9.48
1/2" PVC gray conduit, 10 ft, #72808	Lowe's	2	$4.10
1/2" PVC conduit box adapter, #72855	Lowe's	8	$3.84
1/2" PVC conduit transition, #75783	Lowe's	3	$7.44
1/2" PVC gray conduit Tee connection, #115991	Lowe's	1	$2.48
1/2" PVC gray conduit 90-deg bend, #50916	Lowe's	6	$3.90
1/2" PVC gray conduit coupling, #115930	Lowe's	6	$1.50
16-AWG copper wire, 25', #89173	Lowe's	2	$9.74
SPDT toggle switch, on-off-on, #543137	Lowe's	1	$4.94
AC relay LB2-110A-S-R, #175573	Jameco	1	$5.95
Relay socket PTF08A-A-R, #175531	Jameco	1	$5.09
SJE Pumpmaster WPS float switch, 1047427	SJE Rhombus	2	$189.38
1" PVC plug, #22707	Lowe's	2	$1.80
1 1/2" hose clamps, #80887	Lowe's	2	$1.88

5.2 Pressure Pump Runtime Monitor

Another electronic safety device that should be considered essential for rainwater harvesting is one that will automatically shut off the water pressure pump and sound or display an alarm if pressurized plumbing fails and results in a major water loss. Since the pressure pump automatically turns on with a drop in pressure, a major leak or pipe rupture could potentially drain your storage tanks while you are asleep or away from home. This has always been a concern of mine since implementing rainwater harvesting. Rainwater is very aggressive on metal fittings and leaks can occur without warning. When going on vacation I would usually shut off power to the pressure pump at the breaker panel to prevent it from draining our storage tanks if a major leak occurred.

This problem actually occurred one night due to my own negligence and cost us nearly 2000 gallons of stored water (about 2/3 of total storage). After cleaning the prefilter and shutting off water at the garden hose nozzle, I forgot to turn off the house faucet, leaving the hose pressurized for several days. Pressure eventually blew the hose fitting off the hose sometime during the night and our pressure pump proceeded to empty our storage tanks! (It was an old garden hose that I had repaired a few times.) Fortunately, my wife woke up during the night, hearing the pressure pump in the basement running continuously when no water was being used, and woke me up. I immediately jumped out of bed and turned off the outdoor faucet, stopping the water loss, but not before we had lost about 2000 gallons of water dumped to ground that night.

I could not sleep the rest of the night, both from worry over the costly mistake and trying to figure out how to prevent this problem. My fears were realized and this problem demanded a solution. Rainwater harvesters budget water usage and generally keep it below a certain predetermined level. All pressure pumps have a specific average flow rate (GPM) that is known or can be easily determined. So an electronic sensor that detects when a pressure pump runs too long during any 24-hour period and automatically shuts off pump power could have prevented this water loss. Such a safety device would need to permit manual overriding during rare times of intentional high usage of water, such as tank cleaning or pressure washing. This electronic safety switch could also be used to train teenagers (or adults) that like to take long showers!

This sensor must distinguish between normal short-term high flow water usage (e.g. showers and washing machine) and an abnormal continuous water leak. Normal water usage may occur at either high or low flow rates, but averages out to about 50 gallons per day for our home. So pressure drop or flow rate cannot be used to sense an abnormal leak. Even a low flow continuous leak could result in hundreds or thousands of gallons of water lost in a day. However, in the case of a catastrophic rupture we do not want this sensor waiting for 24 hours before determining that a serious rupture has occurred. Rather, we want it to take action as quickly as possible before we loose too much water, but not sound false alarms during normal short term high flow water usage. It needs to recognize both catastrophic leaks and significant slow leaks, and take action to shut down pump power as quickly as possible.

An electronic flow meter installed between pressure pump and house would make this task easier, but I did not have one on my system. My flow meter is a traditional mechanical flow meter typically used by water utilities. However, pump runtime can be converted to approximate gallons of water pumped knowing the pump's average flow rate. Therefore, a device that accumulates pump runtime every

24-hour period and shuts down pump power if total runtime exceeds a preset maximum number of minutes provides the safety feature we are looking for. After every 24-hour period the runtime counter is automatically reset back to zero if no alarm has been triggered. This runtime monitor displays total number of minutes the pump has run during the current 24-hour period.

In order to relate pump runtime directly to total gallons pumped, you must determine average pump flow rate in gallons per minute (GPM) for both high water usage and low water usage. This is due to the fact that flow rate varies with back pressure, which decreases as a pressure pump recharges a pressure tank. Pressure tank recharging also takes longer if water is being consumed at the same time. My pressure pump is a 0.5 HP cast iron shallow well pump rated for 9 GPM maximum, which could potentially pump 100 gallons in 11 minutes with no back pressure. Obviously, I want pump power shut down in a matter of minutes if a catastrophic rupture occurs. However, a slow continuous leak of only 0.07 GPM will also result in a 100 gallon loss in 24 hours. This slow leak can only be determined by monitoring over a long period of time. My pump is connected to a 52 gallon pressure tank (a bit oversized). Drawdown on this tank is between 14 to 19 gallons, depending on the on/off pressure switch settings. So for a slow 0.07 GPM leak my pump would turn on between 5 and 7 times over a 24 hour period. As the pressure tank is filled, pump flow rate decreases, which causes the pump to run longer than its specified 9 GPM would indicate.

Average pressure pump runtime can be mathematically determined from pump flow rate versus back pressure curves and pressure tank volume versus pressure curves. But an easier way is to simply measure it. This assumes a flow meter has been installed between your pressure tank and house distribution plumbing. Connect an adjustable nozzle to a garden hose and discharge water from an outdoor faucet back into your raw water tank so as not to waste any water. Use the nozzle to simulate various usage flow rates. Record the number of minutes your pressure pump runs to pump your specified maximum daily water allotment, for both high flow and low flow discharge rates. For example, if your maximum daily water usage is 100 gallons then record the total runtime required to move 100 gallons through your flow meter. Total runtime is most easily recorded with the electronic runtime monitor. You can then set maximum allowed runtime on your runtime monitor to this value, or a slightly higher value to minimize false alarms. Your measured average pump flow rate under normal usage then equals 100 gallons divided by total runtime.

Once you determine normal usage average flow rate (GPM) for your pressure pump and maximum pump runtime before power is shut off, calculate your maximum water loss under a catastrophic rupture with zero back pressure on pump. Maximum potential water loss equals maximum allowed runtime times maximum flow rate of pump. Decide if this is an acceptable loss for this low-probability event. If it is not then reduce maximum allowed runtime (which will increase probability of false alarms) to reduce maximum water loss during a catastrophic event. For example, say normal-usage average flow rate of my 9 GPM pump is actually 5 GPM and I set the maximum allowed water usage in any 24-hour period to be 100 gallons. Then maximum allowed total runtime in any 24-hour period equals 100 gal/5 GPM = 20 minutes. Maximum water loss during a catastrophic pipe rupture before pump power shuts down equals 9 GPM × 20 min = 180 gallons. I decide this can be tolerated, so I use 20 minutes as my maximum allowed total runtime programmed into the runtime monitor. This can always be adjusted up or down later if I find it to be non-optimal.

Fig. 5-3. Inside and outside of pressure pump runtime monitor.

My original pressure pump runtime monitor, shown in the first edition of this book, was a custom-built digital circuit of discrete ICs on a custom PC board. This circuit is still in operation on my system. We also showed a software implementation of a runtime monitor in that first edition using a BASIC Stamp microcontroller. However, subsequent developments centered on the Arduino microcontrollers due to their popularity. These developments demonstrated a huge cost savings over my original custom circuit. So this Arduino-based runtime monitor is the one we show here in this second edition.

Figure 5-3 shows my Arduino-controlled pump runtime monitor. An Arduino Leonardo is mounted to a Sparkfun solderable breadboard (PRT-12070) with 1/2" nylon standoffs. A 2-digit LED display is soldered to the bottom of this breadboard and is visible through a rectangular hole cut in the lid of a 6"x6"x4" gray PVC junction box. The pump-on sense relay (shown in figure beside lid) is mounted separately inside the box. Its coil is tied to the built-in pressure switch on the pump so that when the pressure switch contacts are open this relay coil is energized. This provides a dry contact signal to the Arduino to indicate when the pump is running or turned off.

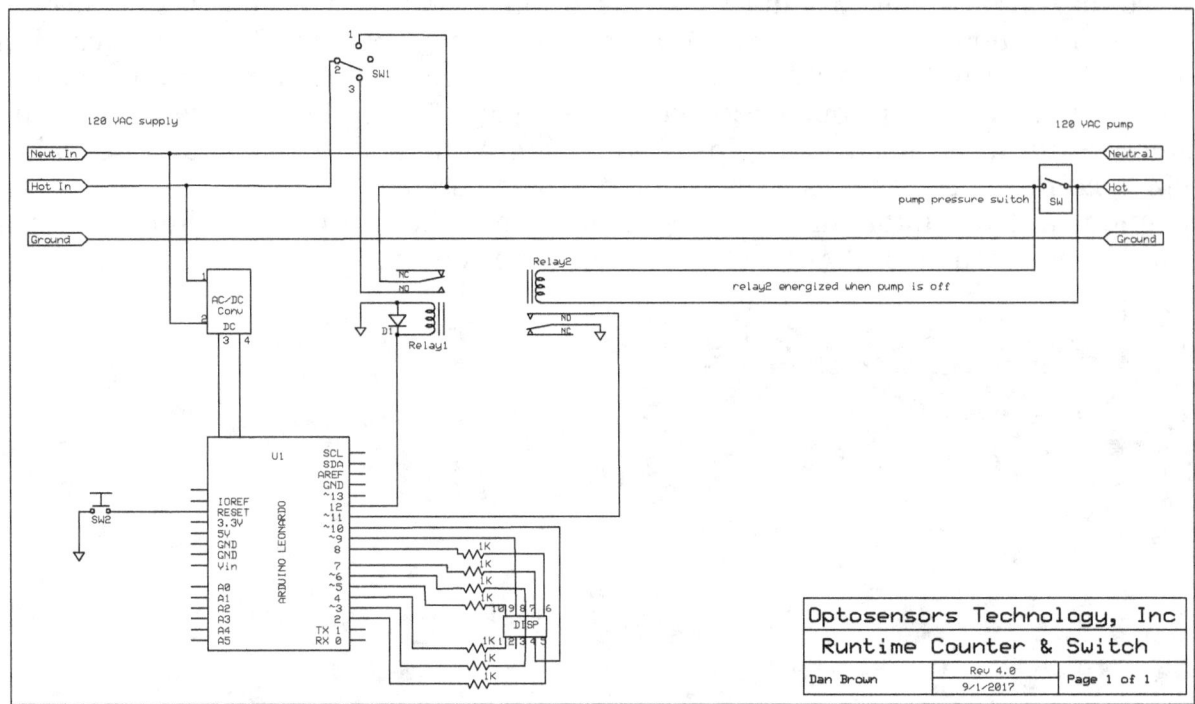

Fig. 5-4. Electronic schematic of pump runtime monitor.

The power relay that cuts power to the pump is soldered to the top of the Sparkfun breadboard near the 3-position toggle switch. Sparkfun technical staff said the traces on this prototyping breadboard are 0.032" wide with a thickness of 1 oz/ft^2, which should handle a maximum current of 2 amps. The maximum current draw of the 120VAC pressure pump we planned to use with this controller is 1.75 amps. But just to be safe I soldered a short piece of wire between the connection points on the underside of the breadboard so that the trace is not carrying the full current load.

A 3-position SPDT toggle switch (Lowes #543137) allows the user to either; 1) bypass this controller and turn on pump power, 2) turn off pump power without shutting down power at the breaker panel, or 3) let the runtime monitor automatically shut off pump power if an alarm condition is triggered. Normally this switch stays in the AUTO position. But there are special maintenance conditions when the other two positions are required. A pushbutton switch allows resetting the Arduino if an alarm condition is triggered. The Arduino sketch keeps the alarm condition set (flashing display and no power to pump) until the user manually resets this controller. If the user subsequently puts the toggle switch in ON position without pushing the reset button, the Arduino will continue counting and displaying total runtime while flashing the display. This feature allows a user to experimentally determine optimal maximum runtime before triggering an alarm condition.

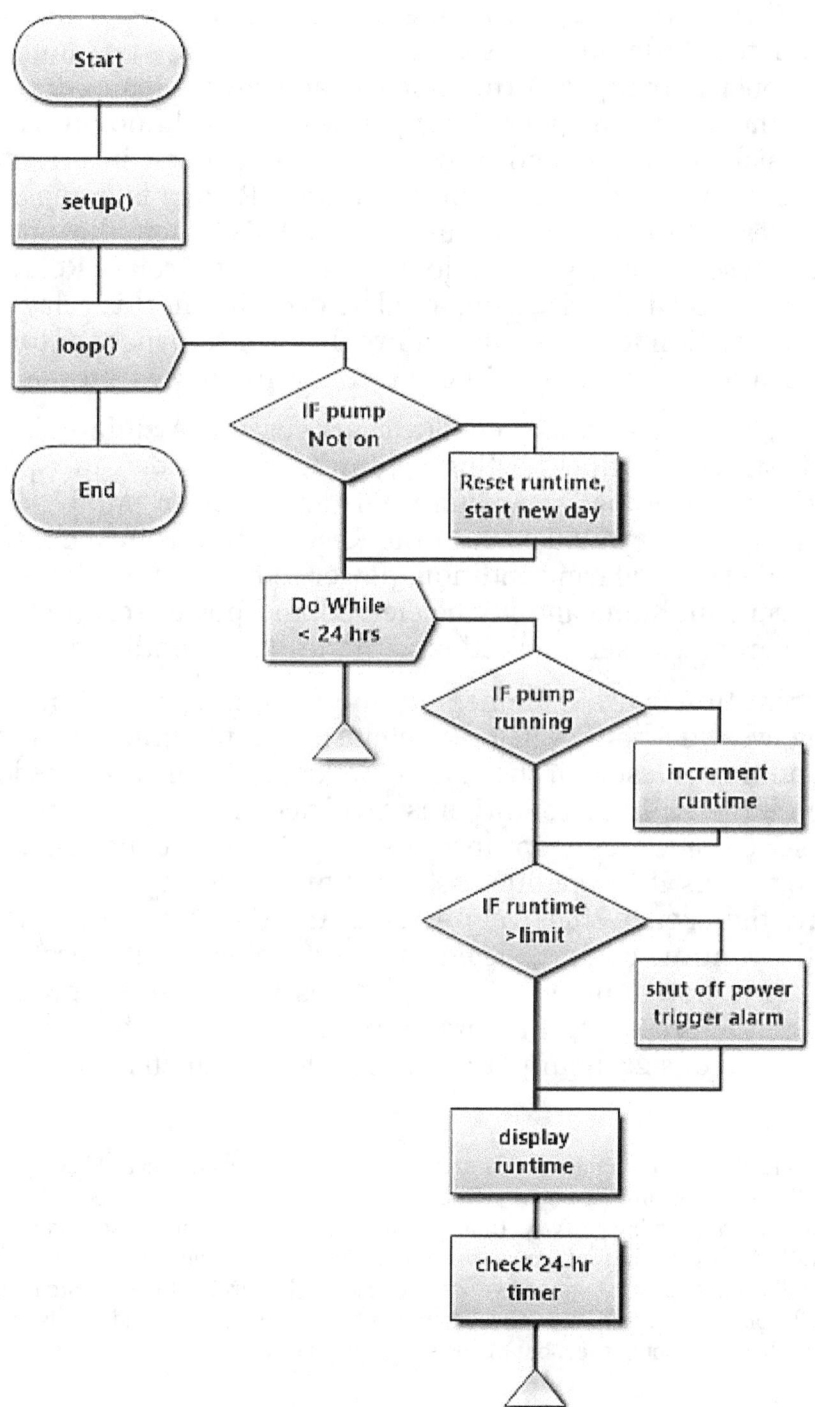

Fig. 5-5. Structured flowchart of pump runtime monitor.

Figure 5-4 shows the electronic schematic for this pump runtime monitor. We use a low-cost AC relay (Relay2) across the pump's built-in pressure switch to isolate AC voltage from the Arduino and signal when the pump is running.[35] When the pressure switch opens (pump off) this relay is activated, closing the normally open (NO) contact on the relay. This pulls the signal wire on Arduino pin 11 to ground. The relay coil has a high resistance and draws very little current in this state. When the pump pressure switch closes, turning on the pump, Relay2 is de-energized. Arduino pin 11, which uses internal resistor pullup, is HIGH when the pump is running. Arduino pin 12 is used to energize the low-voltage power relay, Relay1. If maximum allowed runtime is exceeded, triggering an alarm condition, this relay turns off pump power. The three-position toggle switch allows the user to bypass automatic operation for maintenance or testing purposes. Pins 2 through 10 power runtime display.

Figure 5-5 shows the structured flowchart for the Arduino sketch. If you are unfamiliar with structured flowcharting a tutorial is provided in Appendix A. After running setup() once to initialize Arduino pins and turn on pump power, the sketch enters loop() that runs continuously until the Reset button is pushed. If the loop turns off pump power, due to an alarm condition, power remains off until the Reset button is pushed. The alarm condition can only be cleared and power restored by resetting the program. This guarantees user interaction with the alarm condition.

Two different timers are incremented in loop(); a 24-hour timer and a runtime counter. During every 24-hour cycle the sketch runs in the inner Do-While loop where the 24-hour condition is tested at the end of the loop. This insures the loop executes at least once, even if the 24-hour condition is exceeded. The 24-hour timer and runtime counter are reset outside this loop, inside an IF statement that checks to see if the pump is running. Thus if the pump is still running when the 24-hour timer crosses into a new day, the sketch effectively extends the day beyond 24 hours and keeps incrementing the runtime counter. The runtime counter is only reset (after 24 hours) only if the pump is not running. This eliminates the potential problem of a pump running for twice the preset limit before power is shut off if the pump happens to be running near the end of a 24-hour period. The Arduino sketch is provided below.

/*

Records and displays total runtime (in minutes) of pressure pump and shuts down 120 VAC power to pump if runtime exceeds 20 minutes within a 24-hour period. Pump on/off status is monitored with a 120 VAC relay across pressure switch, isolating 120VAC from low voltage Arduino inputs. A second power relay, activated with 5 VDC across coil, shuts off power to pump if limit is exceeded. When limit is exceeded displayed runtime will begin flashing. Pump power can be restored by turning toggle switch on panel to "on," but display will continue to flash. The only way to remove flashing alarm condition is to press Reset button. If Reset button is not pushed, but toggle switch is placed in "on" position, total

---

35 Electrical engineers may notice that this is not the best way to wire the relay. However, due to this particular pump's construction it was not possible to access the motor's hot and neutral terminals without voiding pump warranty. Relay coil resistance is 3600 ohms.

displayed runtime will continue to increase beyond the 20-minute limit, but it will continue to flash. If 20 minutes is not appropriate for your system, change the value for "maxMinutes" in first line of program then recompile and upload to the Arduino.

Wire Connections:
Arduino
12 ------------------- coil of 5V power relay (Omron G6DN)
11 ------------------- NO contact of 110VAC relay (Jameco LB2-110A-S-R)
10 ------------------- display anode pin 4 (Lite-On LTD-4608JR)
9 -------------------- display anode pin 9
8 ------- 1K res ----- display pin 6
7 ------- 1K res ----- display pin 7
6 ------- 1K res ----- display pin 8
5 ------- 1K res ----- display pin 10
4 ------- 1K res ----- display pin 1
3 ------- 1K res ----- display pin 3
2 ------- 1K res ----- display pin 5
Reset ---------------- pushbutton switch

by Dan Brown, September 23, 2017.
*/

```
const unsigned long maxMinutes = 20;  // max allowed runtime in minutes
const int tripPin = 12;
const int pumpPin = 11;
const int ones = 10;   // ones display digit pin
const int tens = 9;    // tens display digit pin
const byte numeral[10] = {B0111111,   //0  segments gfedcba
                          B0000110,   //1
                          B1011011,   //2
                          B1001111,   //3
                          B1100110,   //4
                          B1101101,   //5
                          B1111101,   //6
                          B0000111,   //7
                          B1111111,   //8
                          B1101111 }; //9
const int segPin[7] = {5,6,4,2,3,8,7};  // defines pin numbers for each segment, abcdefg
unsigned long oldTime;     // time at previous execution of loop
unsigned long newTime;     // time at current execution of loop
unsigned long timeInc;     // length of time to execute loop
unsigned long runtime;     // total length of time pump has run in current 24 hr period
unsigned long limit;       // maximum allowed runtime before shutting down power
unsigned long startTime;   // starting time of current 24 hr period
int pumpOn = LOW;          // low when pump is off, high when pump is on
boolean overLimit = false; // becomes true if runtime exceeds limit

void setup()
{
  limit = maxMinutes * 60000;  // runtime limit in milliseconds
  pinMode(tripPin, OUTPUT);    // controls power relay
```

```
  digitalWrite(tripPin, HIGH); // turn on power to pump
  pinMode(pumpPin, INPUT_PULLUP);  // pump on/off sense
  for(int i=0; i<9; i++)        // initialize all display pins
  {
     pinMode(i+2, OUTPUT);
     digitalWrite(i+2, LOW);
  }
  digitalWrite(ones, HIGH);  // test all segments both digits
  delay(1000);
  digitalWrite(ones, LOW);
  digitalWrite(tens, HIGH);
  delay(1000);
  digitalWrite(tens, LOW);
}

void loop()
{
  if (pumpOn == LOW)      // reset runtime and start new day only if pump not running
  {                   // otherwise extend old day and keep adding to old runtime
     startTime = millis(); // start of new 24 hour period
     runtime = 0;        // reset runtime to zero
     oldTime = startTime;  // initialize variable assuming pump is off in first pass through loop
     timeInc = 0;        // becomes nonzero after first pass through loop
  }
  do  // execute this loop at least once even if beyond 24 hours
  {
     pumpOn = digitalRead(pumpPin);  // determine if pump is running
     if (pumpOn == HIGH)
     {
        runtime = runtime + timeInc;
     }
     if (runtime > limit)
     {
        // shut off pump power and flash runtime in display
        overLimit = true;   // alarm condition can only be made false by pushing RESET
        digitalWrite( tripPin, LOW );
     }
     if (overLimit) // flash display runtime
     {
        for(int i=0; i<10; i++) {displayTime(byte(runtime/60000));}
        delay(500);
     }
     else
     {
        displayTime(byte(runtime/60000)); // convert to minutes and continuous display
     }
     newTime = millis();
     timeInc = newTime - oldTime; // time to execute loop and increment runtime
     oldTime = newTime;         // set previous time to current time for next pass through loop
  } while ((newTime - startTime) < 86400000);   // while less than 24 hours continue looping
```

```
}
void displayTime(byte num)
{
  byte tensDigit;
  byte onesDigit;
  tensDigit = num / 10;
  onesDigit = num % 10;
  for(int seg=0; seg<7; seg++)
  {
    if (bitRead(numeral[tensDigit],seg)==1)
    {
      digitalWrite(segPin[seg], LOW);  // turns on segment
    }
    else
    {
      digitalWrite(segPin[seg], HIGH);  // turns off segment
    }
  }
  digitalWrite(tens, HIGH);
  delay(10);
  digitalWrite(tens, LOW);
  for(int seg=0; seg<7; seg++)
  {
    if (bitRead(numeral[onesDigit],seg)==1)
    {
      digitalWrite(segPin[seg], LOW);  // turns on segment
    }
    else
    {
      digitalWrite(segPin[seg], HIGH);  // turns off segment
    }
  }
  digitalWrite(ones, HIGH);
  delay(10);
  digitalWrite(ones, LOW);
}
```

Table 5-2 show the components and hardware costs to build this pressure pump runtime monitor and controller. Total hardware cost is about $75.

Table 5-2. Pressure pump runtime monitor hardware costs.

6"x6"x4" gray PVC junction box, #10030	Lowe's	1	$11.98
SPDT toggle switch, #543137	Lowe's	1	$4.94
Arduino Leonardo, #2163778	Jameco.com	1	$20.49
DPDT relay, LB2-110A-S-R, #175573	Jameco.com	1	$5.95
Relay socket, PTF08A-A-R, #175531	Jameco.com	1	$5.09
Solderable breadboard, PRT-12070	Sparkfun.com	1	$4.95
9VDC wall adapter, TOL-00298	Sparkfun.com	1	$5.95

USB micro-B cable, CAB-10215	Sparkfun.com	1	$4.95
5V power relay, Z5439-ND	Digikey.com	1	$1.79
2 digit 7-seg display, 160-1538-5-ND	Digikey.com	1	$2.10
1K resistor, CF14JT470RCT-ND	Digikey.com	7	$0.70
Push-button switch, EG2025-ND	Digikey.com	1	$1.15
6-ft extension cord, #70291	Lowe's	1	$1.57
4-40 nylon screw 3/8" long, 36-9328-ND	Digikey.com	4	$0.56
4-40 nylon screw 1/4" long, 36-9427-ND	Digikey.com	5	$0.70
4-40 nylon hex standoff, 36-1902C-ND	Digikey.com	3	$2.10
4-40 nylon nut, 36-9605-ND	Digikey.com	3	$0.51

5.3 First Flush Controller

Figure 5-6 shows a picture of the front panel of our electronic first flush valve controller and inside circuit boards. Swiches and PC boards are easily mounted to the lid of a large gray PVC junction box. An Arduino microcontroller is at the heart of this controller.

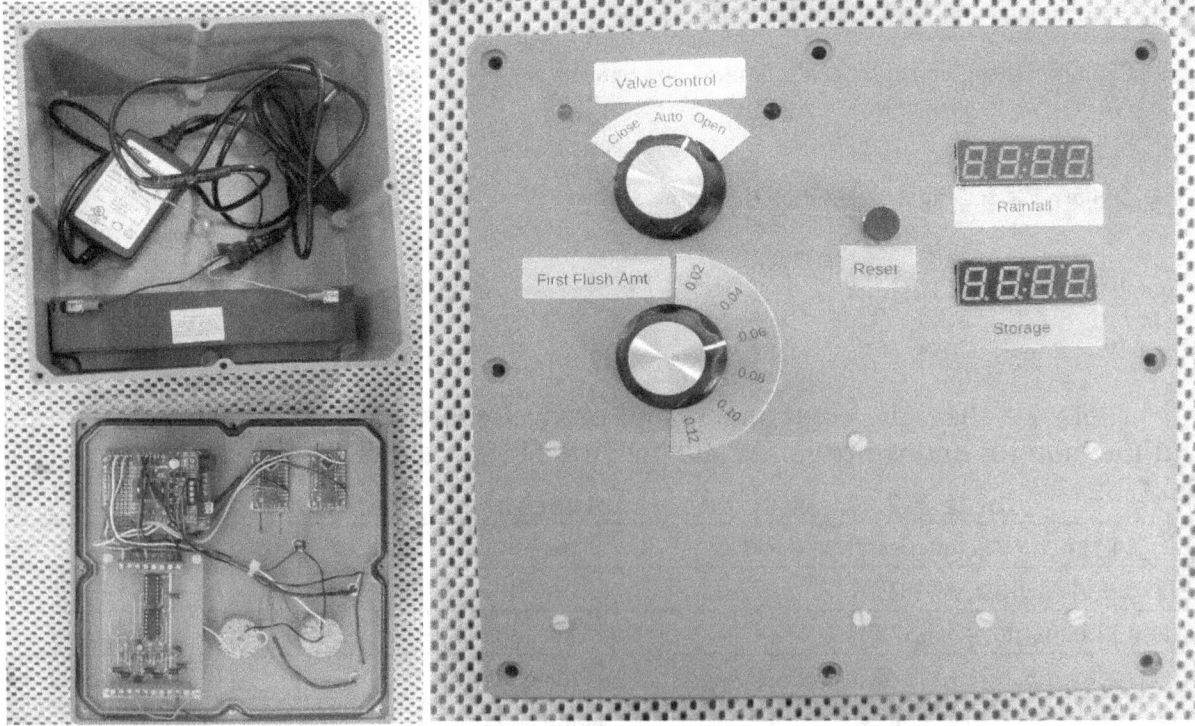

Fig. 5-6. Inside and outside of first flush controller.

The first flush controller primarily controls opening and closing of the motor-driven first-flush valve. This controller closes the first-flush valve (directing rainwater into raw water storage) when two conditions are met: 1.) a specified amount of initial rainfall has been dumped to ground to wash roof AND 2.) raw water storage is not full. Otherwise the valve remains open and rainwater is dumped to ground if either one of these conditions is false. A 6-position rotary switch allows the operator to select six different first-flush amounts; 0.02, 0.04, 0.06, 0.08, 0.10, and 0.12 inches of rainfall. This set of first-flush amounts are fixed in software and can be easily changed. Simply change a constant in the code, recompile and upload to the Arduino. A user-selectable first-flush amount allows adapting the system to seasonal water quality changes (pollen or leaf matter) and to climate changes (drought or heavy rain). For example, during heavy spring rains and heavy pollen season I set the first-flush amount to 0.08 or higher. As summer approaches, pollen disappears, and rainfalls become lighter, I move the first-flush amount to lower settings.

A 3-position rotary switch allows the operator to manually open or close the valve, or place it under automatic control by the Arduino. Normally this switch stays in the "auto" position. The open position is used for occasional maintenance purposes. The closed position is selected during severe droughts when I wish to bypass first-flushing altogether and collect every bit of rainfall. This controller also displays total current rainfall amount and total stored water level on 4-digit LED displays. Pressing the reset button (after recording data) restores rainfall amount to zero.

Figure 5-7 shows an electronic schematic for this controller. All circuit boards in this controller are low-cost off-the-shelf PC boards, except for one, PCB3. This custom interface board provides surge protection for the Arduino and necessary low-pass filtering of input sensor signals. Since we use high frequency capacitive water level sensors and the Arduino is capable of only measuring one high frequency at a time, this interface board also provides a 2×1 multiplexer (MUX) to select measuring raw water or clean water levels. After measuring both frequencies and converting them to gallons of water, the two storage levels are added together and displayed on front panel of controller.

Sparkfun.com's Ardumoto shield (PCB2) allows driving two linear actuators. Generally we only need to drive one for the first flush valve. However, if roof area or maximum rainfall rate requires it, this shield can drive two first flush valves. This board stacks on top of the Arduino after soldering header pins onto it. Default pins required by this shield are pins 2, 3, 4, and 11. Alternate pins 7, 8, 9, and 10 can be used if certain traces are cut and bridges are soldered. But Fig. 5-7 assumes the default pin assignment. Pin 7 is an interruptible pin which we use for input from the digital rain gauge. The rainfall counter is incremented with each pulse input on pin 7 from a digital rain gauge. Rainfall is displayed on front panel with resolution in hundredths of an inch.

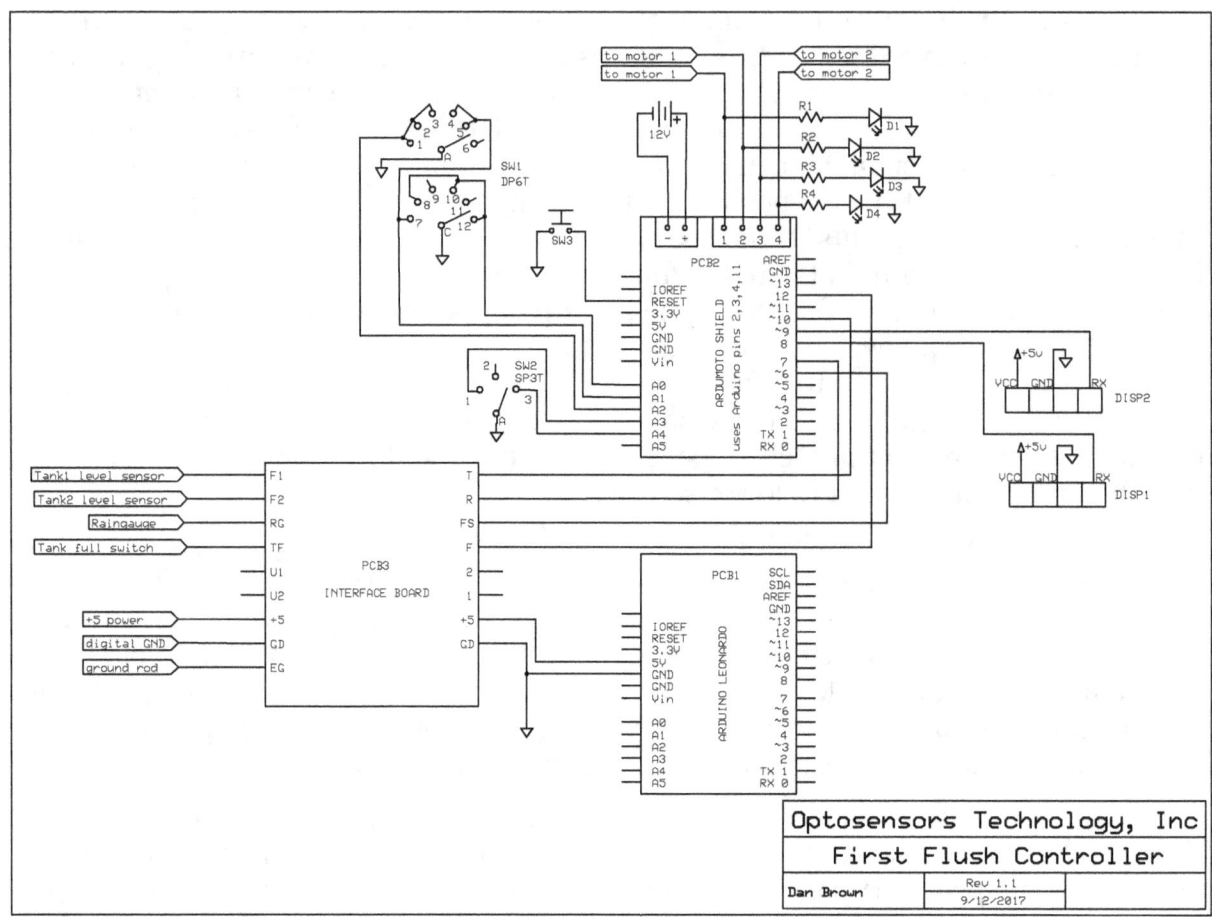

Fig. 5-7. Schematic of first flush controller.

If a tipping bucket style rain gauge is used, it should be placed on an interruptible pin since its signal generally occurs infrequently. Pins that allow inputs to an interrupt service routing (ISR) on an Arduino Leonardo include pins 0, 1, 2, 3, and 7. We avoid using pins 0 and 1 since we often use a serial monitor for program debugging. Pins 2 and 3 are occupied by the Ardumoto shield. That leaves pin 7 for the rain gauge. We use pin 6 as an output frequency select (FS) to the MUX to select between one of two frequencies from tank level sensors. Only one pin on the Leonardo allows direct access to a timer register or T1 pin on the ATmega32u4 for high frequency input, and that is pin 12. So this pin is tied to frequency out from the MUX. Pins 8 and 9 are serial output pins for the two displays. Pin 10 is an input pin for the tank full sensor, a small magnetic reed float switch at top of raw water tank. I originally implemented this controller with an Arduino Uno and different pin assignments were required for that board. I later chose to use the Leonardo due to its lower cost.

If a capacitive type rain gauge is used, as discussed in section 5.5.2, then the simple 2×1 MUX will need to be replaced with a single-chip MUX that provides addition inputs, such as an SN54HC151. In this case two Arduino pins are needed to switch the MUX and select up to one of four frequencies for reading. We can use pins 5 and 6 for this, reserving interruptible pin 7 for other hardware.

Five analog pins are used as digital input pins for the two rotary switches. These pins are pulled HIGH (or 1) with internal pullup resistors. So connecting pins to ground makes them LOW or 0. Two pins are needed to sense between one of three switch positions on the SP3T rotary switch (SW2). A3 and A4 are used by the valve control switch (open, close, auto) to read three binary numbers: 01, 10, and 11. Green and red LEDs on the motor output pins change with position of this switch. The DP6T six position switch (SW1), used to select first-flush amount, requires three digital pins to read six different binary numbers: 001, 010, 011, 100, 101, and 110. This switch, wired as shown in Fig. 5-7, provides these six different binary numbers to pins A0, A1, and A2 on the Arduino. Push-button switch (SW3) is used to reset the Arduino, zero out rainfall display, and open the first-flush valve. Since a power failure during rainfall can reset the Arduino and cause the first-flush valve to open (which is undesirable), a battery backup is required on this controller to prevent that from happening.

Figure 5-8 shows the structured flowchart for the Arduino program or "sketch" as commonly referred to. After setting up pin assignments, displays, and an interrupt service routine, the program enters into a continuous loop of reading and displaying water storage levels, reading rotary switch positions, reading and displaying rainfall amount, reading tank full float switch, and opening or closing the first-flush valve. A boolean variable, firstFlushMet, is initially set to false after a program reset. This variable becomes true when recorded rainfall equals or exceeds first-flush amount setting on the DP6T switch. The ISR sets this variable and the main loop reads it. Once set to true, firstFlushMet remains true until a program reset is executed. This allows rainfall to be collected over multiple days without having to repeat another first-flush cycle. With the valve control switch in "auto" position, and the reset button has not been pushed, the valve will open when the tank-full switch is activated and re-close when raw water drops (from household use) and deactivates this switch. So the reset button need not be pushed until rainfall has completely ended and user wishes to zero out rainfall display and repeat a new first-flush cycle. The tank-full switch and firstFlushMet variable are ignored if the valve control switch is in either "close" or "open" positions. These two positions are used for drought conditions and system maintenance.

Normally the first-flush amount setting is not changed during a rainfall event. However, since this switch is continually read in the main loop it is possible to change its setting during a rainfall if firstFlushMet has not yet been set to true. After the ISR changes this variable, subsequent changes to first-flush setting are ignored until the reset button is pushed. So if rainfall is lighter than expected and collection is essential a user can reduce first-flush amount and close the valve earlier. Alternatively, a user can ignore first-flush amount altogether and simply manually close or open the first-flush valve with the valve control mode switch. The Arduino sketch for this first-flush controller is provided below.

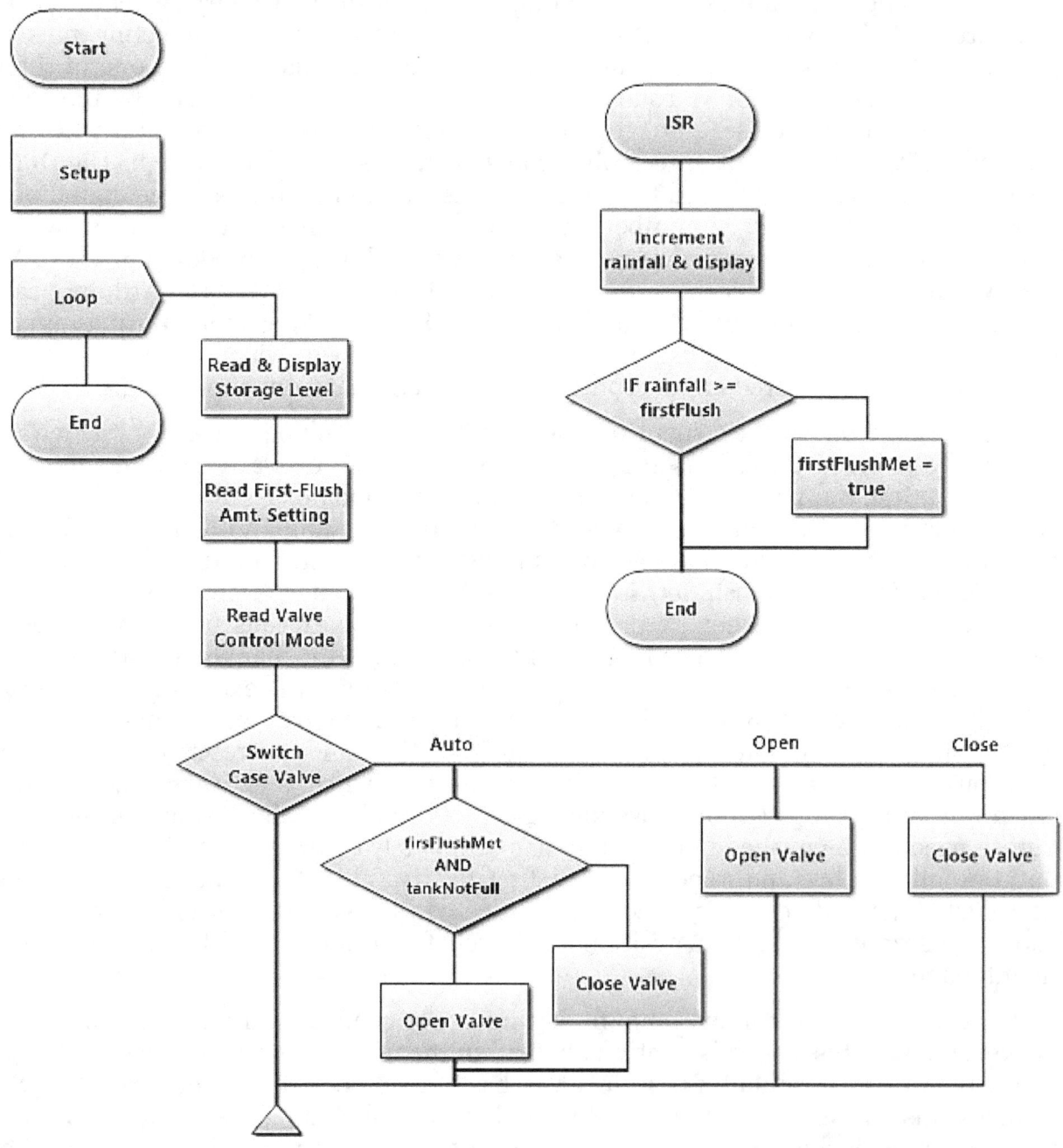

Fig. 5-8. Algorithm for electronic first-flush diverter using an Arduino.

/*

This sketch is written for Arduino Leonardo. Some pin assignments will be different on Uno or other boards.

Reads tipping bucket rain gauge and display inches rain on 4-digit serial 7-Segment Display 1.

Rain gauge has two-wire reed switch. Schmidt trigger debounce circuit essential with high speed microcontrollers. Closing of raingauge switch pulls "debounce in" to ground. Debounce circuit out goes to Arduino.

Debounce circuit with larger RC time constant used with float switch (tank full sensor). Read storage tank levels via custom capacitive water level sensors. Output frequency inversely proportional to water level.

Display total gallons on 4-digit serial 7-Segment Display 2 (Displays: COM-11441 from Sparkfun Electronics)

Control motor-driven first flush valve based on rain gauge and tank full sensor. Use Ardumoto shield from Sparkfun to drive motors. Pins D2, D3, D4, D11 used by Ardumoto board (DEV 14129).

Two power leads from Ardumoto shield to 12 VDC linear actuator with limit switches. Interface circuit contains debounce, MUX, and surge protectors.

Wire Connections:
Arduino/Ardumoto --- Interface Board
5V ----------------------------- +5
GND ---------------------------- GD
D12 ---------------------------- F
D6 ----------------------------- FS
D7 ----------------------------- R
D10 ---------------------------- T
Interface Board --------- Sensors
F1 ----------------------------- F of level sensor1
F2 ----------------------------- F of level sensor 2
GD ----------------------------- G of level sensors 1 and 2
+5 ----------------------------- +V of level sensors 1 and 2
RG ----------------------------- Rain gauge reed switch wire 1
GD ----------------------------- Rain gauge reed switch wire 2
TF ----------------------------- Tank full switch wire 1
GD ----------------------------- Tank full switch wire 2
EG ----------------------------- Earth grounding rod
Arduino/Ardumoto -------7 Segment Serial Displays (COM-11441)
5V ----------------------------- VCC of displays 1 and 2
GND ---------------------------- GND of displays 1 and 2
D8 ----------------------------- RX of display 1
D9 ----------------------------- RX of display 2
Arduino/Ardumoto ------- First flush selector switch
A0 ----------------------------- switch pins 8, 10, 12
A1 ----------------------------- switch pins 4, 5, 7
A2 ----------------------------- switch pins 1, 2, 3
Arduino/Ardumoto ------- Valve control switch
A3 ----------------------------- switch pin 1
A4 ----------------------------- switch pin 3
Arduino/Ardumoto ------- Pushbutton switch (NO)
 RESET ------------------------- pin 2, pin 1 tied to ground

2-pole 6-throw (DP6T) rotary switch on panel wired to provide numbers 1, 2, 3, 4, 5, 6 on pins A0, A1, A2 for first flush amount selection. 2-pole 3-throw (DP3T) rotary switch on panel wired to provide numbers 0, 1, 2 on pins A3 and A4 for automatic or manual valve control. In automatic mode first flush valve is closed IF (raincount > firstflush) AND (raw_tank_not_full), otherwised valve is open. Reset button returns rainfall display to zero and opens first flush valve. 7-Seg1 displays total rainfall in inches 7-Seg2 displays total storage amount in gallons

by: Dan Brown
date: December 29, 2016 revised Nov 21, 2017
to be published in first revision of "Modern Potable Rainwater Harvesting."
*/

```
#include <SoftwareSerial.h>
// first flush amount selection is minFlush times 1, 2, 3, 4, 5, 6 in hundredths of inch
// change minFlush value and recompile if different values are required
const byte minFlush = 2;      // minimum amount in 0.01 inches to dump to ground
const float C11 = 5.845E-11;   // C1 capacitance for level sensor 1
const float C12 = 3.405E-11;   // C2 capacitance per foot for level sensor 1
const float RT1 = 6.6E+5;     // resistance total for level sensor 1
const float C21 = 3.197E-10;   // C1 capacitance for level sensor 2
const float C22 = 2.99E-11;   // C2 capacitance per foot for level sensor 2
const float RT2 = 6.6E+5;     // resistance total for level sensor 2
const float k1 = 26.515;      // gallons per foot for level sensor 1
const float k2 = 51.96;       // gallons per foot for level sensor 2
const byte T1Pin = 12;        // frequency input pin to Timer/Counter 1.
const byte rainTx = 8;        // Arduino transmit pin 8 to rainfall display 1
const byte storeTx = 9;       // Arduino transmit pin 9 to storage level display 2
const byte unusedRx = 13;     // not used except for initialization
const byte tankSel = 6;       // tank select pin output to MUX
const byte raingaugePin = 7;  // raingauge on interruptible pin 7
const byte floatswitchPin = 10; // tank full float switch on pin 10
const byte openvalve = 0;     // opens first flush valve, exchange if motor wires reversed
const byte closevalve = 1;    // closes first flush valve
const byte PWMA = 3;          // speed control motor A on pin 3
const byte PWMB = 11;         // speed control motor B on pin 11
const byte DIRA = 2;          // direction control motor A on pin 2
const byte DIRB = 4;          // direction control motor B on pin 4
const byte normal = 3;        // automatically controlled first flush valve
const byte opened = 1;        // manually opened first flush valve
const byte closed = 2;        // manually closed first flush valve
char storString[6];           // string to write storage level to display
char rainString[6];           // string to write rainfall amount to display
unsigned int firstflush;      // rainfall amount (hundredths of inch) to dump to ground
boolean firstFlushMet = false; // true when raincount > firstflush, set to false on reset
volatile int raincount = 0;   // number of bucket tips, accessed only by ISR after setup
unsigned long freq;           // frequency from tank level sensors
unsigned int level1;          // water level in tank 1 in gallons
unsigned int level2;          // water level in tank 2 in gallons
volatile boolean countDone;   // flag for ending getfrequency
volatile unsigned long timer1OVF; // overflow count on timer1

SoftwareSerial RainfallS7S(unusedRx, rainTx);   // declare procedure names for displays
SoftwareSerial StorageS7S(unusedRx, storeTx);

unsigned long getFrequency ()
```

```
// Use timer1 to count pulses and timer3 to set gate time
{
 countDone = false;
 timer1OVF = 0;
 cli();  // stop interrupts to get things set up
 // temporarily stop timer 0 interrupts
 byte oldTCCR0A = TCCR0A;
 byte oldTCCR0B = TCCR0B;
 TCCR0A = 0;
 TCCR0B = 0;
 // reset timer/counters 1 and 3
 TCCR1B = 0;  // stop timer1
 TCCR1A = 0;  // disable PWM to use all 16 bits
 TIMSK1 = (1 << TOIE1);  // enable overflow interrupt on timer1
 TCCR3B = 0;  // stop timer3
 TCCR3A = 0;  // disable PWM to use all 16 bits
 OCR3A = 15624;  // set timer3 output compare register to value for calling interrupt
 // timer/counter3 counts 15625 (64 us) pulses to get 1 second of time,
 //  use 31250 for 2 sec, 46875 for 3 sec, 62500 for 4 sec
 TIMSK3 = (1 << OCIE3A);  // enable output compare interrupt on timer3 when OCR3A value is hit;
 // zero counts just before starting counters
 TCNT1 = 0;  // zero count on timer1
 TCNT3 = 0;  // zero count on timer3
 // loading TCCRxB registers starts clocks
 TCCR3B = (1 << CS32) | (1 << CS30);  // prescale timer3 to clock on 16 MHz system clock divided by 1024
 TCCR1B = (1 << CS12) | (1 << CS11) | (1 << CS10);  // set timer1 to clock on T1 pin (D12), rising edge
 sei();  // enable interrupts
 while (!countDone) { }  // wait here till timer/counter3 is done
 // timers 1 & 3 are stopped, now restart timer 0 interrupts
 TCCR0A = oldTCCR0A;
 TCCR0B = oldTCCR0B;
 return ((timer1OVF << 16) + long(TCNT1));
}

ISR (TIMER1_OVF_vect)
{
 ++timer1OVF;  // increment timer1 overflow counter
}

ISR (TIMER3_COMPA_vect)
{
 TCCR1B = 0;  // immediately stop timer1
 TCCR3B = 0;  // immediately stop timer3
 TIMSK1 = 0;  // disable timer1 interrupt
 TIMSK3 = 0;  // disable timer3 interrupt
 countDone = true;
}

boolean tankNotFull()
// float switch is closed when tank not full, Schmitt trigger inverts
```

```
{
 int val = digitalRead(floatswitchPin);
 if (val == HIGH)
 {
  return true;
 }
 else
 {
  return false;
 }
}

void raingaugeISR()
{
 raincount++;
 sprintf(rainString, "%4d", raincount);  //  integer to 4-digit string.
 RainfallS7S.print(rainString);   // upper LED display
 if( raincount >= firstflush )    // don't test raincount outside ISR
 {
  firstFlushMet = true;
 }
}

void setup()
{
 pinMode(T1Pin, INPUT);    // frequency input to T1 pin
 pinMode(floatswitchPin, INPUT);  // tank float switch
 pinMode(raingaugePin, INPUT);  // raingauge on pin 7
 attachInterrupt(digitalPinToInterrupt(raingaugePin), raingaugeISR, RISING);
 pinMode(tankSel, OUTPUT);  // pin for selecting tank level (frequency) to read
 pinMode(PWMA, OUTPUT);    // designate Arduino pins for motor control
 pinMode(PWMB, OUTPUT);
 pinMode(DIRA, OUTPUT);
 pinMode(DIRB, OUTPUT);
 digitalWrite(PWMA, LOW);    // initialize all motor control pins as low
 digitalWrite(PWMB, LOW);
 digitalWrite(DIRA, LOW);
 digitalWrite(DIRB, LOW);
 pinMode(A0, INPUT_PULLUP); // pins for selecting first flush amount
 pinMode(A1, INPUT_PULLUP);
 pinMode(A2, INPUT_PULLUP);
 pinMode(A3, INPUT_PULLUP); // pins for open, close, auto valve control
 pinMode(A4, INPUT_PULLUP);
 RainfallS7S.begin(9600);  // Set baud rate of 7-seg displays
 StorageS7S.begin(9600);
 RainfallS7S.write(0x76);  // Clear displays
 StorageS7S.write(0x76);
 RainfallS7S.write(0x7A);  // Set brightness command
 RainfallS7S.write(200);  // brightness: 0 minimum, 255 maximum
 StorageS7S.write(0x7A);
```

```
StorageS7S.write(200);
RainfallS7S.write(0x77);  // decimal point command
RainfallS7S.write(0x02);  // display decimal point 2nd digit
delay(1500);  // Delay before starting readings
sprintf(rainString, "%4d", raincount);  //  display zero rainfall
RainfallS7S.print(rainString);  // upper LED display
}

void loop()
{
 // get tank water levels and display
 digitalWrite(tankSel, HIGH);  // select first tank to read
 freq = getFrequency();
 level1 = int(k1/C12*(1.443/float(freq)/RT1 - C11));
 digitalWrite(tankSel, LOW);   // select second tank to read
 freq = getFrequency();
 level2 = int(k2/C22*(1.443/float(freq)/RT2 - C21));
 sprintf(storString, "%4d", level1 + level2);  //  integer to 4-digit string.
 StorageS7S.print(storString);  // display storage level on lower LED display

 // read first flush amount selection
 byte pinA0 = digitalRead(A0);
 byte pinA1 = digitalRead(A1);
 byte pinA2 = digitalRead(A2);
 firstflush = minFlush * ( pinA2 << 2 | pinA1 << 1 | pinA0);
 // read valve control mode selection, either normal (auto), opened, or closed
 byte pinA3 = digitalRead(A3);
 byte pinA4 = digitalRead(A4);
 byte valve = pinA4 << 1 | pinA3;
 switch (valve)
 {
  case normal :
    if( firstFlushMet && tankNotFull() )
    {
    // close valve and fill tank
      digitalWrite(DIRA, closevalve);
      analogWrite(PWMA, 255);
    //   digitalWrite(DIRB, closevalve);  // use only motor A if one motor installed
    //   analogWrite(PWMB, 255);
    }
    else
    {
    // open valve, dump to ground
      digitalWrite(DIRA, openvalve);
      analogWrite(PWMA, 255);
    //   digitalWrite(DIRB, openvalve);
    //   analogWrite(PWMB, 255);
    }
   break;
  case opened :
```

```
      // open valve, dump to ground
        digitalWrite(DIRA, openvalve);
        analogWrite(PWMA, 255);
      //   digitalWrite(DIRB, openvalve);
      //   analogWrite(PWMB, 255);
     break;
   case closed :
     if( tankNotFull() )
     {
     // close valve and fill tank
       digitalWrite(DIRA, closevalve);
       analogWrite(PWMA, 255);
     //   digitalWrite(DIRB, closevalve);
     //   analogWrite(PWMB, 255);
     }
     else
     {
     // open valve, dump to ground
       digitalWrite(DIRA, openvalve);
       analogWrite(PWMA, 255);
     //   digitalWrite(DIRB, openvalve);
     //   analogWrite(PWMB, 255);
     }
    break;
   default : break; // unrecognized selection
  } // end of switch-case
 delay(1000); // delay before reading tank levels again
}
```

The first nine constants in the above sketch may need to be changed for your system, particularly the eight constants used by the water level sensors. The first constant sets the minimum and maximum first-flush amount to 0.02 inches and 0.12 inches respectively. If you live in a very low rainfall area and want to reduce this to 0.01 and 0.06 inches, change this constant from 2 to 1. Recompile the sketch and upload to the Arduino.

The getFrequency() function uses two 16-bit timer/counters in the ATmega32U4 to measure frequency at high speed. Input pulses are counted on Timer1 via pin 12 (internally wired to T1 pin) and the gate time is controlled with Timer3. Timer3 uses the internal 16 MHz clock scaled down by 1024 to 15.625 kHz in order to keep its count within 16 bits. It receives a pulse every 64 microseconds. Both timers are zeroed and started at the same time. Timer3 starts at zero and counts to 15624 for a 1 second gate time, triggers an output compare ISR, and stops both counters. The count on Timer1 is the frequency on pin 12. Frequency is converted to water level and then displayed on the first-flush controller.

This first-flush controller is a primary tool for system maintenance. It will be viewed by your entire household as everyone learns to conserve water. So it should be located in a convenient spot where anyone can easily view it. A hallway, pantry, or walk-in closet are good possibilities. Table 5-3 shows components and their costs for this first-flush controller, including interface board and digital rain gauge. Your conduit requirements may be different, depending on length of runs. Total hardware cost is about $367, including the digital rain gauge.

Table 5-3. First-flush Controller Costs.

4-digit 7-segment serial display, COM-11441	Sparkfun.com	2	$29.30
Ardumoto motor driver shield, DEV-14129	Sparkfun.com	1	$22.57
Arduino Leonardo, #2163778	Jameco	1	$20.49
8"x8"x4" PVC junction box, #145145	Lowe's	1	$21.98
1/2" PVC conduit bend, #50916	Lowe's	4	$2.60
1/2" PVC conduit 45-deg bend, #115969	Lowe's	1	$2.48
1/2" PVC conduit clamps, #75748	Lowe's	1	$3.08
1/2" PVC conduit Tee, #115991	Lowe's	1	$2.48
1/2" PVC conduit transition, #75783	Lowe's	2	$4.96
1/2" PVC conduit, #72808	Lowe's	7	$14.35
22 AWG shielded cable, W506-50-ND	Digikey.com	1	$44.78
18 AWG shielded cable, CE2204SN-41-50-ND	Digikey.com	1	$25.28
2.3AH 12V battery, 522-1008-ND	Digikey.com	1	$19.42
Battery charger, 62-1233-ND	Digikey.com	1	$25.53
Switch knob, 450-1735-ND	Digikey.com	2	$5.16
6-pos rotary switch, EG1954-ND	Digikey.com	1	$5.50
3-pos rotary switch, EG1958-ND	Digikey.com	1	$5.50
Pushbutton switch, EG2025-ND	Digikey.com	1	$1.15
Custom interface PC board	Optosensors Technology, Inc	1	$16.17
Green & red LEDs, COM-09592 & COM-09590	Sparkfun.com	2	$0.70
10uF capacitor, 445-8465-ND	Digikey.com	1	$1.06
1uF capacitor, BC1168CT-ND	Digikey.com	3	$0.99
4.7uF capacitor, BC1168CT-ND	Digikey.com	1	$0.33
0.1uF capacitor, BC2665CT-ND	Digikey.com	2	$0.36
4700pF capacitor, BC2683CT-ND	Digikey.com	2	$0.36
22K resistor, 22KQBK-ND	Digikey.com	6	$0.60
100K resistor, 100KQBK-ND	Digikey.com	2	$0.20
55 ohm resistor, PPC54.9XCT-ND	Digikey.com	2	$0.40
TVS diode, SA5.0ALFCT-ND	Digikey.com	2	$3.68
TVS diode, SA5.0CALFCT-ND	Digikey.com	6	$3.06

Schottky diode, BAT46CT-ND	Digikey.com	4	$1.80
74HC14 IC, 296-1577-5-ND	Digikey.com	1	$0.54
74HC00 IC, 296-1563-5-ND	Digikey.com	1	$0.54
14-pin DIP socket, AE9989-ND	Digikey.com	2	$0.46
8-pin terminal block, ED2615-ND	Digikey.com	1	$1.31
10-pin terminal block, ED2616-ND	Digikey.com	1	$1.84
Hex standoff, 4-40, 1/2", 36-1902C-ND	Digikey.com	7	$4.90
4-40 screw, 36-9427-ND	Digikey.com	7	$0.98
4-40 flathead screw, 36-9528-ND	Digikey.com	7	$0.98
Breakaway headers, PRT-00116	Sparkfun.com	1	$1.50
2-pin terminal blocks 3.5 mm pitch, PRT-08084	Sparkfun.com	2	$1.90
Rainwise digital rain gauge	AmbientWeather.com	1	$65.66

5.3.1 Interface Boards

Figure 5-9 shows a schematic of the interface board with a simple 2×1 MUX. This board has capability of receiving six inputs, but only four are used; two frequency inputs from tank level sensors, an input from a tipping bucket digital rain gauge, and an input from the raw tank full switch. Surge protectors (bidirectional or unidirection zener diodes) were originally placed on all inputs as shown in the figure. I later discovered that surge protector capacitance caused a capacitive coupling between the two frequency inputs from tank level sensors, causing unstable readings. Removing D7 and D8 solved this problem. Lower capacitance zener diodes might also have solved this problem. An oscilloscope is helpful for isolating problems of this nature.

Following the surge protectors, simple low-pass RC filters and Schmitt triggers (U1) condition input signals for the Arduino. Long signal wires can add both capacitance and inductance which can cause ringing at pulse edges. Test your signal channel with an oscilloscope to determine filtering requirements. The smaller resistance and capacitance for the two frequency inputs were chosen to provide an RC time constant that passes the highest output frequency from the tank level sensors. The four NAND gates (U2) form a 2×1 MUX that selects one of two frequencies to read on Arduino pin 12. Arduino digital pin 6 tells the MUX which frequency to read. A 2×1 MUX can be implemented with either four NOR gates or four NAND gates, both of which come in single ICs, but I chose to use NAND gates here since I had several of these in stock.

The two frequency inputs do not have pullup resistors since these inputs are driven HIGH or LOW by active capacitive sensors (discussed in 5.2.3 below). The other four inputs do have pullup resistors, which assumes an input from a switch being opened or closed. The RC time constant (22K times 1uF) was chosen after testing it with my selected rain gauge. I wanted a time constant long enough to filter out switch bouncing but not so long that it prevented voltage dropping to zero during the momentary reed switch closure. The rain gauge signal is a quick momentary closure of a small magnetic reed switch and if the RC time constant is too long then the Schmitt trigger will not change states. A digital storage oscilloscope can easily capture this switch bouncing and help set the correct RC time constant.

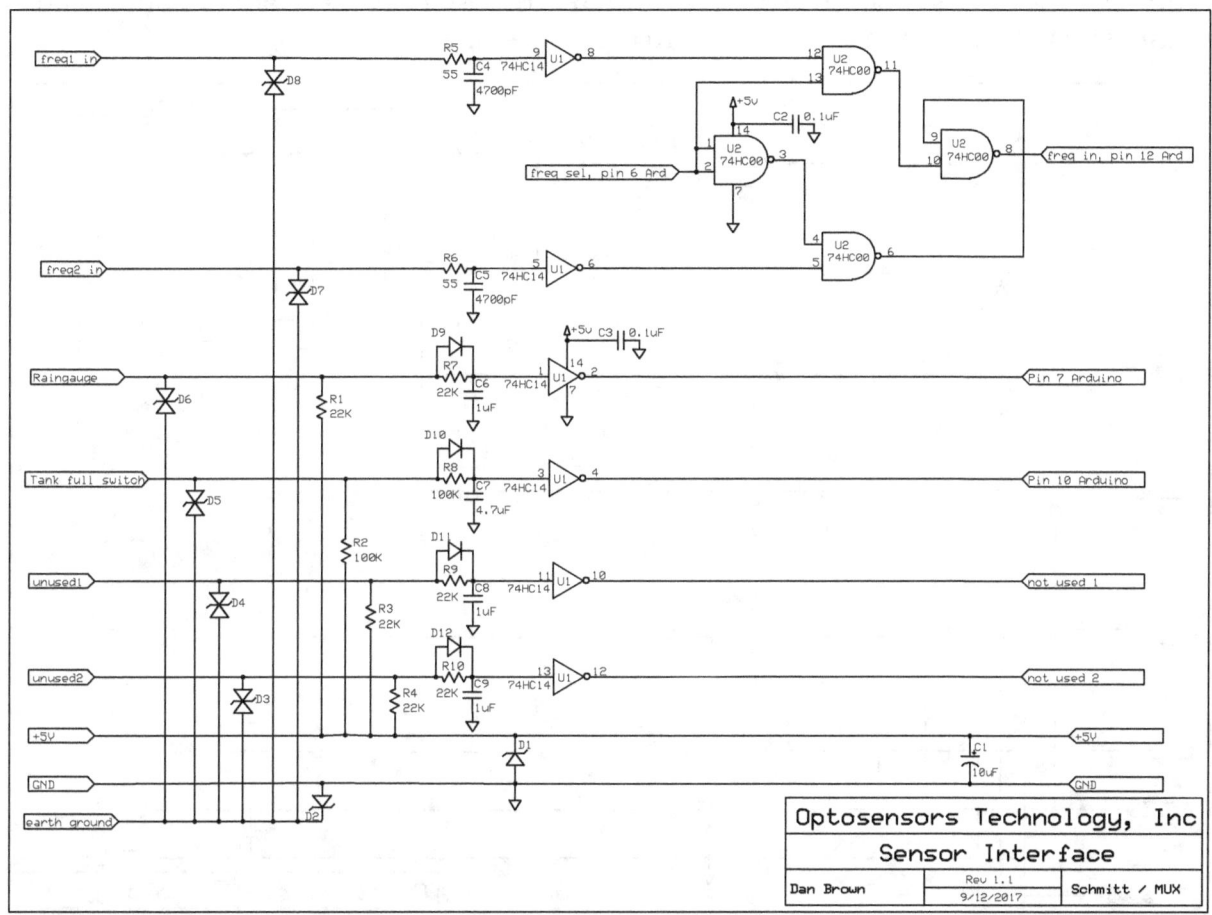

Fig. 5-9. Schematic of interface board for two frequency inputs.

The tank full switch is also a low-cost magnetic reed type float switch, rather than a large expensive float switch with built-in hysteresis. In this case we want to filter out rapid opening and closing of the switch due to water turbulence and only determine if the water level is above or below the threshold switch point. So its RC time constant can be quite large, on the order 0.5 seconds or longer. The Schmitt trigger following the filter adds hysteresis. Note that frequency inputs do not have pull-up resistors and do not require diodes on their low-pass filters, whereas the four switch inputs do require diodes so that both charging and discharging RC time constants are identical.

Long signal wires on high impedance digital inputs can act like antennas for lightning-induced transients. Include transient voltage suppression on signal inputs and use shielded cable in buried conduit to reduce lightening-induced transients. Zener diodes D1 through D8 are Littelfuse transient voltage suppression diodes to shunt voltage transients to earth ground. Digital ground is tied to earth ground through D2. If your household AC ground is noisy due to typical ground loops with utility pole grounds (measure with voltmeter or oscilloscope), add another metal ground stake near your tanks and tie it to digital earth ground.

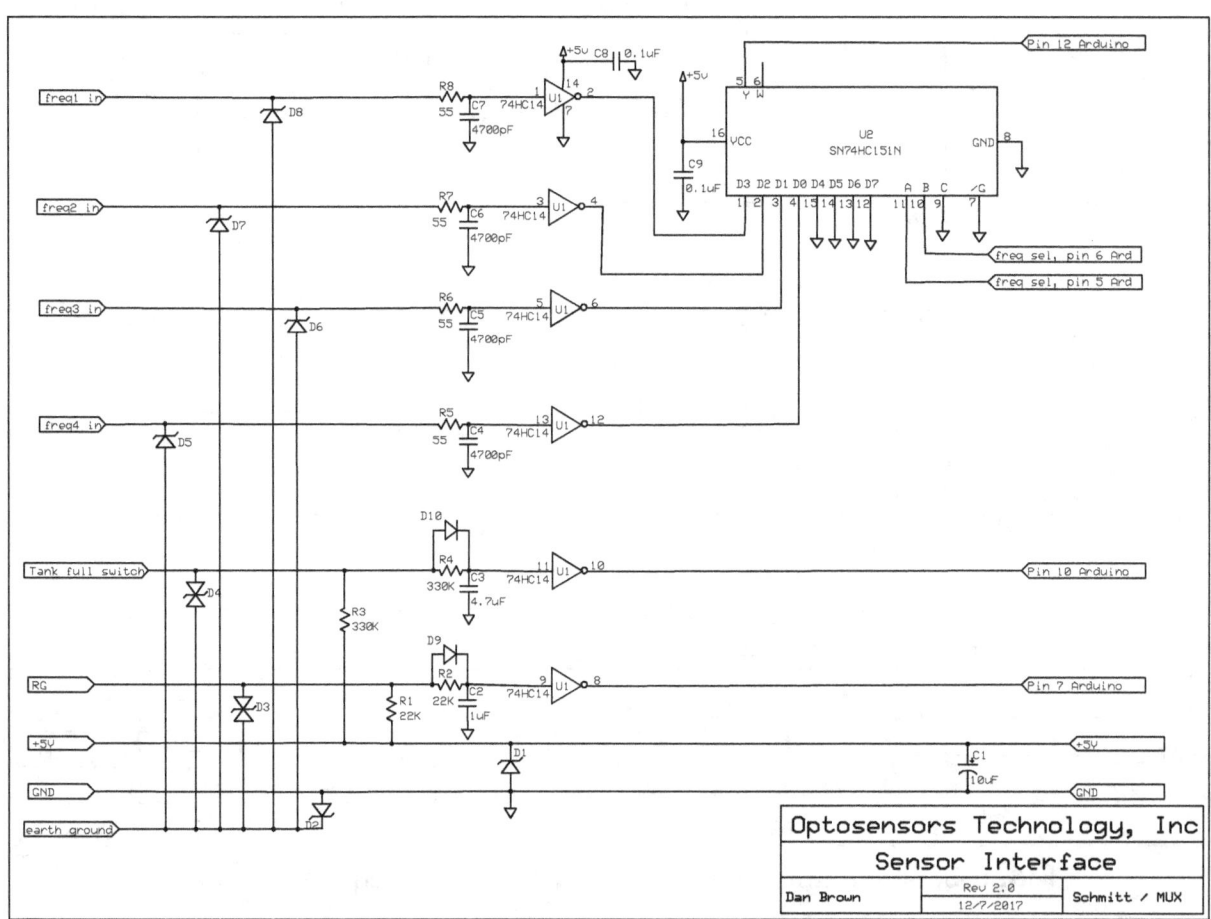

Fig. 5-10. Schematic of interface board for four frequency inputs.

If a capacitive type rain gauge is used, or three water level sensors are required for raw water tanks at different levels as in Fig. 4-13, a MUX IC must be substituted for the NAND gate IC as shown in Fig. 5-10. Arduino pins 5 and 6 can be used to select between four different input frequencies. Pin 7 is still available for hardware that requires an ISR. As with the previous interface board, adjust RC time constants on the low pass filters for the range of frequencies on your sensors and make sure surge protector capacitance is not too large.

Instead of running shielded cable in buried conduit to storage tanks located a long distance from house, you can install a wireless channel incorporating a pair of XBee 2.4GHz transceivers (e.g. XBP24-AWI-001) tied to an Arduino on each end. The increased hardware cost of an extra Arduino, a pair of XBees, and two XBee shields may be offset by cost savings of eliminating long shielded cables in buried conduit. Since a potential exists for heavy rainfall to degrade wireless signal strength, run the main control program on a hardwired Arduino at the tanks. The Arduino in the house only sends switch positions and displays data from the hardwired Arduino. If wireless communications are lost the system continues to operate automatically. Alternatively, you could run a terminal program, such as CoolTerm, on a computer in your house to interface with the hardwired Arduino and eliminate an extra Arduino. You still need a pair of XBees plus one XBee shield and a XBee Explorer (Sparkfun WRL-11697) to interface the wireless link to your computer.

5.4 Water Level Sensor

Providing an indoor display of actual stored water levels to all household members is a valuable system management tool. Rather than depending on you to go outside and physically measure water levels in tanks, any household member can immediately see remaining stored water levels and adjust his/her water usage accordingly. This helps everyone in your household improve water conservation practices. We show here a simple, low-cost, electronic water level sensor that provides a continuous display of water storage level on your first flush controller. This sensor can also be used to calculate the collection efficiency constant in your water storage tracking spreadsheet, WaterStorage2, discussed in more detail in Chapter 6. Collection efficiency is assumed to be a constant in this spreadsheet. Using a good estimate of actual collection efficiency in this spreadsheet gives better water storage tracking and prediction. However, collection efficiency can gradually degrade over time as gutter screens become clogged or the first-flush valve starts leaking water to ground. So periodically calculating actual collection efficiency helps determine required system maintenance.

There are many different types of electronic water level sensors. The one shown in this book's first edition was based on measuring pressure at the bottom of a tank. These proved to be somewhat noisy and unreliable. The one we show here is a capacitive type level sensor with much greater reliability and noise immunity. An astable NE555 oscillator generates a frequency based on the variable capacitance of an insulated conductive rod immersed in the storage tank water. Frequency is read by an Arduino and converted into gallons of water. Long signal wires between tank level sensors and microcontroller have a finite capacitance that can attenuate high frequencies. So we must keep generated frequencies low enough that signal wire capacitance does not seriously degrade the high frequency signal.

Fig. 5-11. Low cost capacitance water level sensor.

Development of a capacitive sensor that had zero leakage current went through several iterations before I achieved success. My initial prototype consisted of two parallel copper foils mounted on structural fiberglass angle bars and separated by about 1/8". The foils and bars were coated with multiple coats of epoxy to electrically insulate them. However, testing still showed a very small leakage current between the two electrodes when the capacitor was immersed in water. The next couple of attempts focused on applying an insulating coating to a copper pipe with a cap soldered on its bottom. Again, tiny invisible pinholes in the coating yielded a leaky capacitor. Perhaps better coating methods would solve this problem, but the time required to apply these coatings was unacceptably high. I wanted capacitors that were easy and cheap to build.

Success finally came when I realized that standard 3/4" copper pipe OD is less than 1/16" smaller than the ID of 3/4" thin-wall PVC pipe. PVC pipe is easy to seal up with no pinhole leaks whatsoever. Figure 5-11 shows a picture of our very inexpensive water level sensor before being inserted through a hole in top of tank. A 3/4" copper pipe is inserted inside a 3/4" thin-wall SDR-21 PVC pipe and standard schedule 40 PVC fittings are used at the ends. The inside of the 3/4" pipe is filled with cheap 80 proof vodka to add weight, eliminate the air gap between copper pipe and inner surface of the PVC pipe, without risking freezing during winter. Vodka is a good "food grade" antifreeze with an electrical conductivity five to ten times higher than stored rainwater. So it is actually better than water for extending the copper pipe diameter to the full ID of the PVC pipe. Testing with and without water showed adding water provided a slightly higher capacitance per foot. But water can freeze if the tank water level gets low and temperatures go well below freezing. Vodka solves this problem.[36] A PVC cap on bottom and copper cap at top completely seal the vodka inside and electrically insulates the copper pipe from tank water.

This sensor rod is easy to construct. Glue a 3/4" PVC cap on one end of a 3/4" thin-wall PVC pipe. Drill a 1.25" hole in the top center of your tank, just large enough for the PVC cap to fit through. Lower the capped end of the PVC pipe through the hole until it touches the bottom of the tank, keeping the pipe vertical. Mark the top of the pipe flush with the top of the tank, remove it, and then cut the pipe at the mark. Note, when the pipe is glued into the adapter the pipe will be raised about an inch off the bottom, insuring that it hangs free. Insert a 3/4" copper pipe into the PVC pipe and cut it about 1 inch longer, just long enough for a soldered copper cap to rest on top of the PVC pipe. Drill and tap a copper cap for 1/8" MIP threads and then solder the cap onto the copper pipe. This hole is used to later fill the rod with vodka and then seal with a brass 1/8" MIP plug. Solder a signal wire to the outside of the copper cap. Drill a couple of small vent holes in the side of the copper pipe about 1.5" below the cap. These will allow the 1/16" space between copper pipe and PVC to be filled with vodka and increase the sensor's capacitance per foot. Cement the PVC pipe into a 2" x 3/4" PVC bushing. After the glue dries, insert the copper pipe into the assembly and seal the gap around the copper cap with epoxy. Let this assembly dry for 24 hours before filling it with vodka.

Using a small funnel fill the electrode rod with cheap 80 proof vodka (40% ethanol by volume). This will protect the sensor from freezing down to -11°F (-24°C) and increase the sensor's capacitance per foot. Unlike other antifreezes, this presents no health risks to your potable water if it should leak out. Apply thread compound or glue to a 1/8" brass plug and screw it into the copper cap to seal the vodka inside. Cement a 2" PVC thread-to-slip adapter to the PVC bushing to form a housing for the electronic circuit. Drill a small hole through the side of this housing for inserting shielded signal cable from first-flush controller. Also drill a small hole for inserting

---

36 So if you are a Baptist you now have a good excuse to go to the liquor store!

14Ga copper ground wire from a stainless steel plate electrode inside the tank. Insert the sensor into the hole in top of tank and seal around the edges with silicone sealer. After soldering the sensor circuit to five wires (copper pipe, stainless plate ground, +5V, GND, and F from Arduino) stuff the PC board and wires inside the housing and screw a 2" PVC cap on top. If necessary apply a short length of electrical tape to the copper cap to prevent it from shorting the electronic circuit. The reason we have brought the copper pipe up into the circuit housing is to help stabilize the electronic circuit temperature. Most applications do not require complex thermal stabilization like we show in section 5.5.2 below. But if you need greater water level accuracy you can easily access probe temperature by cementing a thermistor or other sensor to this copper cap.

The 3/4" copper pipe serves as one terminal of the capacitive probe and the storage tank water itself serves as the other terminal, essentially forming a cylindrical capacitor. A one square foot stainless steel plate inside the tank electrically connects the water to digital ground. Capacitance increases linearly with water height on the probe rod. The capacitance of a cylindrical capacitor formed by two concentric conductive cylinders is given by:

$$C = \frac{2\pi\kappa\epsilon_0 L}{\ln(r_2/r_1)} \qquad (5\text{-}1)$$

where κ is the dielectric constant, ϵ_0 is the permittivity constant, L is the length of the cylinders, and r_2 and r_1 are the radii of the outer and inner cylinders respectively. Equation (5-1) shows that as the ratio of the two radii approach unity the capacitance per unit length goes to infinity (natural logarithm of one equals zero). So for a given dielectric thickness ($r_2 - r_1$) a larger diameter center rod will give a higher capacitance per foot. This is why we add vodka to our probe, in addition to helping hold the sensor down under the water. We also see from Eq. (5-1) that capacitance is linearly proportional to the common length of the two concentric cylinders, which equals the water height.

Pure water has very low conductivity, which can show up as equivalent series resistance (ESR) on the capacitor's grounded side and change oscillator timing. Since ESR prevents the capacitor from fully charging to the trigger point of the NE555 oscillator, it causes the output frequency to be slightly higher than expected. A large steel plate area in contact with the water and very low current traveling through it to the probe surface (due to high timing resistance values) help reduce effects of ESR. If in doubt you can measure ESR with an oscilloscope and signal generator. Drill a hole in the center of the stainless steel electrode and bond insulated solid 14 gauge THHN copper wire to this plate with a stainless screw and nut, with the screw head on the bottom and wire on top. Seal this connection point with blob of epoxy to minimize corrosion between the two dissimilar metals. Bend the pointed plate corners upward so that they do not wear holes in your plastic tank if water turbulence or tank cleaning moves the electrode around. Bring the 14 gauge wire out through a small hole in tank top to the sensor housing and tie it to digital ground.

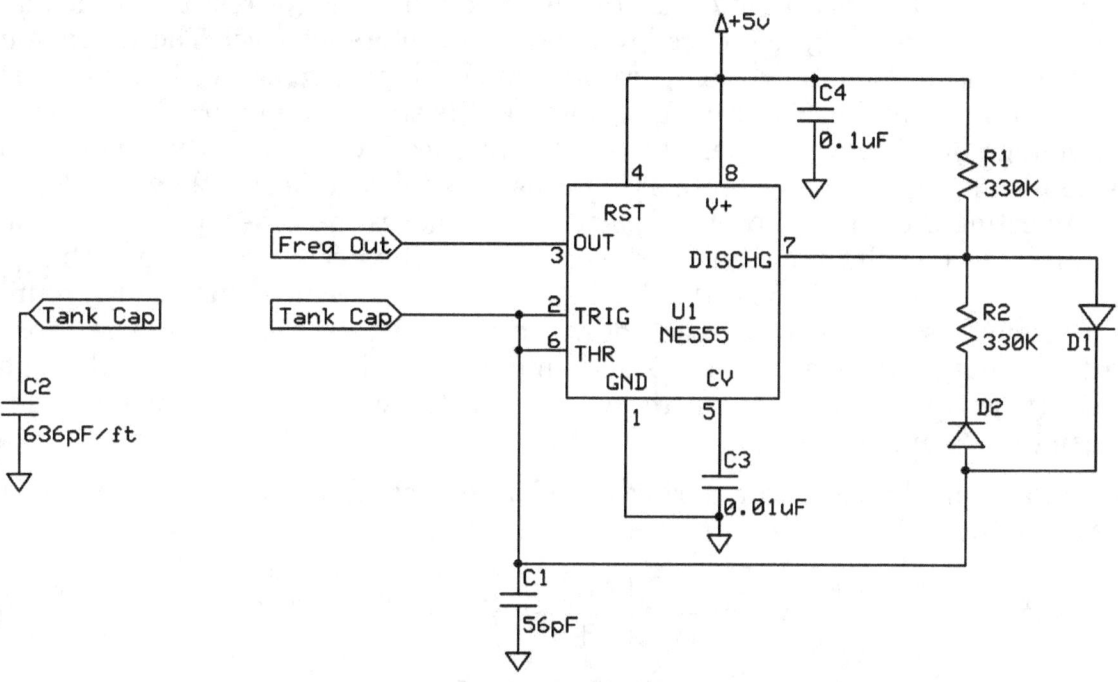

Fig. 5-12. Electronic schematic for capacitive level sensor.

Figure 5-12 shows the simple NE555 oscillator circuit for our capacitive water level sensor. The two diodes, D1 and D2, are added to make charging and discharging RC time constants equal and thus provide a 50% duty cycle. Since long signal wires have inherent capacitance which attenuates high frequencies, we want to minimize high frequency harmonics in our signal and a 50% duty cycle helps achieve that. I avoided simpler 50% duty cycle circuits that use output pin 3 to charge and discharge C2 due to RC loads on output. Note that tank prob capacitor, C2, is in parallel with a small 56pF capacitor, C1. This small capacitor provides an upper limit on frequency output from the oscillator. The constant capacitance, C1, is actually the sum of a physical 56pF capacitor and residual capacitance in the probe with zero water level. Both C1 and C2 are experimentally determined during water level sensor calibration.

From Eq. (5-1) and Fig. 5-12 we see that input capacitance tied to this oscillator is a linear function of water height, h :

$$C = C_1 + h C_2 \qquad (5-2)$$

Output frequency is given by:

$$f = \frac{1}{0.693 C (R_1 + R_2)} \tag{5-3}$$

The constant 0.693 in Eq. (5-3) comes from the way the NE555 works. This integrated circuit (IC) has an internal voltage divider of three 5K resistors which divide IC supply voltage to a 1/3 reference level and a 2/3 reference level. These two voltage reference levels are tied to two comparators which compare reference levels with input voltages on pins 2 and 6. These comparators flip the state of an internal flip-flop when input voltage on pin 2 drops below the 1/3 reference level or input voltage on pin 6 rises above the 2/3 reference level. Tying pins 2 and 6 together as shown in Fig. 5-12 and connecting these to the probe capacitor then causes the flip-flop to change states depending on the voltage charge on the capacitor. When capacitor voltage charges to 2/3 reference level and changes the flip-flop state, pin 7 becomes internally grounded, which discharges the capacitor back down to the 1/3 reference level. The capacitor repeatedly charges through R1 and D1 (when pin 7 is high) and discharges through R2 and D2 (when pin 7 is low) with the voltage on the capacitor continually swinging between the 1/3 and 2/3 levels.

Voltage on the capacitor during the charge cycle follows a simple exponential curve given by:

$$V(t) = \frac{V_{cc}}{3} + \left(\frac{2 V_{cc}}{3} - v_f \right) \left(1 - e^{-t_1 / R_1 C} \right) \tag{5-4}$$

where v_f is the forward bias voltage of the diode. During the discharge cycle, capacitor voltage is given by:

$$V(t) = \left(\frac{2 V_{cc}}{3} - v_f \right) e^{-t_2 / R_2 C} \tag{5-5}$$

Setting diode bias voltage to zero in Eq. (5-4) and solving for time, t when capacitor voltage equals $2 V_{cc}/3$ gives time $t_1 = 0.693 R_1 C$. Similarly solving for time in Eq. (5-5) when capacitor voltage equals $V_{cc}/3$ gives $t_2 = 0.693 R_2 C$. The period is the sum of the two times, which gives frequency as:

$$f = \frac{1}{t_1 + t_2} = \frac{1.443}{(R_1 + R_2) C} \tag{5-6}$$

which is identical to Eq. (5-3).

When diode forward bias voltage in included in the above equations, calculated period is slightly longer and frequency is slightly lower than that given in Eq. (5-6). Diode forward voltage is actually a nonlinear function of current, rather than a constant, which further complicates Eqs. (5-4) and (5-5). These effects are easily modeled in simulation software such as LTSpice.[37] Use identical germanium diodes with minimum forward bias voltage to reduce these effects.

For cylindrical water tanks with constant water surface area, volume of water is simply a constant, k, times water height:

$$Volume = k\,h \tag{5-7}$$

This becomes nonlinear at the domed tank top, but we are primarily interested in water volume for partially filled tanks. This is where we want highest accuracy in our formula for volume versus frequency. Combining Eqs. (5-2), (5-3), and (5-7) provides water volume as a function of frequency:

$$Volume = \frac{k}{C_2}\left(\frac{1}{0.693f\left(R_1 + R_2\right)} - C_1\right) \tag{5-8}$$

This is the equation of a straight line as a function of reciprocal frequency. We program this into the Arduino sketch to convert frequency into water volume. Then we calibrate the sensors by slightly adjusting C_2 (or k) and C_1 to obtain a good fit between displayed water volume and measured volume.

Only two of the four parameters in Eq. (5-8) are independent. Resistance $R_1 + R_2$ is fixed by measured values. Constant, k, can be determined from manufacturer's volume markings on tanks. Simply measure distance between volume markings on tank and divide volume difference by distance in feet (or inches). Errors in tank markings will be corrected by adjusting C_2. The two independent parameters, C_2 and C_1, are adjusted to achieve a good linear fit of measured frequency versus volume data. Use consistent units for C_2 and k, either pF/ft and gallons per foot respectively, or pF/in and gal/in respectively. Initial values of 636pF/ft for C_2 and 56pF for C_1 are reasonable starting values for a probe and circuit as shown above.

To calibrate tank level sensors make a temporary change to the first-flush controller program to display measured frequencies and water volume levels on a serial monitor on your computer. For example, immediately after the calculation of level1 in the main loop of the first-flush controller sketch insert:

```
Serial.print("f1=");
Serial.print(freq);
Serial.print(" level1=");
Serial.print(level1);
```

Then after the level2 calculation enter:

```
Serial.print("        f2=");
Serial.print(freq);
Serial.print(" level2=");
```

---

37 Freeware developed by Linear Technology. Can be downloaded at: www.analog.com/en/design-center/design-tools-and-calculators/ltspice-simulator.html

Serial.println(level2);

Printing out both frequency and water level is redundant, but it allows you to choose either one for a curve fit of the two capacitance constants via either Eq. (5-3) and (5-2) or Eq. (5-8). The latter allows you to also adjust the constant k.

Turn off the transfer pump and begin filling the raw water tank from the clean water tank through a garden hose connected to an outdoor faucet. Record data points of water meter reading, displayed frequency and volume, and distance between water surface and manhole opening on both tanks. The latter is converted into water height from base of tank for curve fitting. Record several data points over a range of water levels. Alternatively, if you know rainfall will not occur for the next several days then you can record data points over several days with normal water consumption. You want data points with water level changing in both tanks. After several days turn the transfer pump back on to make water level in raw water tank change. Use these changes in distance to water surface and changes in water meter reading to determine a more accurate value for constant, k, in Eq (5-7). Then use water meter readings and displayed volume to curve fit C_1 and C_2 using Eq (5-8). Alternatively, use water height and displayed frequency to curve fit C_1 and C_2 using Eq (5-3) and (5-2). This is a slow process that will take several hours (or several days), but it is the most accurate way to calibrate your sensors since it makes your water meter the standard for water volume in tanks.

A quicker but less accurate calibration method is to simply raise the probe by various amounts on the tank and record data points of frequency and height that the probe is raised. This height equals the reduction of sensor penetration depth into the water. When frequencies are converted to capacitance and plotted against these heights in a spreadsheet, you should get a straight line. The slope of this line equals the capacitance per foot value of C_2. Use the top volume mark on tank side and divide it by its height above tank bottom to calculate an initial value for k. You then fine tune this parameter to make Eq (5-8) more accurately track water levels. Collect more data points over the next several weeks as water levels fluctuate through use, recording frequency, distance of water surface to manhole, and water meter reading. Insert data into a spreadsheet and slightly adjust the k parameter for each tank until calculated volume changes from frequencies equal measured volume changes determined by distance from manhole or water meter reading. Note that either k or C_2 is adjusted, but not both, since Eq (5-8) slope is set by either parameter. The C_1 parameter moves the straight line up or down to agree with an absolute reference volume. After calibrating these two parameters for both tanks restore the program to its normal operation with these new adjusted constants.

Figure 5-13 shows an example calibration of my raw water tank level sensor using the second method above. Frequency was measured as a function of raised height of probe. Frequency was then converted to capacitance using Eq. (5-3). Manufacturers markings of 200 gallons every 8 inches were assumed to be correct. This constant was later fine tuned when both tanks were completely full after a heavy rainfall so that storage level readings matched maximum storage tank capacity. A straight line nicely fits the data with a slope of 1.4 pF/gal and y-intercept of 77.2 pF.

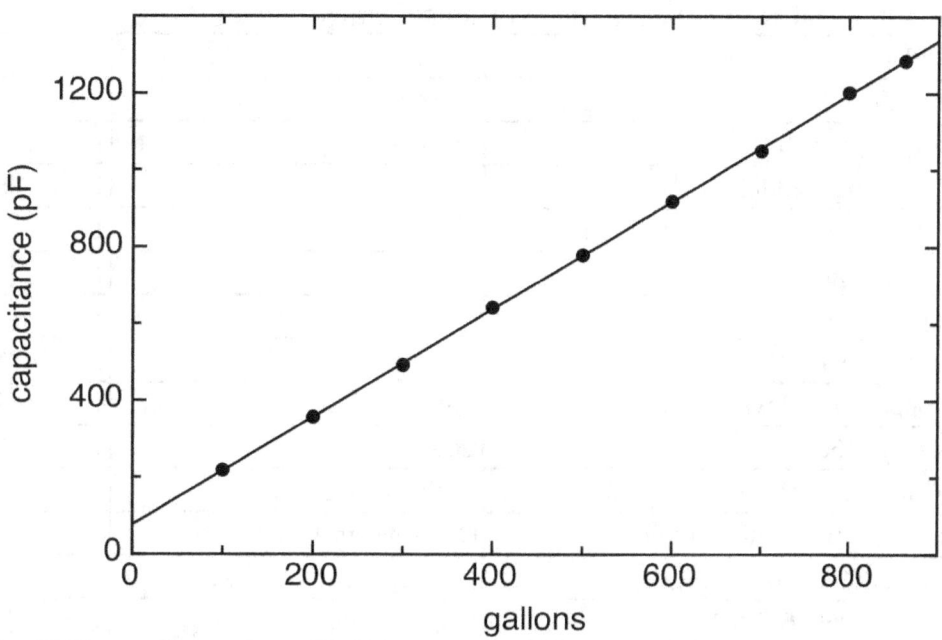

Measured Capacitance vs Water Level

Fig. 5-13. Example calibration of 1650 gallon tank sensor.

The C_1 capacitor restricts maximum oscillator frequency when water level falls to zero. For example, using the values shown in Fig. 5-12 and assuming a maximum water height of 5 feet, Eq (5-3) shows frequency will vary from 39 kHz to 676 Hz. If C_1 is increased to 200 pF then the frequency range is reduced to 10.9 kHz to 647 Hz. So C_1 primarily affects the upper frequency when the tank is empty. This smaller frequency range still provides adequate resolution of water volume, but the larger C_1 value helps reduce circuit noise when the tank is near empty.

The next section shows that dielectric constant of plastic insulators generally changes with temperature. To a lessor extent it is also dependent on frequency. Storage tanks in contact with ground will tend to stay near average ground temperatures, which helps minimize temperature-dependent changes of dielectric constant. Fortunately, dC/dT is reasonably flat for PVC near typical ground temperatures. However, if you require greater accuracy on your tank level sensors then you can implement the temperature compensation methods discussed in the next section. Due to low conductivity water, ESR is also a concern with capacitive sensors. A comparison of sensor noise and ESR between a stainless steel ground plate in bottom of tank and a stainless steel wire coil wrapped around the probe rod is on my R&D "to do" list, but I have not yet had time to investigate this.

Table 5-4 provides hardware costs for building a tank level sensor. Look in any industrial catalog of RF sensors or capacitive sensors and you will see that these typically cost thousands of dollars. This one costs you a little time and about $94!

Table 5-4. Water level sensor hardware costs.

3/4" thinwall PVC pipe, #23990	Lowe's	1	$2.57
3/4" copper pipe, #23791	Lowe's	1	$16.97
Brass 1/8" MIP plug, #748358	Lowe's	1	$2.69
2"x3/4" PVC bushing, #23003	Lowe's	1	$1.38
2" PVC MIP thread-to-slip adapter, #23904	Lowe's	1	$1.32
3/4" PVC cap, #23896	Lowe's	1	$0.53
3/4" PVC plug, #22697	Lowe's	1	$0.98
2" PVC threaded cap, #22994	Lowe's	1	$2.18
14 Ga solid copper THHN wire, #383318	Lowe's	1	$9.97
3/4" copper cap, #21664	Lowe's	1	$1.00
12"x12"x0.048" T316 stainless steel plate	Onlinemetals.com	1	$29.75
Custom PC board for oscillator circuit	Optosensors Technology	1	$15.00
0.01uF capacitor, 399-9858-1-ND	Digikey.com	1	$0.26
0.1uF capacitor, BC2665CT-ND	Digikey.com	1	$0.18
56pF capacitor, BC1039CT-ND	Digikey.com	1	$0.21
Diode, BAT46CT-ND	Digikey.com	2	$0.90
330K resistor, RNV14FAL330KCT-ND	Digikey.com	2	$0.52
NE555 timer IC, 296-1411-5-ND	Digikey.com	1	$0.44
8-dip IC socket, AE9986-ND	Digikey.com	1	$0.19
750 ml 80 proof vodka	Liquor store	1	$6.99

5.5 Digital Rain Gauges

We discuss three types of digital rain gauges in this section. A digital rain gauge is required for one of the inputs to the first-flush valve controller. A common tipping bucket rain gauge (TBR), discussed in the first subsection below, can easily provide this electronic digital input of initial rainfall. These gauges are mechanically simple and cheap to build, but they suffer from erratic under-reporting error. We discuss more accurate approaches to rainfall measurement, including capacitive and acoustic gauges, in the second and third subsections.

5.5.1 Tipping Bucket Rain Gauge

A TBR gauge incorporates a pair of buckets that rock back and forth on a seesaw under a drip stream of water from a rain catchment funnel. The seesaw's pivot is vertically offset so that it stays in the tilted position until the bucket under the drip becomes filled. When this high bucket is sufficiently filled the weight imbalance tilts the seesaw in the opposite direction and places the empty bucket under the water stream, while the full bucket dumps its water. These gauges incorporate a magnetic reed switch that momentarily closes when the seesaw passes through the halfway point (seesaw is horizontal). TBR gauges are typically calibrated to provide one bucket tip for every 0.01 inches of rainfall, or 0.2 mm of rainfall for European gauges.

All mechanical switches, including small reed switches, have switch bounce that appears as a very rapid oscillation of on-off cycles to a high speed microcontroller. You can see this switch bounce oscillation with a digital storage oscilloscope. If you hook a digital rain gauge up directly to an Arduino it will record multiple bucket tips for every single actual bucket tip and produce erroneous rainfall readings. To avoid this we put a simple switch debounce circuit between the rain gauge and Arduino. This consists of a low-pas filter followed by a Schmitt trigger on the interface board previously discussed in section 5.3.1.

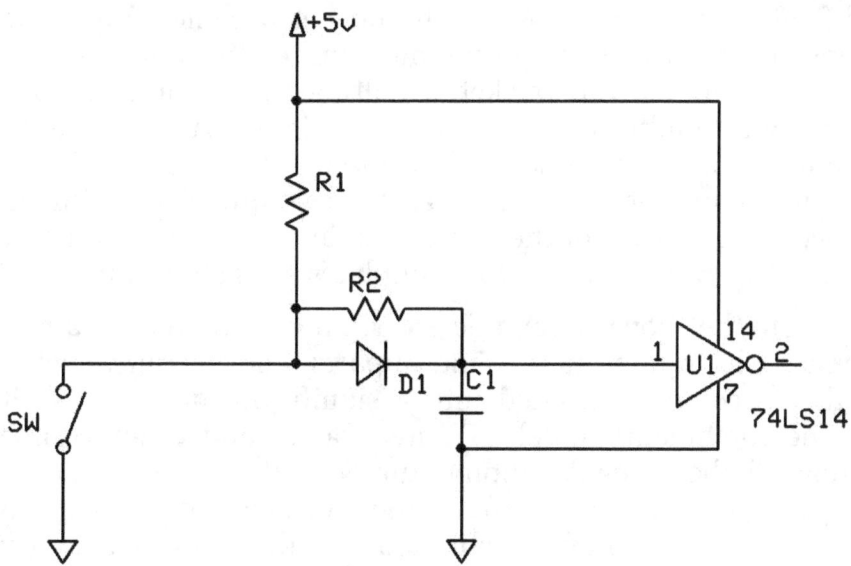

Fig. 5-14. Switch debouncing circuit.

Although software algorithms exist for removing switch bounce, debouncing a switch in hardware is much faster and more reliable. We have more important tasks for our Arduino to do than to tie it up doing simple things like debouncing a switch! Figure 5-14 shows a simple switch debounce circuit implemented on our interface board. When switch SW is open (reed switch on rain gauge), capacitor C1 charges through resistor R1 and diode D1 up to the 5VDC supply. When SW closes the capacitor is discharged through R2 to ground. The diode allows the RC time constant for charging and discharging to be identical. If the diode were not present the RC time constant would be longer for charging. This resistor/capacitor/diode network forms a low pass filter that removes high frequency switch bouncing. An inverting Schmitt trigger, U1, adds hysteresis and outputs a clean square wave pulse on pin 2 for the Arduino to read.

Tipping bucket rain gauges are notorious for under-reporting rainfall during extremely heavy rainfall. I have often seen more than 20% error on my Davis Instruments TBR, especially during heavy rainfall. I am certainly not the only researcher to observe this. Numerous other scientists have observed similar errors on TBR gauges (e.g. Habib et. al.[38], Lanza and Stagi[39], Hodgkinson et. al.[40]). My backup manual rain gauge is a calibrated large catchment area device which practically always shows higher rainfall amounts than my TBR. Sadly, many climatologists around the world blindly trust these TBR devices to provide accurate measurements of rainfall, when in fact they are seriously under-reporting rainfall. Error increases nonlinearly as rainfall rate (inches per hour) increases.

The reasons for these huge errors on TBR gauges are actually quite simple (although difficult to eliminate). When the bucket becomes full and starts to tip, it does not tip instantaneously. During the time required to accelerate the seesaw rainfall continues to drip into the full bucket, resulting in a rainfall recording loss. TBR manufacturers try to minimize this acceleration time by using lightweight materials, but they cannot completely eliminate it because all real materials have mass. And the mass of water in the bucket further increases acceleration time. This under-reporting error due to acceleration time of the bucket seesaw is best determined experimentally using a constant drip-rate calibrator that simulates a constant rainfall rate.

In addition to the above systematic error, TBRs also produce a random error that tends to increase with rainfall rate. Wind imparts a horizontal velocity component to rain drops, that becomes translated into a significant angular velocity as the drop slides down the catchment funnel. (Figure skaters and other competitive athletes intuitively know all about angular momentum!) As the bucket seesaw passes through the halfway point, dripping water from the overhead funnel can spill into either bucket. Random lateral motion fluctuations of the drip stream (due to angular

---

38 Emad Habib, Witold Krajewski, Anton Kruger, <u>Sampling Errors of Tipping-Bucket Rain Gauge Measurements,</u> p. 159, Journal of Hydrologic Engineering, March/April 2001.

39 Luca Lanza and Luigi Stagi, <u>On the Quality of Tipping-Bucket Rain Intensity Measurements, University of Genova</u>, Dept. of Environmental Engineering, Genova, Italy.

40 R A Hodgkinson, T J Pepper and D W Wilson, <u>Evaluation of Tipping Bucket Rain Gauge Performance and Data Quality</u>, Science Report: W6-084/SR, Environment Agency, Bristol, UK.

momentum imparted by the funnel and horizontal velocity of rainfall) prevent knowing exactly which of the two buckets receives the water during this narrow region about the balance point. Therefore, this balance point region becomes a region of uncertainty. Only when the bucket is fully tipped to one side or the other can one be certain about which bucket water falls into (assuming angular momentum is not too large). The speed which the bucket passes through this region of uncertainty does not change with rainfall rate. In other words, the pulse width of the momentary reed switch contact closure does not change with rainfall rate. What changes with rainfall rate is the pulse repetition rate. Therefore, a first-order estimate of random errors is simply the pulse width divided by time between pulses, which increases as rainfall rate increases. Or more accurately, it is the ratio of lateral movement in the drip stream to the lateral velocity of the bucket divider, divided by the time between bucket tips. As time between bucket tips decreases (increasing rainfall rate) this ratio, which is proportional to random errors, increases exponentially.

In addition to lateral motion of the drip stream, friction in the pivot point from dirt or normal wear can increase random errors. A TBR gauge is a mechanical device whose accuracy depends on detecting a very small imbalance of forces. Ideally, that imbalance should be near zero for maximum accuracy, but friction prevents that. Unlike systematic error, this random error cannot be removed through calibration. It can only be minimized by making the water drip stream as thin and stable as possible (zero angular momentum), maximizing the speed of the bucket tip, minimizing rainfall rate, and minimizing friction.

These rainfall rate-dependent errors prevent calibrating a TBR gauge by simply pouring a measured amount of water into it. A measured amount of water must be slowly dripped into it so that the time between bucket tips is typically 30 seconds or more. All TBR gauge manufacturers (at least the reputable ones) specify a maximum rainfall rate for which their gauge is considered calibrated. Calibration is performed through a very time-consuming process of slowly dripping water into the gauge with a constant drip-rate calibrator. For example, if a manufacturer specifies their gauge is calibrated for a rainfall rate of 1.5 inches per hour and each bucket tip represents 0.01 inches of rainfall, then you know that calibration was performed with a time between bucket tips equal to: (1 hr / 1.5 in)(3600 sec / hr)(1 in / 100 tips) = 24 seconds per tip. If the volume equivalent of 2.53 inches of rainfall is dripped into this gauge (253 bucket tips), this test will take 101 minutes to complete. Even at that slow rate typical random errors are more than +/-2% and multiple runs must be taken at various simulated rainfall rates to obtain a statistical average of systematic error. Obviously, proper calibration is something very few people will take the time to do!

I kept seeing large errors in my Davis Instruments TBR when compared with my manual gauge, especially at high rainfall rates. Errors were often around 20% or more. After numerous calibration tests (requiring literally hundreds of hours), I discovered that average under-reporting errors approached 2% only for slow rainfall rates of less than 1.5 in/hr (greater than 24 s/tip) but increased to an average of 23% at a rainfall rate of 7.2 in/hr. During ten calibration runs at 7.2 in/hr (5 s/tip) I saw a maximum error of 30%! Random errors were unacceptably huge. Short-term high rainfall rates are not uncommon. A thunderstorm cloud burst can dump a majority of its contents in a matter of minutes, producing double-digit rainfall rates. According to Christopher C. Burt, author of <u>Extreme Weather: A Guide and Record Book</u>, Unionville, Maryland holds the record for highest rainfall rate in which 1.23 inches was dumped in 1 minute in 1956, producing an equivalent rainfall rate of 73.8 in/hr! Obviously, a TBR gauge did not capture that rainfall rate. After collecting manual and TBR gauge data on 51 actual rainfall events spanning about a year, I found the average under-reporting error on my Davis Instruments TBR to be 14% with a standard deviation of 7.3%. This was way outside the +/-2% range of the calibration screw on the device. In other words, the error was too large to correct.

I repeatedly tried to obtain basic calibration information for my Davis Instruments TBR, but they claimed this information was proprietary. The Davis Instruments customer service representative (who will remain nameless since he was only saying what his boss told him to say) told me, "We know the standard drip method does not work. ... We actually use a laser test fixture to calibrate the gauges." I told him that I operated and programmed laser interferometric surface profilometers for many years for a micro-optics fabricator and that I knew of no possible way to calibrate rain gauges without water. There is a huge difference between component quality assurance testing during fabrication and instrument calibration testing after assembly. The first assures components are built according to engineering specifications. The latter proves an assembled instrument accurately measures what it was designed to measure. When I asked exactly what their laser system was measuring he told me it was proprietary! He essentially implied my calibration method was wrong but refused to tell me what their calibration method was, claiming it was proprietary! He admitted to having equations to correct measured rainfall based on rainfall rate, but refused to tell me what they were, because they were proprietary. They even refused to tell me what rainfall rate their gauge was calibrated at, even after I told them that Rainwise freely told me their gauge is calibrated at 1.5 in/hr (which I proved to be true with my own calibration tests on their gauge).

No doubt any company has every right to withhold any and all performance data from their customers, including proof of instrument calibration. Companies also have every right to obfuscate and hide performance issues on their products. But I also have a right to not buy equipment from such companies. Look in any Edmund Optics catalog, or an Analog Devices catalog, or a typical data sheet for any integrated circuit, and you will find a thousand times more performance data on these high technology devices than Davis Instruments was willing to provide for their simple TBR! I personally believe that basic characterization data and proof of calibration should never be proprietary on any instrument. When someone claims basic characterization and calibration data is proprietary, I begin to suspect they are trying to hide obvious performance issues. So if you want a digital TBR rain gauge for your system, I recommend RainWise, Inc gauges (www.rainwise.com).

We rainwater harvesters require accurate rainfall measurements in order to make sure we do not run out of water and to calculate collection efficiency to determine whether or not our systems are working properly. A low collection efficiency is a sure indicator that something is not working properly. Gutters and gutter screens may need pressure washing. Or the first-flush valve may not be holding during heavy rainfall. Or some other system component is not working properly, resulting in rainfall not being captured as expected. We also need accurate rainfall measurements to know when to implement rigorous water conservation measures. Accurate rainfall measurements are an essential part of maintaining a rainwater system. Therefore, we need a way to easily calibrate these TBR gauges for ourselves, since some TBR manufacturers will not provide this data, for whatever reasons.

Having failed to obtain basic calibration data on my Davis Instruments TBR, I was forced to calibrate it myself with a constant drip rate calibrator. Since water flow rate (gal/min) out of a hose nozzle increases as water pressure in the hose increases, you cannot simply fill a container with water and drip it into the TBR from a small hole in the container's bottom. Drip rate will gradually slow as water height in the container reduces. We want to drip water into a TBR gauge at a constant drip rate. In order to do this we use a special device that maintains constant pressure at the drip hole (and thus constant drip rate) regardless of water column height. This device is easily fabricated with PVC fittings.

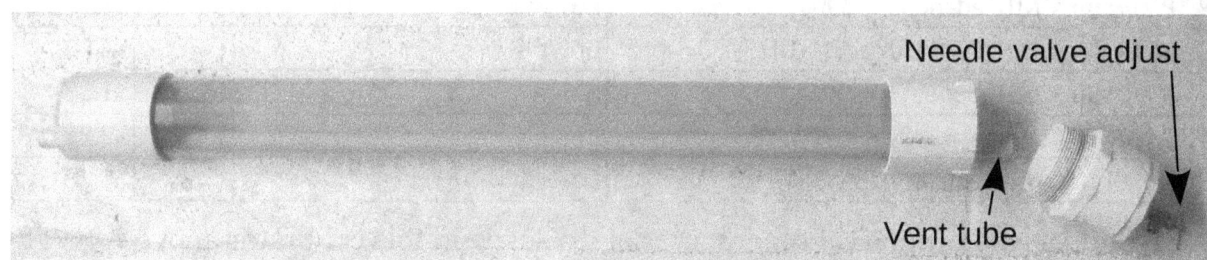

Fig. 5-15. Constant drip rate digital TBR gauge calibrator.

The way to maintain a constant pressure at the drip hole is to bring a vent tube down to the bottom of the water column next to the drip hole, as shown in Fig. 5-15. The air space above the water column is sealed off. Therefore, as the water column falls a partial vacuum is formed in this air space which pulls air down through the vent tube. Since outside air pressure must push water inside this vent tube down to the bottom in order to bubble air into the sealed air pocket above the water column, the reduction in air pressure above the main water column equals the head pressure of the water column, thus maintaining constant pressure at the drip hole. Or another way of looking at this is that the air pressure everywhere inside the vent tube equals atmospheric pressure, and since the vent tube opens up next to the drip hole at the same level then the pressure at the drip hole also equals atmospheric pressure.

This calibrator is built with standard PVC fittings. A clear body helps determine when all the water has dripped from the calibrator. I used 2" diameter clear thinwall PVC pipe from FlexPVC.com, who sells it by the foot. Two feet of this with installed fittings will hold about 1400 ml or 85.4 cubic inches, which is plenty adequate for a good calibration. You can use 1/2" CPVC for the vent tube or purchase clear rigid 1/4" pipe from FlexPVC. The top of the calibrator (left side in picture) has a coupling and plug glued to it. The vent tube is glued into an offset hole drilled in the plug. The offset aids in filling from the bottom. Glue a 2" slip x FIP adapter to the bottom of the 2" clear pipe. Glue a 2" slip x 1/2" FIP bushing into a 2" slip x MIP adapter and screw a brass bushing and needle valve into it. Apply some grease to the threads and screw this assembly into the bottom adapter. Measure required length of vent tube with bottom adapter screwed in before gluing the vent tube into the top plug. Table 5-5 shows the hardware components and costs required to build this calibrator. Total cost is about $45.

Table 5-5. Rain gauge calibrator costs.

Clear thinwall 2-inch PVC pipe, 24 inches long	FlexPVC.com	1	$15.86
Brass 1/2" MIP x 1/8" FIP bushing, #748381	Lowe's	1	$4.99
Brass needle valve 1/8" MIP, #748440	Lowe's	1	$5.99
2" PVC plug, #51457	Lowe's	1	$2.48
2" PVC coupling, #23902	Lowe's	1	$0.98
2" PVC slip x MIP adapter, #23904	Lowe's	1	$1.32
2" PVC slip x 1/2" FIP bushing, #51013	Lowe's	1	$2.08
2" PVC slip x FIP adapter, #23906	Lowe's	1	$1.53
1/2" CPVC x 5' pipe, #23811	Lowe's	1	$2.86
Oatey all-purpose cement, #23541	Lowe's	1	$6.48

Build a wood frame (or other material) to suspend this calibrator above your TBR rain gauge so that you can adjust drip rate with the needle valve. To use this calibrator you first fill it with a precisely measured amount of water. Use a graduated cylinder to measure the amount of water poured into the calibrator. I use a 500 ml graduated cylinder to measure out 1400 ml of water. Turn the calibrator upside down, unscrew the 2" fitting with needle valve and pour the water in. Reassemble and suspend it above your rain gauge on the wood frame. Adjust the needle valve so that time between bucket tips simulates your desired rainfall rate (e.g. 24 seconds for a 1.5 in/hr rainfall rate, 10 sec for a 3.6 in/hr rainfall rate, etc.). After the calibrator stops dripping read the total rainfall on your first-flush controller and compare with calculated rainfall total based on measured amount of water and capture area of your gauge.

For example, a Rainwise TBR gauge has a capture diameter of 8.09 inches. This gives a capture area of 51.45 sq. in. Therefore, 85.24 in³ (1400 ml) of water dripped into this gauge equals 85.24/51.45 = 1.66 inches of rainfall. Repeated calibration measurements on this gauge with the above calibrator varied from 1.63 to 1.68 inches, with an average of 1.65 inches at a rainfall rate of 1.5 in/hr. Average systematic error was 0.6% at this slow rate. Maximum observed random error was +/-1.5%. This gauge actually performed much better than my Davis Instruments TBR and is the one I recommend for my customers if they choose to use a TBR gauge.

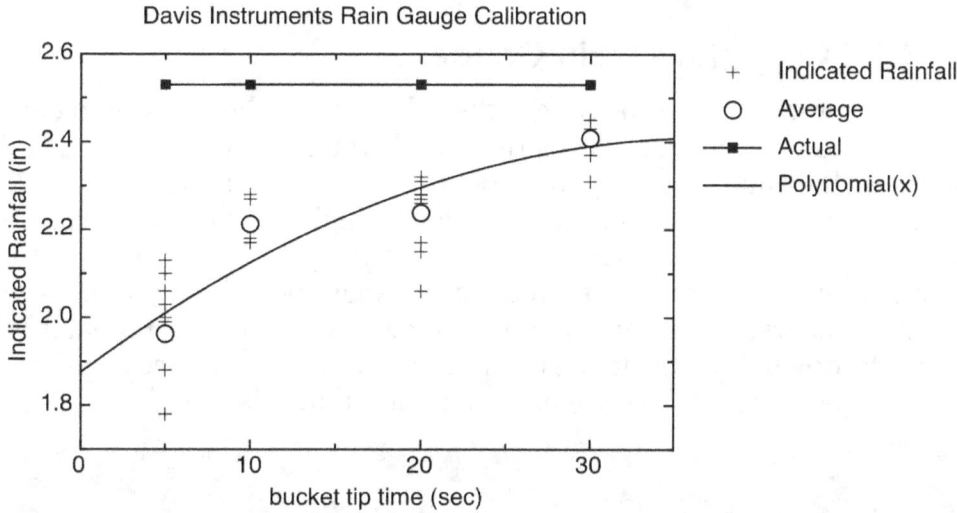

Fig. 5-16. Calibration data points for Davis Instruments TBR.

My Davis Instruments TBR gauge has a capture area of 33.82 in². So 1400 ml of water dripped into this gauge should give 2.53 inches of rainfall on my controller. Figure 5-16 shows actual data points for a very lengthy calibration test of this gauge. Four different bucket tipping times were set on the calibrator and ten measurements were taken at each tipping time. Note how the scattering in data points increases as simulated rainfall rate increases (decreasing bucket tip time). This random error cannot be removed with calibration. The plot also shows a second order polynomial curve fit through the data for correcting systematic error of the form:

$$y = a_0 + a_1 x + a_2 x^2 \tag{5.9}$$

with $\{a_0, a_1, a_2\} = \{1.8754, 2.8858E\text{-}2, -3.897E\text{-}4\}$. This equation can be programmed into the first flush controller Arduino sketch to correct measured rainfall amounts if time between bucket tips is also recorded.

In spite of several days of calibration effort, I am not at all satisfied with this calibration. Random errors are huge and a second order polynomial through the data points is really not much better than a simple straight line through those data points. This calibration data is consistent with the large variability I saw in the year-long data collection for actual rainfall events, where I saw a mean under-reporting error of 14% with a standard deviation of 7%. No doubt Davis Instruments also saw similar huge random and systematic errors during calibration testing and simply concluded that "the standard drip method does not work." So I currently do not use my TBR to measure actual rainfall. I only use it as a rainfall indicator input for my first-flush valve controller and I use a more accurate manual gauge for actual rainfall measurement. I will eventually replace this TBR with a more accurate capacitive rain gauge, which we discuss in the next section.

5.5.2 Capacitive Rain Gauge

With all the problems of TBR gauges there has to be a better way to electronically and accurately read rainfall, regardless of rainfall rate. In fact there is. And you can build it for about the same money you'll spend on a TBR. Furthermore, unlike a TBR gauge, you can even accurately measure extremely high rainfall rates with this gauge. Figure 5-17 shows one of many prototype rain gauges I fabricated with common PVC fittings. This rain gauge incorporates the same capacitive sensor concept we use for our water level sensors, but is much more precise. We want to repeatably measure rainfall amounts down to 0.01 inches resolution. The key to accomplishing this is temperature compensation, which we discuss in detail in this section.

Fig. 5-17. Low-cost capacitive rain gauge

A capacitive rain gauge is easily fabricated from standard plumbing hardware found at Lowes or Home Depot or purchased on line. The prototype shown in Fig. 5-17 consists of a concentric electrode cylindrical capacitor connected to an NE555 oscillator circuit. The center electrode is 1/2" copper pipe insulated with thin-wall plastic tubing. The outer electrode is stainless steel pipe in contact with captured rainwater. Copper or aluminum on this outer bare electrode will corrode and cause capacitor ESR to increase, so we use stainless steel. The capture funnel is 4 inches diameter and the measuring cell is 2-inch clear PVC. An inside bevel is machined on the 3-inch side of a 3"× 2" PVC reducer coupling with a lathe or milling machine to provide a 4-inch diameter knife edge capture area. (The picture shows a 3" coupling and a 2"× 3" reducer bushing.) This reducer is attached to the top of a clear 2-inch PVC pipe. The oscillator circuit is housed in 2-inch fittings at the bottom, similar to the tank level sensor housings. The center copper electrode extends out the bottom into the circuit housing to thermally conduct water temperature to the oscillator for temperature compensation.

Like the tank level sensors, rain gauge probe capacitance is a linear function of water height, x, according to:

$$C = C_1 + C_2 x \qquad (5\text{-}10)$$

where C_1 is residual capacitance of an empty rain gauge and C_2 is the capacitance per inch of water on the center insulated electrode. We locate the oscillator circuit in the base of the rain gauge so that the circuit can sense temperature on the copper probe and compensate for temperature-dependent frequency changes. The oscillator circuit is basically the same as that of Fig. 5-12. Oscillator frequency output as a function of water column height is given by:

$$f = \frac{1.443}{(C_1 + C_2 x)(R_1 + R_2)} \qquad (5\text{-}11)$$

Inverting this equation, solving for x, and including a magnification factor, k, that accounts for a larger capture area than the capacitive cell area gives:

$$x = \frac{1.443}{k(R_1 + R_2)C_2 f} - \frac{C_1}{kC_2} \qquad (5\text{-}12)$$

Magnification constant k is the actual water height in gauge for one inch of rainfall and equals funnel capture area divided by clear cross sectional area inside the sensor cell. For example, if catchment diameter is 4 inches, cell tube ID is 2 inches, and the insulated center electrode OD is 0.75 inches, then magnification factor equals $4^2/(2^2 - 0.75^2) = 4.65$. Thus, water height after 1 inch of rainfall would be 4.65 inches. This magnification constant is most easily determined experimentally by dumping in a measured volume of water equivalent to 1 inch of rainfall and measuring height change in the cell. For example, if capture area is 4.0 inches diameter then you would dump in 12.57 cubic inches of water (or 16.387 ml/in³ × 12.57 in³ = 205.9 ml of water) and measure the change in water height. Place calibration marks on outer surface of cell and continue adding measured amounts of water until the cell is full. These marks will be used for subsequent calibrations.

Using a catchment area larger than the capacitive cell tube ID provides greater gauge sensitivity, but also makes overall gauge length longer. Sensitivity versus length is a tradeoff that one has to make. In my area single rainfall events rarely go over three inches in one day. When they do I can always use my manual gauge to record rainfall data for my WaterStorage spreadsheet. So if I set maximum rainfall capture to 3 inches before having to dump the gauge and magnification factor equals 4.65 then the cell length must be 14 inches.

These capacitive gauges must incorporate an athermalized oscillator circuit, in order to maintain a constant reading to 0.01 inch accuracy over expected temperature ranges. All plastics exhibit changes in dielectric constant with temperature. Thermal expansion of materials also affects capacitance. Capacitance variation with temperature results in frequency variation with temperature. A change in frequency with temperature causes an error in measured rainfall amount. These temperature-dependent changes are quantified by a component's temperature coefficient (TempCo), typically measured in parts per million (ppm) per degree of temperature. This temperature coefficient can be either positive (resistance or capacitance increases with temperature) or negative (resistance or capacitance decreases with temperature). Temperature compensation of electronic circuits is achieved by including circuit components with temperature coefficients that oppose the unwanted temperature-induced change. You cannot completely eliminate all temperature-induced changes, especially if those changes are nonlinear, but if they can be reduced so that they affect rainfall display by less than one hundredth of an inch then we have achieved our goal.

Starting with the basic equation for oscillator frequency, $f = 1.443/RC$, taking the derivative of both sides with respect to temperature and rearranging gives:

$$\frac{1}{f}\frac{df}{dT} = -\left(\frac{1}{R}\frac{dR}{dT} + \frac{1}{C}\frac{dC}{dT}\right) \tag{5-13}$$

The quantity on the left of Eq. (5-13) is the frequency TempCo and the two quantities on the right side are the resistance TempCo and capacitance TempCo respectively. All three temperature coefficients are in units of ppm per degree Fahrenheit (or degree Celsius).[41] Since temperature is in the denominator of these units and $\Delta T_C = 5\Delta T_F/9$, a temperature coefficient in ppm/°C (or ppm/°K) is numerically larger by a factor of 9/5 than the temperature coefficient in ppm/°F (or ppm/°R). Both resistance and capacitance typically change with temperature. Equation (5-13) shows that in order for the change in frequency with temperature to be zero ($df/dT = 0$) the resistance and capacitive temperature coefficients in the oscillator circuit must be equal and opposite. Or in other words, if one temperature coefficient is positive then the other must be negative, according to Eq. (5-14).

$$\frac{1}{R}\frac{dR}{dT} = \frac{-1}{C}\frac{dC}{dT} \tag{5-14}$$

If the rain sensor capacitor TempCo is positive, then we can add a negative temperature coefficient (NTC) thermistor and adjust it to the desired resistive TempCo with other resistors. (We show below that adding a NTC capacitor in parallel will not work.) We can modify the oscillator circuit of Fig. 5-12 to include an NTC thermistor and a trimmer resistance potentiometer as shown in Fig. 5-18. Thermistors usually

---

41 Temperature coefficients are usually given in degrees C or K, but there is nothing magic about SI units. The physics is independent of chosen units and the mathematics is still valid.

have an order of magnitude higher temperature coefficient than we need; the potentiometer allows us to adjust it down to match the capacitive TempCo of the rain gauge sensor. Very low TempCo potentiometers (expensive) should be used for this.

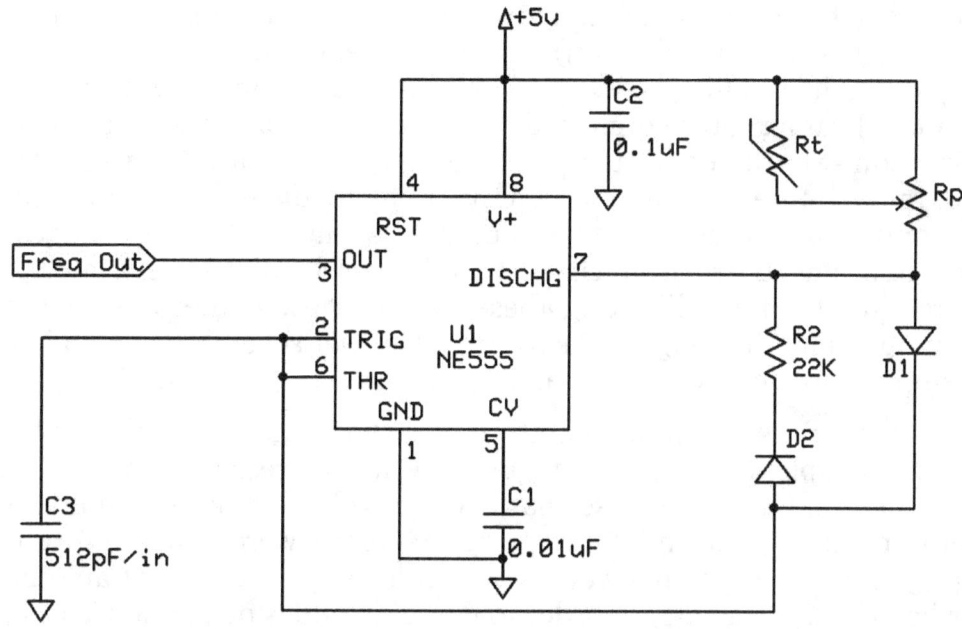

Fig. 5-18. Temperature compensated 555 oscillator circuit.

The capacitive temperature coefficient of a gauge is easy to measure, although it is a fairly lengthy process. This is best done with the oscillator circuit sitting outside the gauge housing in a stabilized room temperature environment so that resistance changes with temperature are near zero, i.e. $dR/dT=0$ in Eq. (5-13), and changes in frequency represents only capacitance changes with temperature. Attach a thermocouple or other temperature probe to the center insulated electrode. Connect a frequency counter to the output of the oscillator or change the Arduino sketch to print frequency. If you use a thermistor as a temperature probe then you can connect it as one of two resistors in a voltage divider. An Arduino can then read voltage on an analog input pin and then immediately read frequency on the digital T1 pin input. This data can then be sent to a computer and analyzed in a spreadsheet to determine the capacitive temperature coefficient.

For my initial measurements of temperature coefficient, I simply dumped a measured amount of ice cold rainwater into the gauge and recorded data points of frequency and temperature as the water slowly warmed up to room temperature. I then repeated the experiment with an identical amount of hot water dumped in and let it slowly cool down to room temperature. However, there was always a discontinuity in the plot of frequency versus temperature at the room temperature point. This is a hysteresis effect due to the different directions of heat flow into or out of the gauge. Measured temperature is not exactly equal to average temperature of the dielectric, since we place the probe on a metal surface in contact with the dielectric but not actually inside the dielectric. For heat flowing into the gauge (warm-up) dielectric temperature will be hotter than temperature readings on the inside electrode. For heat flowing out of gauge (cool-down) dielectric temperature will be cooler than temperature readings. This problem was eliminated by building a test cell wrapped with plumbers heating cable connected to an autotransformer so that heat flows only in one direction over the entire temperature range of about 40°F to 100°F. Although measured temperature is still not exactly equal to dielectric temperature, we are interested only in temperature changes and can assume that these changes at point of measurement are nearly equal to changes in the nearby dielectric. I later purchased a compact refrigerator for calibrating these rain gauges. A measured amount of hot water is dumped into the gauge and slowly cooled over several hours while a computer collects temperature and frequency data.

I used a thermistor for temperature measurements that was calibrated to an Omega thermocouple digital thermometer. Thermistor resistance versus temperature was curve fit to Eq. (5-18) below so that Arduino voltage readings could be converted to temperature in a spreadsheet. Frequency readings were converted to capacitance using Eq. (5-11). Measurements were repeated for several different amounts of water (or water heights, x, in gauge) in order to curve fit values of C_1 and C_2 as a function of temperature. I found that capacitive TempCo actually varied with water height with a concentric tube type capacitor, which is undesirable. We want this temperature coefficient to remain constant so that it can be corrected with a thermistor. I later discovered that a wire coil type capacitive sensor eliminated this variation in capacitive temperature coefficient. The reason for this effect is explained below near the end of this section.

Referring to Fig. 5-18, if R_t is thermistor resistance, R_p is potentiometer resistance, and R_a is the portion of potentiometer resistance in parallel with the thermistor, then charging resistor, R_1, is given by Eq. (5-15).

$$R_1 = \frac{R_t R_a}{R_t + R_a} + R_p - R_a \qquad (5\text{-}15)$$

Equation (5-15) is substituted into Eq. (5-11) to determine frequency dependence on this resistor network. We want $R_1 = R_2$ in order to obtain a 50% duty cycle[42] and we want the TempCo of the combination $R_1 + R_2$ to exactly cancel the TempCo of $C_1 + C_2 x$, giving a stable frequency over the expected temperature range. Taking the derivative of $R = R_1 + R_2$ with respect to temperature and dividing by total resistance gives the

---

42 This only occurs at one temperature point with above circuit. To maintain 50% duty cycle over the entire temperature range requires a similar thermistor-potentiometer network for R2.

temperature coefficient of the total resistance combination in Fig. 5-18:

$$\frac{1}{R}\frac{dR}{dT}=\frac{R_a^2\dfrac{dR_t}{dT}}{2R_2(R_t+R_a)^2}=\frac{R_t R_a^2}{2R_2(R_t+R_a)^2}\frac{1}{R_t}\frac{dR_t}{dT} \tag{5-16}$$

where we have assumed temperature coefficients of other resistances are essentially zero compared to that of the thermistor. Setting Eq. (5-15) equal to R_2 for 50% duty cycle and solving for R_a with the help of the quadratic formula gives:

$$R_a=\frac{-(R_2-R_p)\pm\sqrt{(R_2-R_p)^2-4R_t(R_2-R_p)}}{2} \tag{5-17}$$

Of the two roots for R_a, only the positive root for resistance is meaningful. Since R_a does not vary with temperature, but R_t does, we need to pick an appropriate temperature for calculating R_a. The nominal temperature of 77°F or 25°C is a bit high for our purposes. A temperature closer to the mid-range of outdoor non-freezing temperatures when rain is likely to occur is more appropriate (e.g. 65°F). Of course this is geography dependent.

Equations (5-16) and (5-17) can be placed in a spreadsheet to determine values for R_2, R_p, and R_t that give a resistive TempCo equal and opposite to our measured capacitive TempCo. The thermistor TempCo, the value on the right side of Eq. (5-16), can either be determined from measured data or from a specification data sheet on the thermistor. Thermistor resistance is typically given by:

$$R_t=R_0\exp\!\left(\beta\left(T^{-1}-T_0^{-1}\right)\right) \tag{5-18}$$

where temperature is absolute temperature in either degrees Kelvin or degrees Rankine, R_0 is thermistor resistance at room temperature (25°C or 77°F), and T_0 is room temperature in absolute scale (298°K or 537°R). The beta constant is typically given in units of Kelvins in data sheets, but this is easily converted to degrees Rankine by simply multiplying by 9/5. For example, Digikey.com gives the beta constant for a 47K thermistor, part number 490-14492-1-ND, as 4050°K (B25/50)[43], which is equivalent to 7290°R. I measured temperature and resistance data on this thermistor over a range of 39°F to 104°F and curve fit it to Eq. (5-18) after converting to degrees Rankine. The resulting R_0 and β were 45340Ω and 7925°R respectively. The thermistor TempCo is now easily determined by taking the derivative of Eq. (5-18):

---

43 Beta is normally calculated at two given temperature points with $\beta=\ln(R_{T1}/R_{T2})/(T1^{-1}-T2^{-1})$. A beta of B25/50 means that it was calculated at 25°C and 50°C with nominal resistance given at 25°C. Curve fitting over a different temperature range may yield a slightly different beta since Eq (5-18) is only an approximation to the actual response curve.

$$\frac{1}{R_t}\frac{dR_t}{dT}=\frac{-\beta}{T^2} \tag{5-19}$$

This value is inserted into Eq (5-16) to calculate total resistor network TempCo.

After presetting the potentiometer according to the above equations, we can fine tune it to compensate for smaller temperature effects in the rest of the circuit. This is best done after the gauge has been fully assembled. With two or three different constant water levels in the gauge, we want the output rainfall reading to remain constant within 0.01 inches over a typical outdoor temperature range when rainfall normally occurs (~40°F to 100°F). This is the instrument calibration testing that every gauge should be put through by its manufacturer, proving that the gauge does what it is supposed to do. And there is nothing proprietary about this information!

Frequency vs Temperature

original and compensated

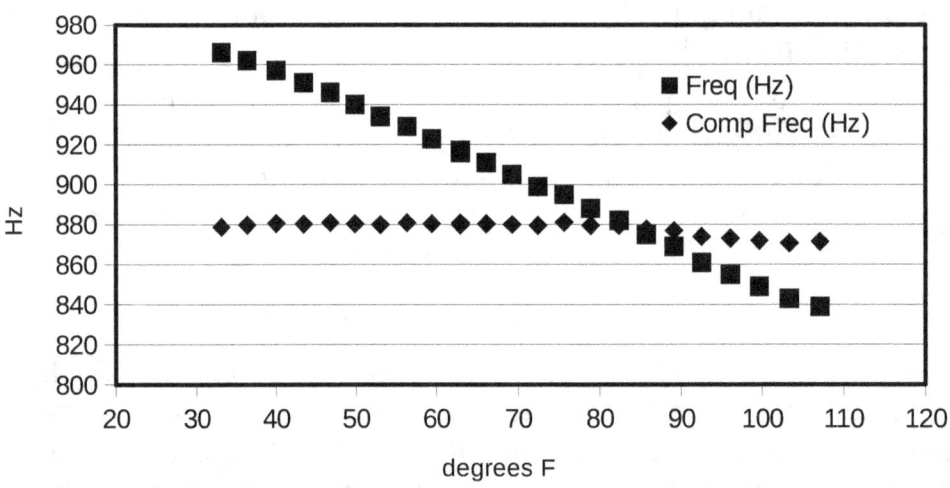

Fig. 5-19. Frequency versus temperature with PVDF and x=16.5 inches water.

As an example to show how we put all this together, consider the Fig. 5-19 plot of measured frequency versus temperature for a fixed amount of water in my test cell for one of the many rain gauge probes I built. Water height on probe is 16.5 inches. With a magnification factor of 4.65 this is equivalent to 3.55 inches of measured rainfall. Frequency decreases with temperature and so capacitance increases with temperature in accordance with Eq (5-13). The slope of this plot divided by frequency at any point gives the frequency TempCo at that point, which equals the negative of the capacitance TempCo, assuming resistance remains constant. Since the plot is reasonably a straight line we can calculate the TempCo using just the two end points: (966 Hz @ 33°F) and

(839 Hz @ 107°F). This gives a frequency TempCo of -1902 ppm/°F at 70°F, which means the capacitive TempCo must be +1902 ppm/°F.[44] So we want a resistor/thermistor combination that gives a resistive TempCo of -1902 ppm/°F at 70°F and matching frequency TempCo at all other temperatures as close as possible, while maintaining a 50% duty cycle.

We first select the thermistor we want to use and calculate its TempCo according to Eq (5-19). For this example we will choose a thermistor with a nominal resistance of 45340 ohms and beta of 7924.56 (previously calibrated thermistor). From Eq (5-19) this gives a thermistor TempCo of -27480 ppm/°F at 77°F (537°R). Then we use Equations (5-17) and (5-16) in a spreadsheet to find the combination of R_2, R_p, and R_t that gives a resistive TempCo of -1902 ppm/°F. After a few educated guesses we find a solution of 100K for R_2 and 130940 ohms for R_p. (We put a resistor in series with a standard 100K trimmer pot to achieve the latter.) This combination of R_2, R_p, and R_t gives a value of 55.993K for R_a.

Lastly, we check the plot of the compensated frequency versus temperature, using Eq (5-18) to calculate thermistor resistance versus temperature, Eq (5-15) to calculate R_1, and Eq (5-11) to calculate frequency with the measured capacitance versus temperature data. Results are plotted in Fig. 5-19, with squares being the original measurements and diamonds being the compensated frequencies. Temperature compensation reduces the frequency variation by a factor of 12.4.

Rainfall is calculated from Eq. (5-12) as a function of frequency, assuming constant values for resistance and capacitance. Taking the derivative of Eq. (5-12) with respect to temperature and assuming $C_1 \ll C_2 x$ gives the temperature coefficient of calculated rainfall:

$$\frac{1}{x}\frac{dx}{dT} = \frac{-1}{f}\frac{df}{dT} \tag{5-20}$$

Equation (5-20) shows that calculated rainfall temperature coefficient has the same magnitude as frequency TempCo. Therefore, if we want rainfall readings to remain constant within 0.01 inches over a 70-degree temperature range (30 to 100°F) at, for example, 2 inches of rainfall, then the TempCo must be reduced to less than 71 ppm/°F.

From Fig. 5-19 and an assumed magnification factor of 4.65 we see that an uncompensated frequency variation of 966 Hz to 839 Hz over the temperature range gives a rainfall error of +/-0.25 inches at 3.55 inches measured rainfall.[45] With temperature compensation the TempCo is reduced to 138 ppm/°F and rainfall error is

---

44 Temperature coefficient at 70 degrees = 2*(966 − 839)/(966 + 839)/(33 − 107). Multiply by 1E6 to get ppm/°F

45 Total error = 0.001902 * 3.55" * (107 - 33)

reduced to +/-0.018 inches at 3.55 inches. At 2.00 inches this error is reduced to +/-0.01 inches which meets our goal. Temperature compensation is necessary to reduce rainfall measurement error.

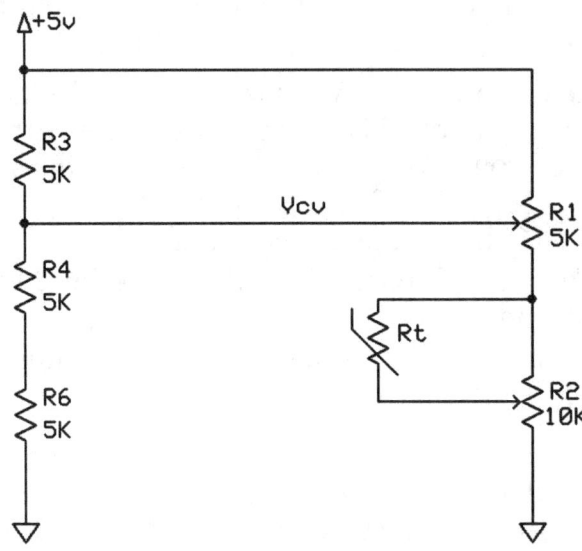

Fig. 5-20. Alternative athermalization means using CV input to NE555.

Figure 5-20 shows an alternative method of temperature compensation using the CV (control voltage, pin 5) input to the NE555. This method has some advantages over that shown in Fig. 5-18. In Fig. 5-18 the capacitor charging current travels through the thermistor, and it is an AC current. In Fig. 5-20 thermistor current is independent of capacitor charging and it is DC. This makes measuring temperature with an Arduino analog input much easier. The three 5K resistors on the left in Fig. 5-20 (R3, R4, and R6) are the voltage divider resistors inside the NE555 IC. CV pin 5 is connected to the 2/3 supply voltage point and nominally equals 3.33V for 5V input supply. When the CV pin is pulled lower than 3.33V frequency increases, when pulled higher than 3.33V frequency reduces. The two external potentiometers, R1 and R2, along with the thermistor, Rt, allow temperature compensation on a capacitive sensor with a positive TempCo when connected as shown in Fig. 5-20. R2 adjusts the effect of the thermistor and overall resistive TempCo. R1 re-balances the network so that control voltage is evenly split above and below 3.33V and 50% duty cycle occurs at a mean temperature (~65°F).

Resistance on the right in Fig. 5-20 below the wiper of R1 is easily written as:

$$R_N = (R_1 - R_{1a}) + \frac{R_t R_{2a}}{R_t + R_{2a}} + (R_2 - R_{2a}) \tag{5-21}$$

where R_{1a} and R_{2a} are resistances above their respective potentiometer wiper. Using the symbol "||" to denote the parallel combination of two resistors, i.e. $R_x || R_y = R_x R_y / (R_x + R_y)$, we can easily write the control voltage in Fig. 5-20 as:

$$V_{CV} = 5 \frac{R_N \| 10\,K}{R_{1a} \| 5\,K + R_N \| 10\,K} \tag{5-22}$$

Equations (5-21) and (5-22) can be inserted into a spreadsheet to determine potentiometer settings for a given thermistor that achieves a small change in control voltage above and below the nominal point at the extreme ends of your temperature range (e.g. 40°F and 90°F). Effects of control voltage change on frequency output from the NE555 is most easily determined with the LTspice software.[46] Simply adjust the two potentiometers until the resistance TempCo exactly cancels your measured capacitive TempCo.

Different plastics produce different capacitance versus temperature curves, but most of the plastics I measured showed a positive TempCo in the temperature region of interest. I measured temperature coefficients of over ten different plastics, with capacitive TempCo varying from a low of 372 ppm/°F for one particular type of polyolefin to over 7000 ppm/°F for vinyl. Even within the same plastic family there was huge variation according to color and thickness. For example, a thin gray polyolefin showed a fairly linear capacitance versus temperature plot, but a thicker red polyolefin gave a highly nonlinear plot. A similar effect showed up with PVC. There was little difference between a stainless steel and copper center electrode, indicating that thermal expansion of the metal probe was not a significant factor in capacitance variation.

Substituting $C = C_1 + C_2 k x$ into the right side of Eq. (5-14), where $k x$ is water height in gauge and x is actual rainfall amount, shows how the capacitive temperature coefficient changes with water height:

$$\frac{1}{C}\frac{dC}{dT} = \frac{1}{C_1 + C_2 k x}\left[\frac{dC_1}{dT} + k x \frac{dC_2}{dT}\right] \tag{5-23}$$

Equation (5-23) shows that the capacitive temperature coefficient approaches that of C_2 for large x and that of C_1 for small x. If the two capacitive temperature coefficients are not identical then the overall capacitive TempCo will change with rainfall amount for small x and Eq. (5-14) would not be satisfied, which is undesirable. We want the combined capacitive TempCo to remain constant over the entire range of water levels so that the thermistor/resistor network can compensate it. So C_1 cannot be a temperature compensating capacitor. Ideally, it should have a TempCo exactly equal to that of C_2. In other words, we want $C_1^{-1} dC_1/dT = C_2^{-1} dC_2/dT$. This is generally not the case with concentric cylindrical capacitors because the dielectric for C_1 is mostly air whereas the dielectric for C_2 is plastic. These different dielectrics make the two capacitive temperature coefficients unequal.

---

46 Freeware developed by Linear Technology. Can be downloaded at: www.analog.com/en/design-center/design-tools-and-calculators/ltspice-simulator.html

As an example, Fig. 5-21 is a plot of Eq. (5-23) with a C_1 temperature coefficient of 143 ppm/°F and capacitance of 105 pF and a C_2 temperature coefficient of 1702 ppm/°F and capacitance of 512 pF/in. Note that for water heights greater than about 3 inches the capacitive TempCo is reasonably flat and approximately equal to the TempCo of C_2. We can correct this region with an equal and opposite resistive TempCo. However, for water heights less than 2 inches the TempCo rapidly drops to that of C_1. This region cannot be corrected with a constant resistive TempCo, resulting in frequency changing with temperature, right in a region where we want the most accuracy. Therefore, the two capacitive temperature coefficients must be nearly identical.

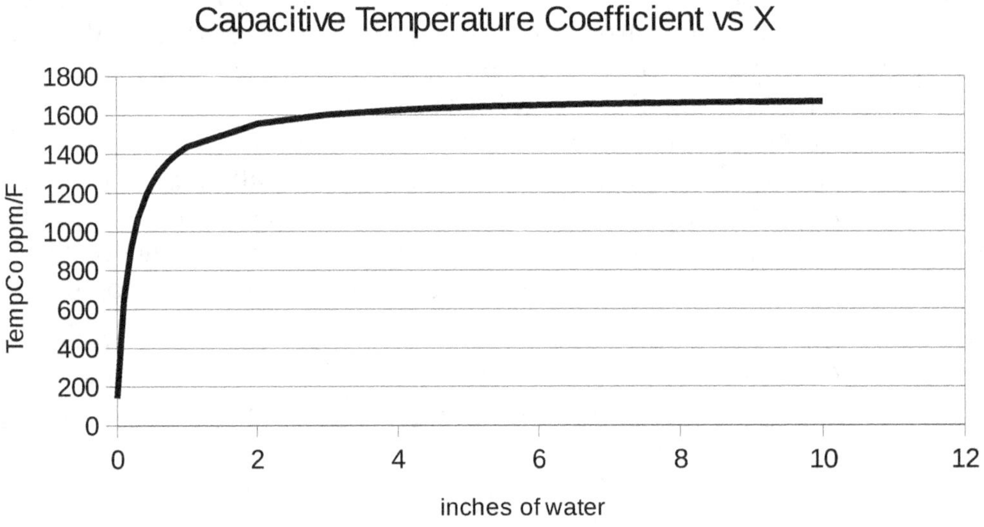

Fig. 5-21. Plot of Eq. (5-21) with C_1 and C_1 having unequal temp. coefficients.

A way to make the TempCo of C_1 more nearly equal to that of C_2 is to wrap a coil of wire around the center electrode on top of the plastic dielectric instead of using a concentric metal cylinder. Although this increases the value of C_1, its dielectric constant is now the same as that for C_2. Temperature coefficient measurements on both types of capacitors (concentric metal cylinders and wire coil) confirmed the above analysis. So I abandoned the concentric cylinder approach. A primary problem with the wire coil approach is that water droplets can cling to the coil, increasing capacitance slightly and causing a reading error. We want the coil to have sufficient surface area to keep ESR low with low conductivity water, but not so large that it reduces C_2 by obscuring surface area of the center electrode.

Alternatively, instead of trying to temperature compensate with hardware, corrections can be mathematically applied in software. An NTC thermistor would simply measure temperature and the Arduino would use this measurement to correct measured frequency. If capacitance changes linearly with temperature for both C_1 and C_2 then you can calculate their values with a pair of linear equations according to:

$$\begin{pmatrix} C_1 \\ C_2 \end{pmatrix} = \begin{pmatrix} C_{10} & dC_1/dT \\ C_{20} & dC_2/dT \end{pmatrix} \begin{pmatrix} 1 \\ T-T_0 \end{pmatrix} \tag{5-24}$$

where C_{10} and C_{20} are capacitances at the reference temperature, T_0. These new capacitance values are then inserted into Eq. (5-12) to calculate rainfall. Or you could use more complex nonlinear equations if capacitance changes nonlinearly with temperature. This basically becomes a 3D curve fitting problem of finding a function, $z(x,y)$, that gives water level as a function of temperature and frequency. For dielectrics with nonlinear temperature curves I have found that shifted inverse hyperbolic functions seem to provide good fits to data.

Note that the above single probe approach depends on a reasonably constant conductivity of rainfall so that ESR remains constant. Rainwater conductivity is generally low, less than 10 micro-Siemens per cm, but it does vary. A change in water conductivity can cause oscillator frequency to change slightly due to ESR. One way to reduce this problem is to use two parallel insulated probes for the capacitive sensor. This makes probe capacitance dependent on the dielectric constant of water rather than its conductivity. Although this approach does bring back the problem of a change in dielectric constant from water to air, as with the concentric cylinders mentioned earlier, water's dielectric constant is about 80 times higher than that of air which helps mitigate the problem. Capacitance of two parallel wires of length h, both of radius r, separated by a center-to-center distance d, is given by:[47]

$$C = \frac{\pi \epsilon_0 \kappa h}{\cosh^{-1}(d^2/2r^2-1)} \tag{5-25}$$

where $\epsilon_0 = 8.854 \times 10^{-12} F/m$ and κ is the dielectric constant of the material surrounding the wires. The inverse hyperbolic cosine in the denominator can be converted to its logarithmic equivalent for numerical calculation purposes. The surrounding material is either air with a dielectric constant of 1 or water with dielectric constant of about 80. The dielectric constant of water does change with temperature, as well as with dissolved air (carbon dioxide, oxygen, and nitrogen) in it. So temperature correction either in hardware or software is still necessary.

Of the three types of capacitive gauges discussed above, this dipole capacitive gauge is the easiest to fabricate. Winding a stainless steel coil around an insulated probe and holding it in place without damaging the insulation requires care. As of this writing I am still in the process of comparing performance between these two latter

---

47 See www.emisoftware.com/calculator/wire-pair-capacitance

types of capacitive gauges. Dielectric constant of water does appear to change slightly with conductivity, making the dual probe sensor also slightly sensitive to conductivity. The two-probe sensor also has a lower capacitance per inch than the wire coil-wrapped single probe. Not surprisingly, tradeoffs exist between these two types of capacitive sensors. On-going research may ultimately show that a water conductivity sensor is also required with these capacitive rain gauges in order to achieve desired accuracy. With a wire coil probe this means including one additional small stainless steel probe in bottom of cell (using the wire coil as the second electrode).

Finally, although Fig. 5-18 and Fig. 5-20 serve to explain temperature compensation with a thermistor, neither figure shows the best way to incorporate trimmer potentiometers. A fixed resistor with a low TempCo should be placed in series with a trimmer potentiometer to reduce potentiometer TempCo effects and protect it from over-current conditions. Low TempCo potentiometers tend to be rather expensive. Determine the range of resistance adjustment you actually need and size the potentiometer for this range. Then select a fixed series resistor to set total resistance equal to R_p. Also, including a second potentiometer and series resistor in place of R_2 in Fig. 5-18 will make calibration adjustments much easier. Increasing R_a will increase the effect of the thermistor and raise the high temperature frequency relative to the low temperature frequency, rotating the frequency versus temperature plot counterclockwise. This makes adjustment of R_2 necessary in order to restore a 50% duty cycle at the mid-range temperature. A second trimmer potentiometer at this point facilitates this re-balancing, just as we use two potentiometers in Fig. 5-20.

5.5.3 Acoustic Resonance Rain Gauge

An entirely different approach for a digital rain gauge is to incorporate acoustic resonance. A tube of length L with one closed end and one open end forms a quarter wavelength resonator that resonates at a fundamental frequency given by:

$$f = \frac{c}{4L} \tag{5-26}$$

where c is the speed of sound in air. The water surface forms the closed end of the resonator tube. As water level rises in the gauge this tube effectively becomes shorter and resonates at higher frequencies, providing a simple means to measure rainfall. We locate a small speaker and microphone in the open top end of the tube. A self-starting oscillator is formed, oscillating at the natural resonance of the tube as soon as power is applied, when we feed the microphone output to an amplifier and drive the speaker with the same amplifier. The tube can also resonate at the third harmonic given by:

$$f = \frac{3c}{4L} \tag{5-27}$$

as well as higher odd harmonics. Bandpass filtering can help reduce unwanted resonances. The electronic schematic of this acoustic oscillator is shown in Fig. 5-22. An Arduino reads resonant frequency using the same algorithm shown earlier for capacitive sensors. The Arduino applies a temperature correction and converts resonant frequency into rainfall amount.

Since speed of sound in air is a function of temperature in accordance with:[48]

$$c = 331.6 + 0.6T \qquad (5\text{-}28)$$

where speed is in units of m/s and temperature is in degrees Celsius, an acoustic resonant gauge must be corrected for temperature. Equation (5-28) shows we can expect a frequency TempCo on the order of 1809 ppm/°C or 1005 ppm/°F. As shown in the previous section, this TempCo is sufficiently large so as to require temperature correction. Interestingly, even a TBR should require temperature correction to obtain desired accuracies, but its random errors are so huge that temperature correction makes no sense at all with a TBR. A small correction to cavity length as a function of frequency and pipe diameter should also be applied to Eq. (5-26) and (5-27). This correction is most easily determined experimentally.

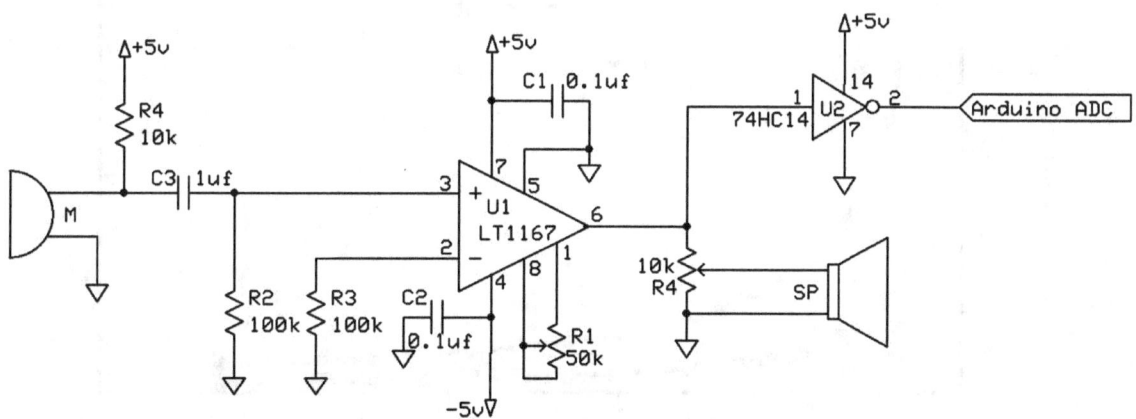

Fig. 5-22. Amplifier circuit for acoustic resonance rain gauge.

One advantage an acoustic resonant gauge has over capacitive gauges is that it is not dependent on water conductivity or dielectric constant. It has other issues, such as annoying emission at audible frequencies that must be handled without destroying the resonant condition. It also has a tendency to suddenly hop into an undesired oscillator mode. Limiting the range of frequencies and careful positioning of speaker and microphone can help prevent this. Also the microphone and speaker must be weatherproof. Placing a waterproof transducer at the bottom of the water column and driving it at higher frequencies (since speed of sound is higher in water) might help

---

48 Lawrence Kinsler, Austin Frey, <u>Fundamentals of Acoustics</u>, John Wiley & Sons, 1950.

mitigate noise, but this also increases transducer cost. Audible range speakers and microphones are much cheaper than ultrasonic transducers.

Figure 5-23 shows resonator air column length as a function of frequency based on Eq. (5-26) and (5-27) at 25°C. Note that a resonator tube of given length has multiple resonant frequencies, all odd-numbered harmonics. Only the fundamental and 3rd harmonic are plotted in Fig. 5-23. This plot shows that the effective range of water heights is shorter for 1st harmonic resonance. The speaker I used had a natural resonance around 450 Hz, so I wanted to stay above 500 Hz. Resonating in the 3rd harmonic allowed this and still provided over 10 inches of water heights. The 3rd harmonic is 180 degrees out of phase with the 1st harmonic. Therefore, if you find your acoustic gauge resonating at 1st harmonic simply reverse the leads on the speaker and it will automatically switch over to 3rd harmonic oscillation.

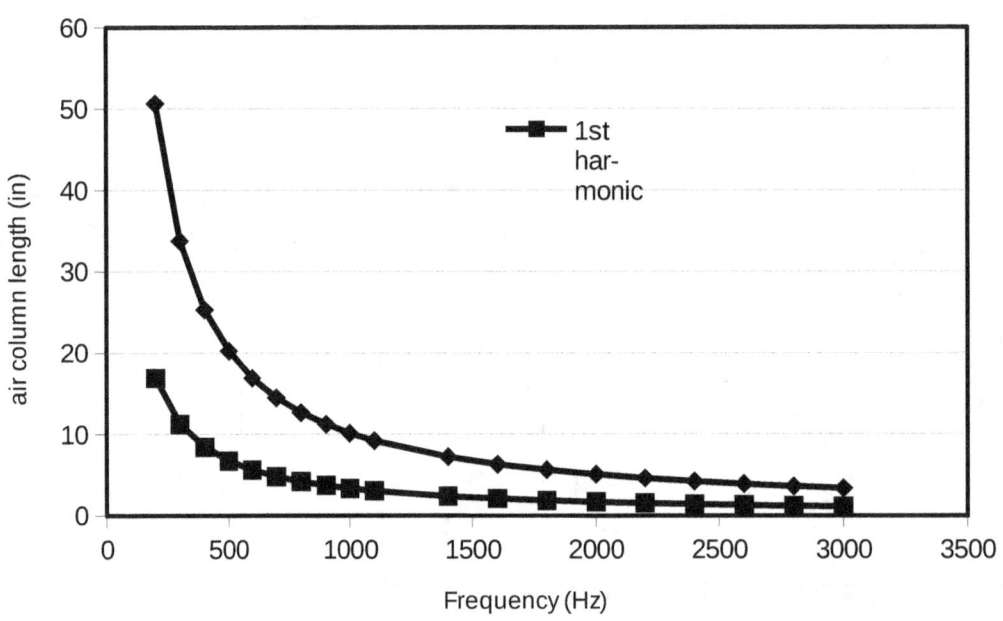

Fig. 5-23. Resonator length versus frequency at 25°C.

The resonance condition of a pipe with one closed end is derived from the complex acoustic impedance at its open end, which includes tube diameter and wavelength, in addition to speed of sound. Resonance occurs when the imaginary part of complex impedance goes to zero. Equations (5-26) and (5-27) are derived from these expressions by making some simplifying approximations and assuming pipe diameter is much smaller than wavelength. As resonating column length becomes shorter and resonant frequency increases this assumption becomes less valid and a more complicated resonant condition can be established, including transverse modes. This becomes obvious with a sudden jump to a lower frequency, which is still a third harmonic (as can be verified by reversing the speaker leads) but does not obey the simple form of Eq. (5-26) or (5-27). On my gauge this occurred at about 2.5 inches with ¾-inch CPVC pipe. To avoid this sudden mode hoping point simply allow sufficient air column length (e.g. 3 inches) when the gauge is full.

My early rain gauge prototypes used clear 2" PVC pipe in the sensor section because this pipe was readily available and low-cost. Constant sensor section diameter gives a constant resolution at all rainfall amounts. This is ideal if one's goal is rainfall rate measurement. However, we rainwater harvesters are mainly interested in high resolution at low rainfall amounts since initial rainfall determines first flush valve control. After a few inches of rainfall when tanks are full and excess rainwater is being dumped to ground we are less interested in resolving rainfall amounts to hundredths of an inch. A simple way to increase resolution for initial rainfall and gradually reduce it as rainfall accumulates is to use a cone-shaped collector with a small diameter at the bottom and large diameter collector. Volume of water in a right circular conic frustum of height, h, with radii, r_1, at the bottom and, r_2, at the top is given by:

$$Vol = \frac{\pi h}{3}\left(r_1^2 + r_1 r_2 + r_2^2\right) \qquad (5\text{-}29)$$

If the collection cone is of height, L, and has a collection diameter, D, then r_1 remains fixed but r_2 varies linearly with water height in this conical collector:

$$r_2 = r_1 + \frac{D - 2r_1}{2L}h \qquad (5\text{-}30)$$

Inserting Eq. (5-30) into Eq. (5-29) and equating the frustum volume of water to captured rainfall volume, $\pi D^2 x/4$, where x is measured rainfall, gives rainfall as a function of water height in the gauge:

$$x = \frac{4}{D^2}\left[r_1^2 h + \frac{D - 2r_1}{2L}r_1 h^2 + \left(\frac{D - 2r_1}{2L}\right)^2 \frac{h^3}{3}\right] \qquad (5\text{-}31)$$

Equation (5-31) shows that magnification factor (previously defined as h/x with gauges of uniform cross section) now varies non-linearly. Magnification factor is the change in water height in gauge per unit change in rainfall amount. For this cone-shaped gauge it equals the reciprocal of the derivative of Eq. (5-31) with respect to h, or reciprocal slope of x plotted as a function of h:

$$\frac{dh}{dx} = \frac{1}{\dfrac{4}{D^2}\left[r_1^2 + \dfrac{D - 2r_1}{2L}2r_1 h + \left(\dfrac{D - 2r_1}{2L}\right)^2 h^2\right]} \qquad (5\text{-}32)$$

Maximum magnification occurs at the bottom of gauge and minimum magnification occurs at the top of gauge. For example, if we have a frustum cone-shaped gauge 10 inches long with a 5-inch collection diameter and 1-inch diameter at the bottom then this gauge will have a magnification factor of 25 when empty and a magnification of 0.89 when full (not including probe volume). Increasing magnification at bottom of gauge is especially helpful with acoustic resonance gauges where resonant frequency is low and only changes by a few Hertz with water height.

Nonlinear magnification helps reduce inherent nonlinearities of frequency versus length, which Fig. 5-23 clearly shows. As the gauge fills (shorter air column lengths) changes in frequency becomes greater. Using Eq. (5-27) to convert Eq. (5-31) into a function of frequency for a given tube length shows a much less nonlinear function, as shown in Fig. 5-24. Although not a perfectly straight line, its slope does not change as rapidly as occurs with a uniform area collection cell (compare with Fig. 5-23).

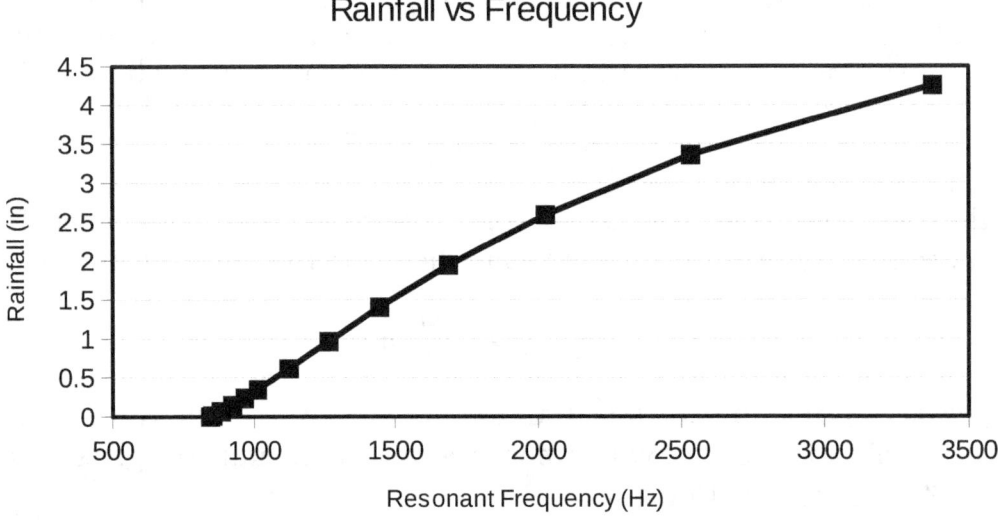

Fig. 5-24. Rainfall versus frequency for 12-inch tube in conical collector at 25°C.

Both capacitive and acoustic resonance rain gauges are storage type gauges. Unlike a TBR, they store rainfall until they are emptied, either manually or electronically. Storage type gauges allow visually checking stored rainfall levels against electronic displayed levels (assuming clear plastic storage section on gauge). A quick visual check immediately verifies sensor working condition. Furthermore, storage type gauges permit repeated measurements of the same sample to overcome random noise. Electronic dumping of storage type gauges is easily accomplished with an inexpensive plastic solenoid valve on bottom of gauge.

With an accurate digital rain gauge and accurate water level sensors, we can now provide immediate and automatic display of system collection efficiency on a system controller. Such a feature would have helped me prevent the loss of over 2 inches of rainfall one night at a time when I desperately needed to collect it. I had noticed that collection efficiency was much lower than normal. After testing gutter screens I assumed they were the problem and cleaned them with chlorine and a pressure washer. The next night we got over 2 inches of rainfall but my system captured very little of it! So I stood in the rain during the next rainfall and watched my first-flush valve. Sure enough water was blowing out the sides of a closed valve and spilling to ground. The rubber plunger had apparently worn to the point that it was no longer holding. I wired it shut with baling wire to capture remaining rainfall until I was able to repair it. This event also prompted development of a custom molded plunger made of food-grade silicone.

5.6 Wireless Control with Raspberry Pi

In section 5.3.1 above I mentioned a wireless connection using a pair of Xbees to overcome the problem of long signal wire runs. Another advantage of a wireless connection includes the ability to display rainwater system data in any format and display it anywhere in your house. With an accurate digital rain gauge that can measure rainfall rate just as easily as total rainfall, a wireless link becomes especially attractive. In fact, you might even find yourself becoming a data collection point for local weather scientists! Rainfall rate as a function of time is important information for weather modeling purposes that very few climatologists and weather scientists have access to. With a Raspberry Pi we can record rainfall and rainfall rate as a function of time into a file and then send it to interested scientists over the Internet.

Instead of using another Arduino for our controller's user-interface in the house, we use a Raspberry Pi single-board computer with a touch screen. The main Arduino-based system controller is hardwired to the system and transmits rainfall, storage levels, and hardware status to a Raspberry Pi at user-selected time intervals. The Raspberry Pi calculates collection efficiency from this data and can sound an alarm if it falls below a user-set value (e.g. 50%). Furthermore, a Raspberry Pi can store rainfall amounts as a function of time and plot both rainfall (inches) and rainfall rate (inches per hour) in whatever time interval a user desires. For example, if two subsequent rainfall readings separated by 5 minutes are 2.0 and 2.5 inches, then the 5-minute rainfall rate at this point in the time record is 0.5 inches / 5 min * 60 min/hr = 6 inches per hour. No doubt, such a high rainfall rate would overwhelm at TBR resulting in huge errors, but a capacitive or acoustic resonance gauge can easily handle it. With a time record of both storage level and rainfall we can then determine how rainwater system collection efficiency varies with rainfall rate.

Implementing a wireless link between the main Arduino controller at storage tanks and a computer in the home running a graphical user interface (GUI) not only provides much greater information to a user, but it also reduces some hardware costs, especially for long runs between storage tanks and house. Controller hardware costs for physical rotary switches, knobs, and LED displays are reduced by about $80 and replaced with software knobs and displays on a computer screen. Much more system information can be displayed on a computer screen and it can be easily changed and organized however a user sees fit. With shorter shielded cable runs between the main Arduino controller and capacitive sensors, we can then run sensors at higher frequencies and reduce high-impedance noise.

Table 5-6. Raspberry Pi controller costs.

Raspberry Pi 3, DEV-13825	Sparkfun.com	1	$39.95
Raspberry Pi LCD 7" Touchscreen, LCD-13733	Sparkfun.com	1	$64.95
Raspberry Pi 16GB MicroSD NOOBS Card, COM-13945	Sparkfun.com	1	$24.95
SmartiPi Touch, PRT-14059	Sparkfun.com	1	$24.95
Wall Adapter Power Supply, USB Micro-B, TOL-13831	Sparkfun.com	1	$7.95
Sparkfun Xbee Explorer Dongle, WRL-11697	Sparkfun.com	1	$24.95
Sparkfun Xbee Shield, WRL-12847	Sparkfun.com	1	$14.95
2.4 GHz XBee Pro 60mW, WRL-08742	Sparkfun.com	2	$75.90
Arduino stackable headers, PRT-11417	Sparkfun.com	1	$1.50

A Raspberry Pi chip computer coupled with a 7-inch touch screen provides an inexpensive powerful hand-held computer that can be placed anywhere in your home. It comes with all the software needed to program GUIs to communicate with an Arduino. You can pick up an inexpensive keyboard and mouse at Walmart to plug into the Raspberry Pi's four USB ports to make programming it easier. I used Python and Tkinter to build a GUI on my Raspberry Pi, but the Pi includes several other programming language choices. This hardware is very easy to assemble and get up and running. Table 5-6 shows hardware components and costs needed to assemble this wireless rainwater system controller, including two XBees. A wireless controller will probably cost slightly more than a completely hardwired controller, depending on savings from not having to run multi-conductor shielded cable in buried conduit to your house, but you also get a fully functional general purpose Linux computer that can do so much more than just display rainfall and total water storage. You can save rainfall versus time data to the Raspberry Pi for later analysis or plotting. And you can do email and internet searches without virus worries, since it's a Linux machine.

The Arduino sketch for the wireless controller is a modified version of that shown in section 5.3. Instead of displaying rainfall and storage amounts, the Arduino stores and continuously updates variables until it receives a command from the Raspberry Pi to transmit current values. Instead of reading physical rotary switch positions (valve control and first flush amount), the Arduino reads software "switches" set by the Raspberry Pi to determine program flow. These switch values remain in Arduino memory until changed by the Raspberry Pi. Therefore, the controller continues to operate if the Raspberry Pi goes off line. Default switch values are assigned in setup( ) in case the Arduino does a hard reboot without the Raspberry Pi. The delay function at the end of loop( ) is no longer needed since the Arduino is not displaying data.

Table 5-7. Raspberry Pi Commands to Arduino.

Byte Command	Action
0x01	Set valve control to automatic
0x02	Set valve control to fixed open
0x03	Set valve control to fixed closed
0x04	Get current valve control setting
0x05	Set first flush to 0.02 inches
0x06	Set first flush to 0.04 inches
0x07	Set first flush to 0.06 inches
0x08	Set first flush to 0.08 inches
0x09	Set first flush to 0.10 inches
0x0A	Set first flush to 0.12 inches
0x0B	Get current first flush setting
0x0C	Get current valve position
0x0D	Get current rainfall reading
0x0E	Reset rainfall to zero, empty gauge
0x0F	Get clean water storage level
0x10	Get raw water storage level
0x11	Reset controller, empty gauge, open valve

The wireless serial link between the Arduino and Raspberry Pi is a "bottleneck" for data flow. Both computers process data much faster than data is transmitted between the two. So we want to minimize amount of data wirelessly transmitted. Two-byte integers have more than enough resolution for both storage levels and rainfall amounts. Rainfall is transmitted in hundredths of an inch as an integer and then converted to inches by the Raspberry Pi. Only seventeen essential commands are needed from the Raspberry Pi, which can easily be handled with single 8-bit bytes. These commands either set software switches on the Arduino, or get current switch positions, or get rainfall and storage levels, or reset hardware. The Raspberry Pi controls data flow across the wireless link, including the rate of transmitting rainfall readings. The Arduino sends no data unless the Raspberry asks for it. If the wireless link is down or if the Raspberry Pi is offline, then the Arduino continues to control the system according to previously set switch positions. The Arduino does not need to be connected to the Raspberry Pi in order to do its job. The Raspberry Pi is simply an interface between the user and the Arduino-controlled rainwater system. Table 5-7 shows the commands I chose to implement.

As Table 5-7 shows, the Raspberry Pi either sets or gets software switches on the Arduino, gets system status information, and resets hardware. The Raspberry Pi also stores rainfall and water level versus time data, if the user so chooses, for subsequent plotting and analysis. Although the Arduino is always running and controlling the system, the Raspberry Pi can be shut down without affecting system control. When it comes back online it first retrieves current settings and displays those to the operator.

The main loop() on the Arduino is shown below. With every cycle through loop( ) the Arduino checks to see if data is waiting in the XBee queue from the Raspberry Pi with the if(Serial1.available()) statement. If there is no data sitting in the Xbee queue then the entire switch(command) case statement is skipped and program flow continues with sensor data collection and processing. The rest of the algorithm is essentially the same as that shown previously in section 5.3.

```
void loop()
{
  byte bh;
  byte bl;
  if (Serial1.available())
    {
    byte command = Serial1.read();
    switch(command)
      {
      case 1:  // set valve control to automatic
          vControl = normal;
          break;
      case 2:  // set valve control to opened
          vControl = opened;
          break;
      case 3:  // set valve control to closed
          vControl = closed;
          break;
      case 4:  // send current valve control setting to Xbee
          Serial1.write(vControl);
          break;
      case 5:  // set first flush amount to 0.02 through 0.12
      case 6:
      case 7:
      case 8:
      case 9:
      case 10:
          firstflush = minFlush * (command - 4);
          break;
      case 11:  // send current first flush setting to Xbee
          Serial1.write(firstflush);
          break;
      case 12:  // send current valve position to Xbee
          Serial1.write(vPosition);
          break;
      case 13:  // send current rainfall reading to Xbee
          bh = highByte(rainfall);
          Serial1.write(bh);
          bl = lowByte(rainfall);
```

```
        Serial1.write(bl);
        break;
    case 14:  // empty gauge and reset rainfall to zero
        emptyGauge();
        break;
    case 15:  // send current clean water storage amount to Xbee
        bh = highByte(cleanStor);
        Serial1.write(bh);
        bl = lowByte(cleanStor);
        Serial1.write(bl);
        break;
    case 16:  // send current raw water storage amount to Xbee
        bh = highByte(rawStor);
        Serial1.write(bh);
        bl = lowByte(rawStor);
        Serial1.write(bl);
        break;
    case 17:  // reset controller, empty gauge, open valve
        emptyGauge();
        vControl = normal;
        firstFlushMet = false;
        break;
    default : // unknown byte value
        break;
    }
  }
rainfall = getRainfall();
if (rainfall >= firstflush)
  firstFlushMet = true;
else
  firstFlushMet = false;
cleanStor = getStorage(clean);
rawStor = getStorage(raw);
switch (vControl)
{
  case normal :
    {
      if( firstFlushMet && tankNotFull() )
      { // then close valve and fill tank, use only motor A if one motor installed
        digitalWrite(DIRA, closevalve);
        analogWrite(PWMA, 255);
        //      digitalWrite(DIRB, closevalve);
        //      analogWrite(PWMB, 255);
        vPosition = closevalve;
      }
      else
      { // then open valve, dump to ground
        digitalWrite(DIRA, openvalve);
        analogWrite(PWMA, 255);
        //      digitalWrite(DIRB, openvalve);
        //      analogWrite(PWMB, 255);
        vPosition = openvalve;
      }
    }
    break;
  case opened :
```

```
      {  // open valve, dump to ground
         digitalWrite(DIRA, openvalve);
         analogWrite(PWMA, 255);
         //    digitalWrite(DIRB, openvalve);
         //    analogWrite(PWMB, 255);
         vPosition = openvalve;
      }
    break;
  case closed :
    {
      if( tankNotFull() )
      { // then close valve and fill tank
        digitalWrite(DIRA, closevalve);
        analogWrite(PWMA, 255);
        //    digitalWrite(DIRB, closevalve);
        //    analogWrite(PWMB, 255);
        vPosition = closevalve;
      }
      else
      { // then open valve, dump to ground
        digitalWrite(DIRA, openvalve);
        analogWrite(PWMA, 255);
        //    digitalWrite(DIRB, openvalve);
        //    analogWrite(PWMB, 255);
        vPosition = openvalve;
      }
    }
    break;
  default : // unrecognized selection
  { }
  }
}
```

A screen shot of the GUI software running on my Raspberry Pi is shown in Fig. 5-25. The dialog box occupies nearly the full 7-inch touch screen so that buttons can be selected with large fingers. The top line provides first flush valve control modes: Automatic, Opened, or Closed. These correspond to the "Switch(vControl)" case statement in the above C code. First flush amount is user selectable with a spin box on the second line. The entry box to the right of this switches between a red "Valve is open" and green "Valve is closed" depending on the variable "vPosition" in the Arduino sketch. The next two lines display measured rainfall and storage. Clean storage should stay nearly topped off at all times. If you see it slowly dropping when sufficient raw water exists, then you know your transfer pump is not working properly. The fifth line allows the user to select whether or not to write rainfall and storage data versus time to a file. File names are automatically generated based on date and time. The last line shows calculated collection efficiency as the ratio of change in storage to water collected by the roof. Currently this drops to zero when there is no rain, as shown in Fig. 5-25, or if storage is topped off and no more water can be added. Although not yet implemented in this software, rainfall and maximum storage capacity could be used to prevent a false alarm when collection efficiency drops to zero.

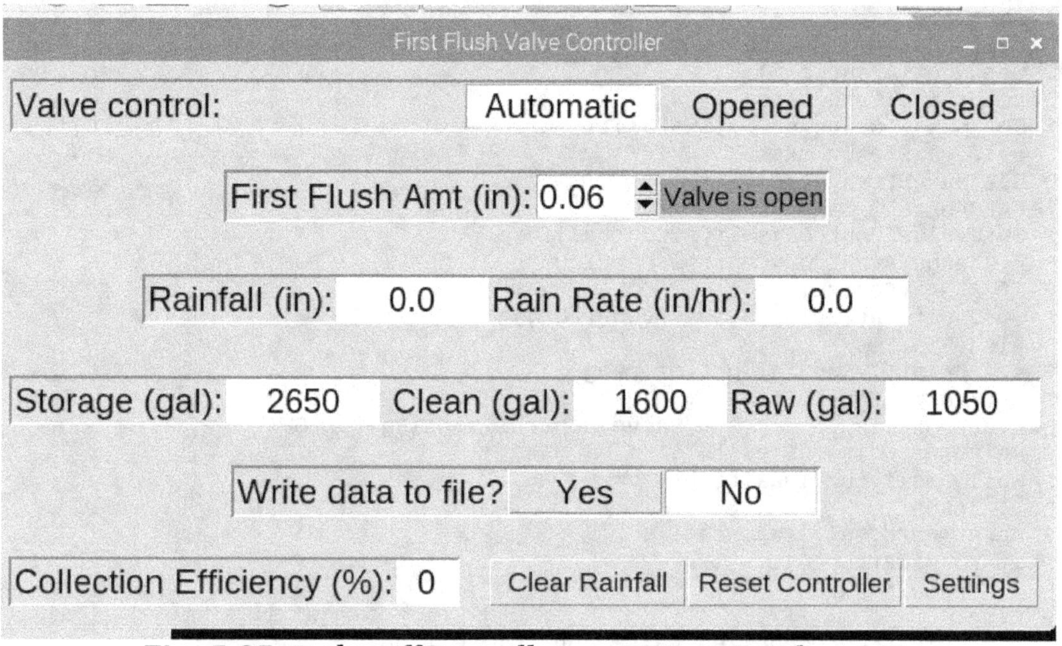

Fig. 5-25. Python ffController.py GUI on Raspberry Pi.

The Python code used to generate the above dialog box is provided below. I am personally not a big fan of Python for a number of reasons, but it is easy to use to make GUIs on the Raspberry Pi and it has a large user base. Nevertheless, those who are more proficient with Python can no doubt improve my code below.

```python
import serial
from time import time, strftime
from tkinter import *
class RainwaterSystem:
    """RainwaterSystem maintains user set controls and continually updates
    key system status data at user-set time intervals (in seconds).
    System is defined in a configuration file (ffconfig.txt)."""
    def _init_(self):
        self.ser = serial.Serial(port='/dev/ttyUSB0', baudrate=9600,
                    bytesize=8, parity='N', stopbits=1, timeout=5)
        filename = '/home/pi/data/ffconfig.txt'
        config_file = open(filename, 'r')
        self.timeInc = int(config_file.readline())# sample time in seconds
        self.roofArea = int(config_file.readline()) # square feet
        self.maxStor = int(config_file.readline()) # currently unused
        config_file.close()
        self.rainfall = 0.0      # current rainfall in inches
        self.rainfallPrev = 0.0  # previous rainfall reading
        self.cleanStor = 1000    # clean storage in gallons
        self.rawStor = 1000      # raw storage in gallons
        self.totStorPrev = 2000  # previous total storage reading
    def Set_TimeInc(self, tinc):
        self.timeInc = tinc
    def Get_TimeInc(self):
        return self.timeInc
```

```python
    def Set_Roof_Area(self, area):
        self.roofArea = area
    def Get_Roof_Area(self):
        return self.roofArea
    def Set_Max_Storage(self, mxstor):
        self.maxStor = mxstor
    def Get_Max_Storage(self):
        return self.maxStor
    def Set_Valve_Control(self, vControl):
        # Change Arduino vControl setting to either 1, 2, or 3
        if (0 < vControl < 4):
            self.ser.write(bytes([vControl]))
        else:
            print('vControl out of range')
    def Get_Valve_Control(self):
        # Get Arduino's current valve control setting for GUI
        self.ser.write(bytes([4]))
        while self.ser.inWaiting() == 0:
            pass
        data = ord(self.ser.read())
        if (0 < data < 4):
            return data
        else:
            print('unrecognized valve control setting')
    def Get_Valve_Position(self):
        # Current valve status, open or closed?
        self.ser.write(bytes([12]))
        while self.ser.inWaiting() == 0:
            pass
        data = ord(self.ser.read())
        if (data == 0) or (data == 1):
            return data
        else:
            print('unrecognized valve position')
    def Set_FF_Amt(self, first_flush):
        # Send Arduino new first flush setting
        cmd = int(float(first_flush) / 0.02) + 4
        self.ser.write(bytes([cmd]))
    def Get_FF_Amt(self):
        # Get current first flush setting on Arduino
        self.ser.write(bytes([11]))
        while self.ser.inWaiting() == 0:
            pass
        data = ord(self.ser.read())
        return data
    def Clear_Rainfall(self):
        # Zero rainfall display and empty gauge
        self.ser.write(bytes([14]))
        self.rainfall = 0.0
    def Get_Rainfall(self):
        # Read rainll gauge connected to Arduino
        self.rainfallPrev = self.rainfall
        self.ser.write(bytes([13]))
        while self.ser.inWaiting() == 0:
            pass
        data = ord(self.ser.read())
```

```python
        self.rainfall = (data * 256 + ord(self.ser.read()))/100
        return self.rainfall
    def Get_Rainfall_Rate(self):
        # Calculate rainfall rate, in/hr
        rainrate = (self.rainfall - self.rainfallPrev)*3600/self.timeInc
        if (rainrate < 0):
            rainrate = 0.0
        return rainrate
    def Get_Clean_Storage(self):
        # Clean water storage level, call this before Get_Raw_Storage
        self.totStorPrev = self.cleanStor + self.rawStor
        self.ser.write(bytes([15]))
        while self.ser.inWaiting() == 0:
            pass
        data = ord(self.ser.read())
        self.cleanStor = data * 256 + ord(self.ser.read())
        return self.cleanStor
    def Get_Raw_Storage(self):
        # Raw water storage level
        self.ser.write(bytes([16]))
        while self.ser.inWaiting() == 0:
            pass
        data = ord(self.ser.read())
        self.rawStor = data * 256 + ord(self.ser.read())
        return self.rawStor
    def Get_Efficiency(self):
        # System collection efficiency
        delStor = self.cleanStor + self.rawStor - self.totStorPrev
        delRain = self.rainfall - self.rainfallPrev
        collect = delRain * self.roofArea * 0.6234
        if (collect > 0.0):
            return delStor/collect
        else:
            return 0.0
    def Reset_Controller(self):
        # Resets controller for new ff cycle, opens valve, empties rain gauge
        self.ser.write(bytes([17]))
#--------------------------------- end of RainwaterSystem ----------------
class DataFiles:
    # Handles I/O files for rainfall & storage as function of time
    def _init_(self):
        self.dfile = open('/home/pi/data/junk.txt','w')
        self.dfile.close()
    def Open_DataFile(self):
        name = '/home/pi/data/' + strftime("%b_%d_%Y_%H%M.txt")
        self.dfile = open(name, 'w')
        s = str(rw.Get_TimeInc())
        self.dfile.write(s + '\n')
    def File_Is_Open(self):
        return not(self.dfile.closed)
    def Close_DataFile(self):
        self.dfile.close()
    def WriteLine(self, str):
        self.dfile.write(str)
#--------------------------------- end of DataFiles ---------------------
def Update_Display():
```

```python
    """ Recursively calls itself every timeInc period to update display.
    Automatically adjusts correction to timeInc so that data sampling
    equals timeInc."""
    global t1  # need these for subsequent calls
    global err
    t2 = time()
    timeInc = rw.Get_TimeInc()
    err = err + t2 - t1 - timeInc
    t1 = t2
    delay = int((timeInc - err)*1000)
    rainfall = rw.Get_Rainfall()
    label5.config(text=rainfall)
    rainrate = rw.Get_Rainfall_Rate()
    if (rainrate < 0):
        rainrate = 0.0
    label6.config(text=rainrate)
    clean = rw.Get_Clean_Storage()
    label12.config(text=clean)
    raw = rw.Get_Raw_Storage()
    label10.config(text=raw)
    storage = clean + raw
    label9.config(text=storage)
    s = u.get()   # get radiobutton rad1 position
    if (s==2):    # data file is open
        s = str(rainfall) + ' ' + str(storage) + '\n'
        df.WriteLine(s)
    vp = rw.Get_Valve_Position()
    if (vp == 0):
        label3.config(text='Valve is open', bg=("red"))
    elif (vp == 1):
        label3.config(text='Valve is closed', bg=("green"))
    else:
        print('unrecognized valve position')
    eff = int(rw.Get_Efficiency() * 100)
    label15.config(text=eff)
    win.after(delay, Update_Display)
def Set_Valve_Cont():
    s = v.get()
    rw.Set_Valve_Control(s)
def Set_First_Flush():
    ff_amt = spin1.get()
    rw.Set_FF_Amt(ff_amt)
def Settings():
    def SaveSettings():
        time_inc = int(entry1.get())
        rw.Set_TimeInc(time_inc)
        roof_area = int(entry2.get())
        rw.Set_Roof_Area(roof_area)
        maxStor = int(entry3.get())
        rw.Set_Max_Storage(maxStor)
        config_file = open('/home/pi/data/ffconfig.txt','w')
        config_file.write(str(time_inc)+'\n')
        config_file.write(str(roof_area) + '\n')
        config_file.write(str(maxStor)+'\n')
        config_file.close()
        root.focus_set()
```

```
            stg.destroy()
    def Cancel():
        root.focus_set()
        stg.destroy()
    time_inc = rw.Get_TimeInc()
    roof_area = rw.Get_Roof_Area()
    max_storage = rw.Get_Max_Storage()
    wfont = ('Helvetica',16)
    stg = Toplevel(root)
    stg.wm_title("Settings")
    Label(stg, text='Readings sample time (s):',font=wfont).grid(row=0, sticky=W)
    entry1 = Entry(stg, width=6, font=wfont)
    entry1.grid(row=0, column=1, sticky=W)
    entry1.insert(0, time_inc)
    Label(stg, text='Roof area (sq ft):',font=wfont).grid(row=1, sticky=W)
    entry2 = Entry(stg, width=10, font=wfont)
    entry2.grid(row=1, column=1, sticky=W)
    entry2.insert(0, roof_area)
    Label(stg, text='Maximum storage (gal):', font=wfont).grid(row=2, sticky=W)
    entry3 = Entry(stg, width=10, font=wfont)
    entry3.grid(row=2, column=1, sticky=W)
    entry3.insert(0, max_storage)
    but1 = Button(stg, text='Save Settings', width=14, font=wfont,
                                            command=SaveSettings)
    but1.grid(row=3, column=0, sticky=E)
    but2 = Button(stg, text='Cancel', width=14, font=wfont, command=Cancel)
    but2.grid(row=3, column=1, sticky=W)
    stg.grab_set()
    stg.focus_set()
    stg.wait_window()
# ----------------------------- Main Program -----------------------------
rw = RainwaterSystem()
rw._init_()
df = DataFiles()
df._init_()
wfont = ('Helvetica',18)
root = Tk()
root.title('First Flush Valve Controller')
root.geometry('750x400')
win = Frame(root)
win.pack(side=TOP, expand=YES, fill=BOTH)
valve = Frame(win, bd=2, relief=SUNKEN)
label1 = Label(valve, text='Valve control:',font=wfont)
label1.pack(side=LEFT)
v = IntVar()
v.set(1)
radio1 = Radiobutton(valve, text='Closed', font=wfont, width=10,
                        indicatoron=0, command=Set_Valve_Cont, variable=v, value=3)
radio1.pack(side=RIGHT)
radio2 = Radiobutton(valve, text='Opened', font=wfont, width=10,
                        indicatoron=0, command=Set_Valve_Cont, variable=v, value=2)
radio2.pack(side=RIGHT)
radio3 = Radiobutton(valve, text='Automatic', font=wfont, width=10,
                        indicatoron=0, command=Set_Valve_Cont, variable=v, value=1)
radio3.pack(side=RIGHT)
valve.pack(expand=YES, fill=X, pady=1, padx=10)
```

```
dat = rw.Get_Valve_Control()
v.set(dat)
dat = rw.Get_Valve_Position()
if (dat == 0):
    valve_status = 'Valve is open'
    color = "red"
else:
    valve_status = 'Valve is closed'
    color = "green"
ff = Frame(win, bd=2, relief=SUNKEN)
label2 = Label(ff, text='First Flush Amt (in):', font=wfont)
label2.pack(side=LEFT)
label3 = Label(ff, text=valve_status, bg=(color), font=('Helvetica',14))
label3.pack(side=RIGHT)
spin1 = Spinbox(ff, from_=0.02, to=0.12, increment=0.02, command=Set_First_Flush)
spin1.config(font=wfont, width=5)
spin1.pack(side=RIGHT)
ff.pack(expand=YES, pady=1, padx=10)
dat = rw.Get_FF_Amt()
ff_amt = dat/100
w = StringVar(ff)
w.set(str(ff_amt))
rainfall = rw.Get_Rainfall()
rainrate = rw.Get_Rainfall_Rate()
rain = Frame(win, bd=2, relief=SUNKEN)
label4 = Label(rain, text='Rainfall (in):', font=wfont)
label4.pack(side=LEFT)
label5 = Label(rain, text=str(rainfall), font=wfont, bg=('white'), width=8)
label5.pack(side=LEFT, anchor=W)
label6 = Label(rain, text=str(rainrate), font=wfont, bg=('white'), width=8)
label6.pack(side=RIGHT)
label7 = Label(rain, text='Rain Rate (in/hr):', font=wfont)
label7.pack(side=RIGHT)
rain.pack(expand=YES, pady=10, padx=10)
raw = rw.Get_Raw_Storage()
clean = rw.Get_Clean_Storage()
storage = raw + clean
store = Frame(win, bd=2, relief=SUNKEN)
label8 = Label(store, text='Storage (gal):', font=wfont)
label8.pack(side=LEFT)
label9 = Label(store, text=str(storage), font=wfont, bg=('white'), width=8)
label9.pack(side=LEFT)
label10 = Label(store, text=str(raw), font=wfont, bg=('white'), width=8)
label10.pack(side=RIGHT)
label11 = Label(store, text='Raw (gal):', font=wfont)
label11.pack(side=RIGHT)
label12 = Label(store, text=str(clean), font=wfont, bg=('white'), width=8)
label12.pack(side=RIGHT)
label13 = Label(store, text='Clean (gal):', font=wfont)
label13.pack(side=RIGHT)
store.pack(expand=YES, fill=X, pady=1, padx=10)
datf = Frame(win, bd=2, relief=SUNKEN)
Label(datf, text='Write data to file?', font=wfont).pack(side=LEFT)
u = IntVar()
u.set(1)
rad1 = Radiobutton(datf, text='No', font=wfont, width=8, indicatoron=0,
```

```
                 command=df.Close_DataFile, variable=u, value=1)
rad1.pack(side=RIGHT)
rad2 = Radiobutton(datf, text='Yes', font=wfont, width=8, indicatoron=0,
                 command=df.Open_DataFile, variable=u, value=2)
rad2.pack(side=RIGHT)
datf.pack(expand=YES, pady=1, padx=10)
eff = int(rw.Get_Efficiency() * 100)
ceff = Frame(win, bd=2, relief=SUNKEN)
label14 = Label(ceff, text='Collection Efficiency (%):', font=wfont)
label14.pack(side=LEFT)
label15 = Label(ceff, text=str(eff), font=wfont, bg=('white'), width=3)
label15.pack(side=LEFT)
ceff.pack(expand=YES, side=LEFT, pady=1, padx=10)
buts = Frame(win)
button1 = Button(buts, text='Clear Rainfall', font=('Helvetica',14),
command=rw.Clear_Rainfall)
button1.pack(side = LEFT)
button3 = Button(buts, text='Settings', font=('Helvetica',14), command=Settings)
button3.pack(side=RIGHT)
button2 = Button(buts, text='Reset Controller', font=('Helvetica',14),
command=rw.Reset_Controller)
button2.pack(side=LEFT)
buts.pack(expand=YES, fill=X, pady=1, padx=10)
t1 = time()
err = 0.0
win.after(3000, Update_Display)
win.mainloop()
```

We point out a key feature in the above code. We want the GUI to remain responsive to user inputs even when it is off taking data and doing calculations. We accomplish this by the last two lines in the above code. The last line, win.mainloop(), makes the GUI operational. The line immediately before this queues up a procedure called Update_Display to be executed 3000 milliseconds later. This allows mainloop() sufficient time to get the GUI up and running. Look up at the procedure or method named "Update_Display" (just under the comment line marked "end of DataFiles") and you will see that this procedure recursively calls itself after a specified "delay." The GUI is still active and responsive while the procedure is sitting in the queue waiting to be executed. The time delay is set by the user under "Settings" (lower right corner of dialog box) and stored in a configuration file, along with roof area and maximum storage capacity. This time increment specifies the sampling time between rainfall and storage readings. Since code execution time is finite, Update_Display determines actual time of subsequent executions of itself using the computer system clock and calculates a correction to the user-specified time increment. These are the first five lines of code under the two global variable declarations in Update_Display. We need these two variables from previous executions but Python does not allow passing in variables with recursive calls. So we have no choice but to use global variables here.

The Arduino can take readings of rainfall and storage levels at much faster rates than the Raspberry Pi would normally request, especially if data is being written to a file. There is no benefit to recording data points at time intervals so short that real values of rainfall and storage change very little. However, random noise always exists in sensor readings causing non-real fluctuations of readings. This random noise can be reduced at the Arduino by averaging multiple readings or by increasing gate time to say five or ten seconds and then dividing the count by gate time to obtain frequency. Alternatively, it could do a weighted moving average on multiple sequential readings to remove noise. A TBR cannot go back and remeasure rainfall since it dumps to ground after every measurement. So it has no possibility of removing random errors through repeated measurements on the same quantity like we can do with capacitive or acoustic resonance gauges. Although complex noise reduction algorithms have not been implemented in the above code, basic noise reduction is easily accomplished by simply increasing rainfall sampling time and frequency measurement gate time.

Chapter 6.
Water System Maintenance and Management

Zero-maintenance water systems do not exist. Either you pay someone to do the maintenance (e.g. a public water utility) or you do it yourself. Indeed, you are more motivated than anyone else to provide clean potable water for your family. Potential sickness or even death from bad water strongly motivate one to properly maintain his system. Employees of public water suppliers are often only motivated by fear of losing their jobs. And if the water supplier is unionized, even that motivation may be lacking! Maintaining a potable rainwater harvesting system is neither complicated nor difficult, it simply must be done. In fact, it should actually be easier than maintaining a swimming pool. If you faithfully execute simple regular maintenance on a well-designed system, your water will undoubtedly be safer, cleaner, and more reliable than water from a typical public utility. I have successfully maintained our own rainwater system for several years and consistently find our potable water is orders of magnitude cleaner and more reliable than typical public water supplies. In fact, I have often found my water having a lower TDS reading than store-bought bottled water.

Having personally maintained a private swimming pool for 17 years in the past, I can honestly say that maintaining a potable rainwater harvesting system is much easier and cheaper. Electricity consumed by low-wattage circulation pumps on rainwater storage tanks is far less than that of big 220VAC circulation pumps on typical swimming pools. Chlorine and chemicals consumed by swimming pools far exceed that of rainwater harvesting systems. Maintaining proper chlorine levels in my swimming pool was a constant challenge because solar UV continually causes chlorine to escape into the atmosphere. Stored rainwater is protected from solar UV and now my main concern is chlorine chemicals going bad from sitting on the shelf too long! For our family, a rainwater system is far more useful than a swimming pool, easier to maintain, and less costly.

You should keep a log book of all tests, inspections, and system maintenance. My log book is a three-ring binder containing three sections. The first section contains weekly log sheets of inspections, tests, and disinfectant maintenance. A sample log sheet developed for my system is provided in section 6.1 below. The second log book section is a detailed record of all special or unusual maintenance and repairs. Tank cleaning, gutter/drain line cleaning, and distribution system cleaning fall under this category. The third section contains written procedures for system maintenance and manufacturers' user manuals. Example written procedures are provided in sections below to assist you in developing your own written procedures.

Although you may be the primary operator/manager of your rainwater system, you must write down all your maintenance activities in simple how-to procedures in case you become incapacitated or unable to perform those duties. You also must train someone else in your family how to maintain your water system. Besides food, water is the most essential and basic resource for survival and you want to make sure your family can continue on without you. Make sure someone else besides you knows how your system works and how to perform regular maintenance tasks. Put all procedures in your maintenance log book. Number valves or show pictures of referenced valves in procedures so others can easily follow your written procedures.

As you develop written procedures for maintenance, testing, and inspection activities, consider the challenges for someone who has not thought as much about how your system works as you have, but must suddenly take over maintaining it. These activities may seem simple to you because you have thought a lot about them. But for someone who has not, they can be very daunting. Every commercial business that seeks ISO 9000 certification must document every single procedure that it performs, no matter how small or simple. This documentation process ensures that the business continues its activities long after its founders and original procedure developers have left the company. You want the same for your water supply. You put a lot of time and effort into developing it. Now make sure it continues to provide clean potable water long after you are gone. The procedures below are provided to help you get started with your own written maintenance, testing, and inspection procedures. If you later modify your system be sure to update your written procedures.

6.1 Weekly Inspections and Disinfectant Maintenance

Maintaining proper disinfectant level in a potable rainwater system is absolutely essential. Water becomes stagnant in distribution plumbing and biofilms begin growing if residual disinfectant gets too low, causing water to develop a musty odor and taste. Redish-brown or pink slim forming around sink drains or in toilet tanks is a good indicator of insufficient residual disinfectant in your clean water. Weekly inspections and tests of disinfectant levels take only a few minutes to accomplish. Special circumstances may require daily testing of disinfectant levels. But normally they should be quite stable. Once automatic chlorine metering is proven stable then chlorine testing can become less frequent, but visual checks should still be performed weekly.

All maintenance and test results should be recorded in a log book so that you can determine trends and quickly spot developing problems. A sample weekly inspections log sheet is provided below in Table 6-1. Certain times of the year, such as springtime and autumn, may require more frequent inspections due to heavier organic matter in rainwater. Your log book and experience will help determine required inspection intervals. My regular testing includes measuring disinfectant levels (free chlorine or hydrogen peroxide), oxidation reduction potential (ORP), pH, and conductivity in both tanks and at kitchen tap. I record these measurements in my log book and quickly compare with previous measurements. Before implementing automatic injection, disinfectant was manually added as needed and amounts were recorded. Small adjustments to a chlorine metering pump may be required initially, until the system stabilizes. Afterward, very little adjustment should be needed. Just check the chlorine reservoir for disinfectant concentrate level and replenished as needed.

Disinfectant testing simply involves taking samples of water from three different test taps and measuring free chlorine ppm, ORP, pH, and conductivity (or TDS). Since I generally use no chlorine in my raw water tank I skip chlorine testing on raw water. I use a standard DPD chlorine test kit that involves adding 5 drops of two different reagents to a measured amount of water and matching resulting color to a chart to determine free chlorine ppm level. ORP is also proportional to free chlorine, but not linearly. ORP depends on pH and residual ozone (or hydroxyl radicals) in addition to free chlorine. I use a Hanna Instruments 98121 hand-held meter to measure ORP and pH and an ExStik II hand-held meter for measuring conductivity. Chlorine disinfectant is actually more effective with slightly acidic water, which rainwater is by nature. The values I want to see are: free chlorine between 0.2 ppm to 1 ppm (preferably closer to the low end of range), ORP of at least 650 mV, pH between 6.0 to 7.0, and conductivity under 25 µS/cm. My raw water tank usually has the lowest conductivity since no residual disinfectant has been added to it. On occasion I'll raise chlorine levels in the clean tank to fight a biofilm problem or add hydrogen peroxide to the raw water tank to solve an odor problem.

If clean water disinfectant levels are too low or too high an adjustment is made to the metering pump, or chlorine is manually added as needed. Since rainwater is practically distilled water its mineral content is zero and it has no buffering capacity. Thus, pH can easily swing over a wide range with very slight changes in acidity. You should avoid adding chemicals to raise pH, especially if AOP is used. Carbonate and bicarbonate ions are strong scavengers of AOP hydroxyl radicals. Getting rid of excess organic matter is usually all that is necessary to restore pH to its desired range. Typically, my raw water tank runs slightly more acidic than my clean water tank.

Under certain conditions, such as making disinfectant adjustments or high pollen conditions, I test disinfectant levels and water chemistry daily. Experience and log book data indicate how often to test stored water. Test as often as needed until you are confident that disinfectant levels will remain stable for over a week, or between scheduled tests. Tank size, water usage, and rainfall all affect chlorine stability. Chlorine test kits can be purchased from a local swimming pool supplier or from online vendors. Minimum and maximum free chlorine levels of 0.2 ppm and 2 ppm (mg/liter) respectively are generally recommended by the Center for Disease Control (CDC) and the World Health Organization (WHO) for disinfecting drinking water.

If you only add chlorine to raw water storage then maximum levels can be exceeded without significantly affecting chlorine level in clean water storage, especially if UV or ozone is installed on the clean water circulation loop. When I initially used chlorine as a primary disinfectant and oxidant with a UV lamp on the clean water tank, I kept residual free chlorine level in my clean water tank close to minimum at less than 0.5 ppm and chlorine level in the raw water tank closer to shock levels at greater than 2 ppm free chlorine. Adding chlorine to clean water storage was rarely needed since a make-up feed of 50 GPD usually supplied sufficient chlorine. Chlorine that is not consumed by organic matter is gradually dissipated by turbulent circulation and exposure to UV light.

The human nose is fairly sensitive to chlorine, but it cannot distinguish between free chlorine and combined chlorine; only a test kit can distinguish between the two. Combined chlorine is chlorine that has been used up in oxidizing organic matter and has little to no disinfecting capacity left. Combined chlorine includes THMs, chloramines, and other byproducts of oxidation reactions. I frequently found that my raw water tank smelled like it had plenty of chlorine in it, only to find that free chlorine levels were inadequate after testing, particularly during early spring when pollen levels were high. So do not rely on your nose as a substitute for proper chlorine level testing with a good test kit or electronic chlorine meter. However, your nose is a good detector of potential biofilm growth. If your water has a musty smell to it then biofilm growth is usually the problem. Pure water is completely odorless and tasteless. With AOP on my raw water tank I look for tasteless, odorless raw water.

Regular Clorox liquid bleach (sodium hypochlorite) is the easiest and quickest way to add chlorine to your water. Amount of chlorine needed depends on free chlorine test results and amount of organic matter in water. Prior to installing micro-mesh gutter screens my water required about 1 to 2 cups of regular Clorox bleach periodically added to 1600 gallons of raw water. After installing micro-mesh gutter screens required amounts decreased to about half that. Reduced chlorine reduces THM formation. Powdered calcium hypochlorite can also be used to add free chlorine to water. I would crush a calcium hypochlorite tablet in a doubled plastic bag with a hammer and store unused powder in a glass jar, measuring it out as needed. My system required 1 level tablespoon of calcium hypochlorite to raise free chlorine level in 1600 gallons of water approximately 1 ppm (parts per million). After measuring an appropriate amount, the crushed calcium hypochlorite is dissolved in warm water in a plastic gallon jug. I let solids settle out and then pour the semi-clear liquid concentrate into the water tank while stirring with a plastic stir stick. I use a half-inch CPVC pipe about 3.5 feet long with caps glued to each end as a stir stick. Stirring while adding chlorine concentrate helps dissipate it more quickly and minimizes possibility of concentrate just getting sucked into the distribution system if the pressure pump happens to turn on while chlorine is being added.

Another method for adding chlorine to raw water is to use a common swimming pool chlorine tablet float. These floats usually have a means for adjusting size of ports to control tablet dissolving rate, but chlorine levels are not nearly as stable with this method as with stirring in measured amounts of liquid concentrate. Nevertheless, chlorine tablet floats can help reduce maintenance time for solving an odor problem in raw water storage. Chlorine tablets are cheaper than hydrogen peroxide, although the latter is preferable for minimizing THM generation. Hydrogen peroxide works well with an AOP system and actually increases its effectiveness, whereas chlorine does not. An AOP purifier will eventually neutralize all the chlorine.

Conductivity, in micro-Siemens per centimeter, is proportional to TDS in the water and is an indicator of water impurities. However, conductivity also increases with disinfectant levels, so this measurement cannot be used in isolation from other measurements. Low-cost electronic TDS meters can be purchased for under $100 from on-line suppliers. I use an Extech ExStik II for measuring conductivity. Conductivity cannot tell you whether or not the water is biologically safe but it does provide a reading on the general purity of water since conductivity usually increases as total dissolved solids (TDS) increase. Conductivity also increases with free chlorine level. So your water will have a certain minimum conductivity even when it is completely free of all other pollutants. A high abnormal reading of conductivity indicates a potential problem, including a possible biofilm problem, that needs further investigation. Store-bought distilled water has a conductivity of less than 10 micro-Siemens per cm according to my TDS meter. The conductivity of my stored rainwater runs around 6 to 15 micro-Siemens per cm during summer and winter months and a little higher in springtime and autumn. Your rainwater should approach distilled water TDS readings, or at least be less than 50 micro-Siemens per cm. Municipal drinking water will usually run in the hundreds of micro-Siemens per cm. Collect several samples of fresh rainfall in a clean jar and measure their conductivities. Average the readings and use this as a standard to compare with stored water conductivity.

Regular periodic water testing is the only way to ensure that your water is biologically and chemically safe to drink. Low-cost bacteria test kits for E.coli can be purchased online, but these will always give a zero reading for bacteria if proper residual disinfectant levels are maintained. So I do not find them all that useful. Monitoring disinfectant level, conductivity (or total dissolved solids), and other chemical factors are much more sensitive measures of drinking water safety. Residual disinfectant concentration is raised during any major tank cleaning or repair that exposes stored potable water to cleaning tools or open air. Depending on extent of repairs, shock chlorination or hydrogen peroxide is used to flush the system after major repairs. After making a major system modification, I may have the water tested by National Testing Laboratories in Cleveland, Ohio. These tests can be rather expensive (~$140) but are necessary for detecting certain pollutants that low-cost test kits cannot detect. With laboratory testing I am mainly looking for disinfection by-products (e.g. THMs) for chemical safety, not biological safety since I always maintain proper disinfectant levels. Farming chemicals and industrial pollutants, often found in ground waters, have always been non-existent in my rainwater. However, you may want to have a laboratory test for these if you live near farms using aerial-sprayed pesticides and herbicides, or a power plant, or anything that generates industrial air pollution. Wind can potentially cause hazardous pollutants to show up in rainwater and laboratory testing is the only way to fully validate water purity. Annual professional laboratory testing is generally recommended by most rainwater harvesting manuals. If your disinfection method includes UV and/or ozone, ensure these are properly working and service according to manufacturer recommendations.

Water flow in both circulation loops is checked at least weekly and filters are backwashed or replaced as needed. A quick look at the UV lamp power supply, ozone generator, and transfer pump control circuit takes only a few seconds to make sure no obvious malfunctions exist. I also record daily rainfall, first-flush setting, water meter reading, and water storage level and enter these in my spreadsheet. This spreadsheet (WaterStorage2) calculates average daily water usage, remaining storage, and date that we run out of water if no rainfall occurs. A sudden jump in water usage or a large discrepancy between calculated and measured water levels could indicate a system water leak, clogged gutter screens, or clogged prefilter. Otherwise, small discrepancies between calculated and measured levels indicate an error in assumed collection efficiency. After major rainfalls I visually check the prefilter and clean it if necessary. I visually inspect the first-flush diverter valve for any obvious malfunctions. All these inspections only take a few minutes to accomplish, but they are a very necessary part of maintaining my system.

Table 6-1 shows a weekly record sheet that I use in my log book. Water is tested at three locations; kitchen tap, clean tank test tap, and raw tank test tap. With AOP in raw water storage I skip "ppm Cl" in raw water but do test ORP since AOP raises ORP readings. ORP measurement of non-chlorinated water is a good indicator of hydroxyl radical generation by AOP. Hydroxyl radicals are short-lived and ORP of water in the test cup will slowly reduce as hydroxyl radicals recombine with hydrogen. Conductivity and pH are also measured in raw water. I want to see a pH of 7.0 (neutral) or slightly acidic (~6.5) and a conductivity of less than 10 micro-Siemens per cm. More important tests are those at the kitchen tap and clean water test tap. These should be nearly the same. If chlorine level or ORP has been fairly stable in clean water storage but the kitchen tap shows a significant drop in chlorine and ORP, then I suspect a biofilm problem in distribution plumbing. Although not indicated on the weekly log sheet, I will periodically check water from the drinking water tap (0.5 micron activated charcoal filter) to insure all chlorine is still being removed. The drinking water from this filter should be completely odorless and tasteless. After performing other visual and odor inspections as discussed above a check mark is placed in the corresponding box on the log sheet. Data to be entered in my WaterStorage2 spreadsheet is recorded on this weekly log sheet. Although, this data is often recorded in my spreadsheet more often than the weekly inspections, particularly during periods of high rainfall. Below is an example written procedure for these weekly inspections.

Weekly Inspections and Maintenance Procedure

Note: Visually inspect system every week. During periods of heavy pollen (April and May) inspect system more frequently. If chlorine level is not stable or is being adjusted, test chlorine levels every day or every other day until stable. Chlorine testing may become less frequent once automatic metering is proven stable and reliable.

1. Use inspections log as a check list for inspections. For chlorine and water quality testing, sample water from faucets at three points; kitchen tap, clean water faucet in pump house, and raw water faucet in pump house. Test for free chlorine, ORP, pH, and conductivity. No need to test for chlorine in raw water tank (except under special circumstances) since chlorine is not normally added to raw water. Record readings in log sheet. Chlorine at kitchen tap and clean water tank should be detectable but less than 1.0 ppm. If chlorine is undetectable add chlorine to tank and slightly increase chlorine metering on Stenner metering pump. If chlorine is too high decrease chlorine metering. Record metering adjustments in log book. If chlorine is sufficient at clean tank but undetectable at kitchen tap, note potential biofilm problem and follow procedures for solving. For chlorine addition: 1 tablespoon calcium hypochlorite increases free chlorine by 1 ppm and 3.5 ounces of 6% bleach increases free chlorine by 1 ppm.

2. Check chlorine concentrate level in pump house reservoir. Add if necessary. Use Clorox Regular Bleach (not splashless or scented). Bleach must conform to ANSI/NSF 60 for drinking water disinfectant.

3. Sample raw water from raw water faucet in pump room and test for ORP, pH, and conductivity. Record readings in log book. During heavy pollen season AOP may become unable to handle organic loading. Musty or unpleasant odor in raw water tank, discoloration of water, or frequent filter changing indicate biological growth. Add a few ounces of hydrogen peroxide (use chlorine if hydrogen peroxide not available) to assist AOP with breaking down organic matter.

4. Check for good circulation in both clean and raw water loops. Raw water circulation is indicated by bubbles in the clear tubing section immediately after ozone injector or AOP purifier. Discharge in raw water tank should be strong, unless transfer pump is running. If circulation is inadequate change filters. Change clean water filter if filter is dark gray or black or circulation is inadequate. Record filter maintenance in log book.

5. Check for water leaks or dripping in pump house plumbing. Record water meter reading in both log book and WaterStorage2 spreadsheet. Periodically record measured water level, as indicated on first-flush controller, in column F of WaterStorage2 spreadsheet, particularly after a rainfall that does not completely fill raw water tank. Note any unusual discrepancy between calculated and measured water levels. These data points are used to calculate collection efficiency and indicate needed maintenance.

6. Inspect pump house for general cleanliness and absence of musty smells. Use shop vacuum to remove any bugs or spider webs. Do not use any insecticides in pump house since you do not want these vapors being sucked up by the AOP purifier and absorbed into your water. If a musty odor is present, spray walls (but not metal hardware) and floor with water-chlorine mixture.

7. If heavy rain occurred during previous week rinse off bag pre-filter with a hose as needed. During heavy pollen season check and clean (if necessary) pre-filter after every rainfall.

8. Periodically check toilet tank for biofilm. If free chlorine level has remained stable in clean water tank for several days, compare with free chlorine level at kitchen faucet. Large difference may indicate biofilm problem in distribution pipes. Record in log book.

9. Check diverter valve for any obvious obstructions. Apply axle grease to pipe where it slides in bracket as necessary. Check screen for buildup of debris and remove as necessary. Periodically check valve during rainfall to make sure it is not dumping rainfall to ground when valve is closed.

Table 6-1. Weekly Disinfectant Tests, Inspections, and Maintenance

Date	Location	ppm Cl	ORP	pH	TDS/cond.
	Kitchen tap				
	Clean tank				
	Raw tank				

Chlorine Reservoir [ ] Circ. Flow [ ] UV & Ozone [ ] Leaks [ ] Filter Color [ ] Prefilter [ ] Raw tank odor [ ]
Diverter valve [ ] Water Meter Reading _____ Measured Storage _____

Rainfall _____ First-flush setting _____ Notes:_____

Date	Location	ppm Cl	ORP	pH	TDS/cond.
	Kitchen tap				
	Clean tank				
	Raw tank				

Chlorine Reservoir [ ] Circ. Flow [ ] UV & Ozone [ ] Leaks [ ] Filter Color [ ] Prefilter [ ] Raw tank odor [ ]
Diverter valve [ ] Water Meter Reading _____ Measured Storage _____

Rainfall _____ First-flush setting _____ Notes:_____

Date	Location	ppm Cl	ORP	pH	TDS/cond.
	Kitchen tap				
	Clean tank				
	Raw tank				

Chlorine Reservoir [ ] Circ. Flow [ ] UV & Ozone [ ] Leaks [ ] Filter Color [ ] Prefilter [ ] Raw tank odor [ ]
Diverter valve [ ] Water Meter Reading _____ Measured Storage _____

Rainfall _____ First-flush setting _____ Notes:_____

Date	Location	ppm Cl	ORP	pH	TDS/cond.
	Kitchen tap				
	Clean tank				
	Raw tank				

Chlorine Reservoir [ ] Circ. Flow [ ] UV & Ozone [ ] Leaks [ ] Filter Color [ ] Prefilter [ ] Raw tank odor [ ]
Diverter valve [ ] Water Meter Reading _____ Measured Storage _____

Rainfall _____ First-flush setting _____ Notes:_____

6.2 Filter and Purifier System Maintenance

Changing, cleaning, or back washing filters is another very important part of regular maintenance. When filters become clogged or overloaded they no longer do their job of keeping water clean, and in fact can even become sources of biological contamination. Clogged gutter screens and prefilter cause rainwater to be spilled to ground. The need for filter maintenance is generally indicated by reduced flow, not by some fixed amount of time in use, since rainwater quality varies during the year and according to gutter and drain pipe cleanliness. This is why inspecting circulation loop flow on a weekly basis (or more frequently if your tanks are dirty) is important. Frequent inspections of gutter screens and prefilter may be required at certain times of the year. During heavy pollen season (April and May) I must clean my prefilter after practically every rainfall. After pollen season has ended I inspect micro-mesh gutter screens and clean as necessary. This is easily done by spraying gutter screens with a 50/50 mixture of bleach and water, lightly scrubbing with a dish washing brush to loosen stubborn pollen buildup, and then rinsing with high pressure water spray from a hose nozzle or pressure washer. Chlorine used in this cleaning activity usually results in cleaner gutters and drain lines as well.

Experience and log book recordings will help you determine about how often to change filters or clean gutter screens and prefilter. One advantage to having circulation loop discharge at tops of tanks is that the sound of splashing water provides a quick clue to filter condition (assuming water level is not above discharge opening). You can also visually inspect water flow by looking inside tanks. Low flow indicates a need to backwash or change filters. If you use an ozone generator or AOP purifier then exiting ozone bubbles through a clear section of hose provides an indicator of flow condition. If little or no ozone bubbles are present then flow is insufficient and the filter needs servicing. A clear plastic cartridge filter housing facilitates quick visual inspection of cartridge filters. I use white polypropylene depth filters or pleated filters. When the cartridge becomes discolored or gray from dirt and biofilm, I change it.

The first filters in the purification process receive heavier dirt loading. As previously discussed, my system incorporates a series of filters, starting with 100 micron gutter screens down to a 0.5 micron activated charcoal filter on the kitchen drinking water tap. Rainwater is first filtered by gutter screens and the prefilter. After entering storage, it is continuously filtered by a 25 micron back-washable filter and a 10-inch cartridge filter on a circulation loop. If raw water circulation is inadequate, I first clean the back-washable filter. A garden hose is connected to the loop, two valves are closed, and the filter is run through a backwash sequence that only takes a few minutes. A clear plastic hose on backwash waste line provides a visual indicator when the filter is clean. Waste water dumped to ground is usually a dark gray and gradually clears. If water is scarce and plants need watering I fill 5-gallon buckets with this waste and water garden plants while waiting for the cycle to complete.

After cleaning my back-washable filter I restore loop circulation and check the flow again. If flow is still insufficient then I change out the second filter in this loop, a 5-micron dual-gradient polypropylene filter, 10 inches long by 4.5 inches diameter, that fits inside a 10-inch "Big Blue" filter housing. Changing this filter usually always restores proper flow. If flow is not restored after cleaning and changing both filters, then I suspect a bad circulator pump and proceed to check its pressure. My raw water test tap is the highest point in this loop and if the circulation pump cannot push water to this point (assuming the tank is not empty) then pump replacement is required.

The bag prefilter on the gutter drain line needs cleaning more frequently, sometimes after every major rainfall if pollen levels are high. All gutters have micro-mesh screens but some fine organic particles get past them. The bag prefilter (the last line of defense before entering storage) captures additional mold, algae, and pollen that is not trapped by gutter screens. Fine particulates can clog the bag filter and cause lost rainfall if the filter is not kept clean. If the bag prefilter is getting clogged with a slimy material after every rainfall, then I suspect biofilms growing in my gutters and/or drain lines that need to be cleaned out. If necessary, a gutter screen can be lifted to inspect inside gutters. Or all the screens can be removed for pressure washing gutters. But I have never had to do this. Drain lines can be cleaned as necessary using procedures discussed in section 6.5. Micro-mesh gutter screens should be periodically checked for pollen or mold build-up by dumping some water on them and visually determining whether water is wicked through the screen or rolls off to ground. If water rolls off screens and falls to ground then first spray screens with a bleach and water mixture to loosen mold. Follow up with a high pressure spray rinse from a garden hose nozzle or pressure washer. This should restore flow through micro-mesh gutter screens. Set the diverter valve control switch to "open" to make sure this valve stays open during cleaning so that gutter wash water does not enter water storage.

UV lamps must be changed each year or according to manufacturer's specifications. UV lamps usually have a timer on its power supply indicating remaining lamp life. During lamp replacement inspect its outer protective quartz tube and clean if necessary. Dirt deposits on this quartz tube can block UV from doing its job of disinfecting. Follow manufacturer's instructions on servicing this tube since it is easy to break if not handled carefully. Wear rubber gloves when handling both the quartz tube and new lamp to prevent leaving finger prints on them. Expended UV lamps contain mercury and should be handled and disposed of properly, similar to fluorescent lamps, in accordance with local and state ordinances. Home building suppliers often have bins to deposit such bulbs. Ozone generators also have a limited lifetime, but it's usually longer. The Prozone AOP system requires a bulb change after 20,000 hours of operation. Keep track of installation date and have it serviced according to manufacturer's instructions. Testing ozone generators is possible using ozone test strips, particularly if the ozone supply line between generator and ozone injector is clear plastic tubing. Temporarily insert an ozone test strip in the tubing at the injector and determine if it adequately changes color while the generator is operating. Below are example written procedures for filter replacement or cleaning and chlorine metering adjustment.

Filter Replacement Procedure

Raw Water Cartridge Filter Replacement (filter on left).
1. Turn off raw water circulation pump (switch on left) and close valve on bottom left on pipe coming from wall.
2. Unscrew clear filter housing. Rinse out housing, check O-ring, install new filter, and re-install housing.
3. Re-open valve and turn on circulation pump. Bleed air from filter housing by pressing red button on top if necessary.
4. Order new filters from www.freshwatersystems.com. Filter is 20" long (be careful not to get 19.5" long filters). Desired filter porosity is 20 microns. Use either pleated filter or polypropylene sediment filter. Part numbers include:
801-20/20, Harmsco pleated filter 20 micron
701-20/20, Harmsco pleated filter 20 micron
PS-27200-20, Neo-Pure pleated filter 20 micron
MB-25200-20, Neo-Pure polypropylene filter 20 micron

Clean Water Cartridge Filter Replacement (filter on right).
1. Turn off clean water circulation pump (switch on right) and close valve on bottom pipe going into pump from wall.
2. Unscrew clear filter housing. Rinse out housing, check O-ring, install new filter, and re-install housing.
5. Re-open valve and turn on circulation pump. Bleed air from filter housing by pressing red button on top if necessary.
6. Order new filter from www.freshwatersystems.com. Filter is 20" long (be careful not to get 19.5" long filters). Desired filter porosity is 1 micron or smaller. Use either pleated filter or polypropylene sediment filter. Part numbers include:
MB-25200-01, Neo-Pure polypropylene filter 1 micron
PS-27200-01, Neo-Pure pleated filter 1 micron
PS-27200-S35, Neo-Pure pleated filter 0.35 micron

Kitchen Tap Cartridge Filter Replacement
1. Under kitchen sink, close valve to filter. Release pressure in lines at kitchen sink.
2. Unscrew filter housing, using strap wrench if needed. Discard old filter and clean out filter housing.
3. Check rubber o-ring for damage and replace if necessary.
4. Insert new filter in housing and screw back in place. Use a strap wrench if needed.
5. Open valve slowly and check for leaks. Press button on top of housing to release air.
6. Order new filter from www.freshwatersystems.com. Desired filter is 9-7/8" long 0.5 micron carbon block filter. Part numbers include:
CTOCL-2510, Neo-Pure 0.5 micron carbon block filter
CTOV-2510, Neo-Pure 0.5 micron carbon block filter
CTOC-2510, Neo-Pure 0.5 micron carbon block filter

Pre-filter Cleaning and Diverter Valve Inspection Procedure

1. Unscrew cleanout plug on pre-filter. If bag does not appear to be seriously clogged, simply spray down with a hose while still mounted inside housing and re-install cleanout plug.

2. If bag is seriously clogged, loosen wing nuts holding bag in place and remove bag filter. Clean bag filter with high pressure spray from garden hose. If mold growth is severe in filter housing wipe off with damp paper towels and small amount of diluted bleach.

3. Reassemble pre-filter and tighten wing nuts. Be careful not to cross-thread cleanout plug when installing.

4. Ensure screen around divert valve is intact and keeping insects out of divert valve. Repair if necessary.

5. If filter bag is torn or damaged, replace with 100 micron, 50 GPM filter bag, Grainger item 1EUE9 or similar.

Chlorine Metering Adjustment Procedure

Chlorine metering adjustment is accomplished by changing either the metering time or metering speed on the Stenner peristaltic metering pump. These two modes of operation are referred to as SECONDS and FLOW SWITCH respectively on page 9 of the Stenner Econ FP Pump Manual. Your system log book should tell you which mode setting the metering pump is currently set to. Refer to pages 10-15 in the Stenner manual for changing pump time in SECONDS mode and page 18 for changing pump speed in FLOW SWITCH mode. In either mode, when the pump receives a dry contact signal from the relay in the transfer pump controller, it turns on and begins pumping chlorine into the transferred water stream.

Ideal free chlorine ppm level is between 0.2 to 0.5 ppm. ORP should be greater than 650 millivolts. Increase free chlorine up to 5 ppm if red or pink biofilm becomes noticeable in toilets or sink drains. Record metering pump settings in weekly log when changes are made to serve as a reference for future changes. Only use chlorine that conforms to the American National Standards Institute and National Sanitation Foundation (ANSI/NSF) Standard 60 for use in potable water. The only Clorox bleach approved for potable water is Clorox Regular Bleach, not splashless type. Do not use any bleach with scenting compounds added. Label should indicate that it meets NSF Standard 60. Contact vendor if uncertain.

If there is a stale musty odor in raw water tank, add 35% food-grade hydrogen peroxide. Normally the AOP system should prevent odor problems, but heavy organic loading (pollen, gutter mold, etc.) can sometimes overwhelm the AOP system. The hydrogen peroxide is like a "booster shot" for the AOP; it works with AOP and makes it more effective. Since chlorine and hydrogen peroxide neutralize each other, you may have to manually add more chlorine to clean water tank until all the hydrogen peroxide is dissipated. Use hydrogen peroxide test strips to test raw water. If food-grade hydrogen peroxide is unavailable, you can use solid NSF approved calcium hypochlorite tablets or liquid bleach in raw water tank. But be aware that this works against AOP and may increase THMs. The UV in the AOP system will gradually dissipate chorine. If chlorine is manually added to raw water tank you may have to reduce chlorine metering in the transferred water to prevent getting the clean water chlorine level too high.

To change chlorine metering in transferred water:

1. Remove plastic cover over peristaltic pump control buttons.

2. Simultaneously press MODE and % buttons for 5 seconds or until display becomes brighter to unlock keypad.

    a. If in SECONDS mode: Display should show a number of seconds and a percentage. Metering time equals the number of seconds displayed times the percentage. For example, if display shows "60 seconds 66% then metering time equals 39.6 seconds.

b. In in FLOW SWITCH mode: Display should show "FLOW SWITCH" and a percentage number. In this mode the pump runs at a set speed the entire time water is being transferred. The percentage number indicates the speed of the metering pump. For example, if display shows "FLOW SWITCH 50%" then pump is running at 50% speed.

3. If display does not show desired mode, SECONDS or FLOW SWITCH, then press and hold down MODE button and repeatedly press up-arrow or down-arrow button until desired mode is displayed.

4. To change metering time in SECONDS mode press and hold % button and repeatedly press up-arrow or down-arrow button until desired percentage is displayed. To change metering pump speed in FLOW SWITCH mode press and hold % button and repeatedly press up-arrow or down-arrow button until desired percentage speed is displayed.

5. Replace plastic cover when done.

6. After making a change to metering pump, test chlorine level in clean water every 2 or 3 days to make sure it is stable at the desired level.

6.3 Tracking With WaterStorage2 Spreadsheet

After making sure your system is working properly and disinfectant levels are correct, the next most important job in water system management (assuming you have no other potable water source) is simply tracking water storage levels, usage rates, and predicted future rainfall. You should track it like a bank account and financial budget. Since your household depends on this water supply for its daily activities you do not want to run your tanks dry due to inattention! A local weather service, such as the National Weather Service (https://forecast.weather.gov), will probably become your most visited internet site. Long range forecasts based on the El Niño Southern Oscillation (ENSO) are also very helpful. The Climate Prediction Center (www.cpc.ncep.noaa.gov) provides long-range probabilistic outlooks of temperatures and precipitation based on historical data and the current state of the ENSO cycle. These sites provide excellent detailed information on weather prediction. Generally, a La Niña means dryer weather in southern states and wetter weather in northern states. El Niño conditions produce the opposite effect. Tracking these long-range predictions, along with water storage and use, will help prevent you from running out of water.

For example, if water storage is slowly getting lower and the National Weather Service and Climate Prediction Center are predicting reduced rainfall for the next several weeks or months, then you should implement rigorous water conservation measures in your household. If drought intensifies, these conservation measures should become more rigorous. Such measures may include delaying clothes washing or washing clothes at a laundry mat, mandatory sea showers, delaying plant watering or watering with gray water, minimizing toilet flushes, using disposable paper plates for meals, etc. Your household will become much more aware of how precious pure water is as you manage a rainwater harvesting system. Over time your household will find all kinds of ways to conserve water, more than you ever thought possible.

We discussed a simple spreadsheet (WaterStorage1) in Chapter 3 for system sizing. We show a more comprehensive spreadsheet here for operational storage tracking and prediction. This system management spreadsheet, WaterStorage2, is your primary tool for determining when to implement rigorous water conservation measures or high water use activities. This spreadsheet also helps you monitor your system's collection efficiency to know when maintenance is required. Data entered in this spreadsheet include: actual daily rainfall amounts, water meter readings, first-flush divert setting, and measured storage levels. This spreadsheet calculates actual water usage based on water meter readings rather than assuming a constant usage rate. It calculates what storage level should be based on rainfall and assumed collection efficiency. If calculated water levels deviate significantly from displayed levels, then you know you have a problem that needs investigating.

I recommend using an accurate manual rain gauge with a resolution in hundredths of an inch to supplement your digital gauge on your controller, at least until you are confident in your electronic gauge's accuracy. As previously discussed TBR digital gauges are notorious for under-reporting actual rainfall. TBR rainfall readings will tend to overestimate collection efficiency unless corrected for rainfall rate. A well-corrected capacitive gauge provides much better rainfall measurement. Use actual rainfall in your WaterStorage2 spreadsheet for maximum accuracy. For weather reporting purposes rainfall is recorded at a specific time (e.g. 6 AM) for the previous 24-hour period. However, this is not important for our purposes. We are only interested in the amount of recent rainfall, not in when it actually occurred. So you can record rainfall in your spreadsheet at whatever time is most convenient for you. Just do not duplicate readings in your spreadsheet.

WaterStorage2 calculates water storage levels based on actual usage from water meter readings, actual rainfall, first-flush divert setting, and assumed collection efficiency. It calculates remaining storage level to end of year if no further rainfall occurs. It is like a bank account tracking random income (rainfall), daily expenses (water usage), and remaining balance (water storage). The spreadsheet is your primary tool telling you when you will run out of water if you do not get rain, and serves as an early warning of when to start rigorous water conservation. With proper resource management there is no reason to ever run out of water. The root cause of letting storage tanks run dry is the same as over-drafting a checking account; slothful negligence! I often reassure my wife during lean water times, "Honey, I'm not going to let us run out of water, we may have to readjust our life style for a season, but we are not going to run out of water." During severe droughts my wife will sometimes do laundry at our children's houses in exchange for babysitting our grandchildren. Our adult children often tell us how much they look forward to us going through droughts!

Actual measured water levels are periodically entered in WaterStorage2 to compare with calculated levels, usually before and after a rainfall event that does not completely fill storage tanks. A large discrepancy between calculated and measured levels indicates a sudden drop in collection efficiency. This is usually due to a clogged prefilter. Clogged gutter screens or a leaky first-flush diverter valve can also reduce collection efficiency. Entering measured water levels after long dry spells, along with water meter readings, can help spot slow leaks on storage tanks. Plastic tanks tend to drip at their bottom tank fittings. Seal with silicone caulk if this occurs.

Table 6-2 shows the first thirteen rows of an example WaterStorage2 spreadsheet, from Jan 1, 2013 to Jan 12, 2013. Cell K2 to the right of collection efficiency (not shown) can be used for total cumulative rainfall to help determine if drought conditions are beginning.

Table 6-2. Operational WaterStorage2 Spreadsheet

Date	Rainfall (inches)	Water Meter Reading (gal)	Consump- tion (gal/day)	Storage Level (gal)	Measured Storage Level	First- flush divert (inches)	Roof Area (square ft)	Max Storage (gal)	Collection Efficiency
Jan 1	0.9	110339	52.5	3200		0.1	1400	3200	0.9
Jan 2	0.7		63.7	3200		0.1			
Jan 3	0		63.7	3136		0.1			
Jan 4	0		63.7	3073		0.1			
Jan 5	0		63.7	3009		0.1			
Jan 6	0.11		63.7	2985		0.06			
Jan 7	0		63.7	2921		0.06			
Jan 8	0		63.7	2857		0.06			
Jan 9	0		63.7	2794		0.06			
Jan 10	0.2	110912	63.7	2840	2800	0.06			
Jan 11	0.04		44.0	2796		0.06			
Jan 12	0.15	111000	44.0	2823		0.06			

Cell E2 contains the starting storage level on January 1st. The following formula is entered in cell E3 (Storage Level for Jan 2) and propagated down to the end of the year:

=IF(B3>G3,MAX(MIN((B3-G3)*0.6234*$H$2*$J$2-D3+E2,$I$2),0),MAX(E2-D3,0))

The formula is similar to that of the basic Waterstorage1 spreadsheet except that it uses actual average water consumption amounts in column D, rather than a fixed daily water consumption amount. If rainfall is captured that day (cell in column B is greater than first-flush amount in column G) then collected rainfall is added to the previous day's storage level, but capped at the maximum storage level allowed by the tanks. Average daily consumption amount is subtracted, but limited by a minimum storage level of zero.

Cell D2 contains the calculated (or assumed) usage rate on January 1st based on the previous year. A water meter reading should be entered in cell C2 for January 1, even if it is an estimated water meter reading. This is needed for the formula that is entered in column D. Thereafter the water meter reading can be entered on any date you want. The following formula is entered in cell D3 (Consumption for Jan 2) and propagated down to the end of the year:

=IF(AND(C2>0,MIN(C3:$C$367)>0),(MIN(C3:$C$367)-C2)/MATCH(MIN(C3:$C$367),C3:$C$367,1),D2)

This complicated-looking formula is really fairly simple. The AND function checks to see if a water meter reading was registered the previous day (the reason we need to enter a reading on January 1 in cell C2) and that there is another reading somewhere in column C from the current day to the end of the year. The MIN function retrieves the minimum nonzero value in the column since blank cells are not zero values. If the AND function is true, then the formula takes the minimum value of the readings from the current day to the end of the year (which presumably is the very next reading recorded) and subtracts the previous day's meter reading. This difference is then divided by the number of days between the two readings to calculate an average daily consumption amount. The MATCH function determines the number of days from the current position to the next meter reading in column C. If the AND is false, then the previous day's consumption amount is copied into the current day.

Note that water consumption (column D) changes when a new water meter reading is entered, which can be entered on any date. The water meter reading does not have to be entered every day, nor on any specific day other than on January 1. When a water meter reading is entered the spreadsheet calculates the average consumption from the previous reading and assumes that average all the way into the future until a new meter reading is added. The consumption and storage levels on Jan 1 are manually entered as carry-overs from the previous year. The meter reading on January 1 is entered to provide a starting point for the formula in D3. The next meter reading taken on Jan 10 in this example recalculates the consumption values as 63.7 gal/day starting with Jan 2. When a new meter reading is entered on Jan 12 the consumption is recalculated to 44 gal/day.

Note that the measured water storage level (2800) is different than the calculated storage level on Jan 10. The rainfall event on this day did not fill storage tanks, so it can be used to recalculate collection efficiency. If collection efficiency in cell J2 is reduced from 0.9 to 0.66 then the calculated and measured storage levels are equal. Normally such a drop would indicate that the prefilter needs cleaning. However, this is in the middle of winter and the precipitation could also be in the form of snow, which most likely blows off roof and accounts for reduced collection efficiency.

The first-flush divert amount was reduced to 0.06 inches on January 6, probably after noticing that rainfall was very light and that most of it was dumped to ground. Rainfall on January 10, 11, and 12 was also very light. This example assumes the controller reset button was pressed after recording rainfall on each of these three days, which forces the controller to execute a new first-flush divert cycle. Thus no rainfall was collected on January 11 because rainfall was less than first-flush divert, and only 0.09 inches was collected on January 12. For these three days 0.18 inches was dumped to ground as first-flush divert and only 0.21 inches was collected. However, if the user were watching his local weather service and noticed on January 10 that rainfall would occur for the next three days then he could wait till rainfall ceased on January 13 before pressing the reset button and simply record total rainfall for the three days. In that case only 0.06 inches would be dumped to ground as first-flush divert and he would have collected 0.33 inches for those three days. This example shows why one should pay attention to local weather forecasts and wait till rain has ceased before recording rainfall and resetting the controller.

There are maintenance situations, such as storage tank cleaning (covered in next section), that can destroy proper tracking of your spreadsheet. This occurs when you transfer water from clean storage to raw storage via pressurized distribution, and then back to clean storage via the transfer pump. The meter reading will indicate a huge loss of water when in fact it was merely transferred. To prevent WaterStorage2 from showing an unreal water loss, edit the column D formula on the day that water was transferred to subtract the amount transferred from the difference between the two meter readings. For example, on September 28th (row 272) I cleaned my tanks and transferred 1500 gallons in the process. So I edited cell D272 so that the inner MIN function just to the left of the MATCH function reads:

(MIN(C272:$C$367)-C271-1500)

This causes subsequent consumption levels into the future to show normal levels and also prevents a spike over 1500 from appearing on the graph of average daily consumption levels. Otherwise the spreadsheet would falsely indicate running out of water in a couple of days. If you copy over your spreadsheet to the following year and delete all the rainfall data and meter readings, as I do at the end of every December, be sure to go back and re-edit this cell and remove the transferred amount. Alternatively, you can simply maintain a clean master copy of WaterStorage2 that you make a copy of for each new year.

Collection efficiency is defined as the change in water storage level divided by the amount of water collected by the roof when the first-flush valve is closed. It does not include rainfall intentionally diverted to ground either through first-flushing or due to a full raw storage tank (when valve is open). Nor does it include water intentionally consumed by the house during a rainfall event. Collection efficiency includes rainfall unintentionally spilled to ground off gutters, leakage at divert valve, or pressure-relief spillage due to a clogged prefilter. A degradation in collection efficiency indicates a degradation in your collection system and shows up as an increasing difference between actual change in total storage and calculated total storage in the WaterStorage2 spreadsheet. With optimally performing gutter screens and first-flush/prefilter, collection efficiency is normally slightly less than 100% due to a small amount of collected rainwater used to wash leaves off gutter screens.

Collection efficiency may vary with rainfall rate due to TBR rain gauge errors and ability of gutter screens to handle water flow. Since TBRs have an increasing under-reporting error with increased rainfall rate, these errors will show an increasing collection efficiency with rainfall rate, which is not realistic. We expect collection efficiency to go down with increased rainfall rate due to increased splashing off roof or water flowing too fast for the gutter screens to handle.

With the Raspberry Pi user interface discussed in section 5.6, collection efficiency is viewed in real time along with rainfall rate (assuming a capacitive rain gauge). If just the hardwired controller of section 5.3 is used then only rainfall events that do not completely fill raw water tanks can be used to calculate this collection efficiency factor. Rainfall events that completely fill raw water tanks cannot be used to calculate collection efficiency since excess water dumped to ground after tanks are full is unknown. The simple linear equation we use in our water storage tracking spreadsheet (WaterStorage2) is:

$$storageChange = efficiency \times RoofArea \times 0.6234 \times (rainfall - firstflush) - consumed \qquad (6\text{-}1)$$

where *efficiency* and *RoofArea* are assumed to be constants. The spreadsheet calculates water storage level changes from rainfall amounts, first-flush settings, and water consumed by the household. However, if we know the actual change in water storage level and the other variables, we can use this equation to calculate actual *efficiency* for every rainfall event that does not completely fill raw water storage because the first-flush valve remains closed the entire time. We then use an average value of these multiple calculations in our spreadsheet.

To calculate actual collection efficiency we need to record certain data with each rainfall event that does not completely fill water storage. This includes recording actual change in water storage as indicated by water level sensors, rainfall in inches, first-flush setting, and amount of water consumed that day. If you suspect that an imminent rainfall event will not completely fill your raw water tank, simply record the current storage level and water meter reading before rain occurs. Then do the same after rain ceases. Then use the first-flush setting on your controller and the above equation to calculate collection efficiency.

Many rainfall events completely fill storage tanks and cause excess rainfall to be dumped to the ground. If your first-flush controller is located inside then you can easily watch water storage levels during a heavy rainfall. If storage levels approach maximum then you can go ahead and take an "ending" reading of rainfall, storage level, and water meter before the diverter valve opens up and begins dumping rainfall to ground and use this data to calculate collection efficiency. Do not press the reset button when you take this rainfall reading, otherwise the first-flush valve will open. Simply note the values and do the calculations outside the spreadsheet. An example written procedure for water storage monitoring is given below.

Water Storage and Usage Monitoring

Although the first-flush valve controller provides current water storage levels, it does not predict when you will run out of water if you do not get rain. That is the job of WaterStorage2 spreadsheet, which monitors water storage and usage. This spreadsheet, along with online weather prediction, is your primary tool for managing water usage so that your household does not run out of water. Instructions are included on spreadsheet. This spreadsheet also provides a tool for monitoring collection efficiency and first flush amount. The spreadsheet automatically calculates average daily water usage based on two subsequent water meter readings. Meter readings can be several days apart and do not need to be done on any specific day or time, but are most conveniently taken during normal weekly inspections. An immediate water usage can also be observed for any household activity (showers, clothes washing, dish washing, etc) by taking before and after water meter readings. Knowing usage amounts for every water-consuming activity helps you to implement water conservation measures during drought conditions. The spreadsheet automatically calculates theoretical storage levels based on rainfall, first-flush setting, and water usage, which should be close to actual measured storage level. A significant difference between theoretical and actual storage levels could indicate either: a) a major storage leak, b) an incorrectly assumed collection efficiency, c) a change in roof collection efficiency indicating micro-mesh gutter screens need cleaning, or d). a clogged pre-filter or leaky first-flush valve resulting in excess water dumped to ground.

1. After every rainfall event record the number of inches of rainfall, as indicated by rain gauge or display on first-flush controller, in WaterStorage2 spreadsheet. For weather reporting, this is done only once per day at a specific time, e.g. 6 AM, so that recording is for previous 24 hours. However, this is not important for your purposes. You simply want an accurate recording of rainfall so that your spreadsheet knows how much water your roof and tanks should have collected. Record in the spreadsheet on the date rain gauge is read, regardless of when rain actually occurred. Dump water from gauge (if using a manual gauge) and press Reset button on first-flush valve controller panel to prepare it for next rainfall event. If rain is predicted the next day (according to online weather prediction site) you may want to skip first flush divert and leave diverter valve closed. In this case do not press Reset button, recognizing that displayed rainfall amount will be the sum of two consecutive day's rainfall with only one first-flush cycle.

2. Periodically record water meter reading in the spreadsheet, on the date you read water meter. Water meter is located in pump house near pressure tank. Water meter reading is often recorded as part of weekly inspections and should also be recorded in weekly log.

3. Observe calculated water usage in spreadsheet and insert notes for any unusual high usage rates, such as pressure washing, clothes washing, etc. If unusual high consumption cannot be explained, look for major leaks in system.

4. Monitor weather forecast from National Weather Service (www.wpc.ncep.noaa.gov) or another service. If no heavy rain is forecast and water storage is approaching 50% capacity, begin water conservation measures in household (delaying clothes washing, sea showers, etc.). Use predicted date when storage runs dry and your expectation of future rainfall to determine needed conservation measures.

If tanks are near full and forecasted rain is imminent, implement heavy water usage activities (clothes washing, cleaning, etc.) to make room in raw water tank to receive new rainfall. In other words, manage and plan water usage in your household so as to maximize efficient use of collected rainfall.

5. Collection efficiency monitoring is performed by comparing theoretical and actual storage levels before and after rainfall events in which rainfall DOES NOT fill raw water tank and open the divert valve. Rainfall events that cause the divert valve to open and dump excess rainfall to the ground cannot be used to calculate collection efficiency since you do not know when valve opened nor how much rainfall was dumped to ground. Adjust collection efficiency (cell J2) in spreadsheet to minimize the differences between predicted and measured storage amounts using only those points where raw water tank is not completely filled. If you must reduce collection efficiency in order to get a good fit between theoretical and actual storage levels then you may need to pressure wash your micro-mesh gutter screens. Alternatively, first-flush valve may be leaking or drain lines may have heavy mold growth causing pre-filter to frequently clog.

6. At end of year make a copy of spreadsheet for following year. Alternatively, make a copy from your clean master copy for the new year. Remove rainfall data, water meter readings, and first-flush divert amounts. Change dates in column A to correspond to new year. Enter measured storage level, water meter reading, first-flush setting, and any rainfall on January 1 of new year to prepare new spreadsheet.

6.4 Storage Tank Cleaning

Keeping storage tank insides clean and free of biofilms, particularly clean water storage, is very important. Sediment buildup on tank bottoms promotes biofilm growth that can harbor dangerous organisms. Biofilms can even form without much sediment, feeding on dissolved organic matter (assimilable organic carbons). Biofilms can grow even in chlorinated water, as every swimming pool owner well knows. Outer layers of biofilms shield bacterial colonies underneath from chlorine. As colonies grow they release large quantities of bacteria into the water through outer layers to either start new biofilms elsewhere or increase chlorine demand and formation of THMs. Biofilms are a problem even for public water utilities, as an internet search will show. Biofilms form inside distribution lines of both public and private water systems. Some public water utilities have excessive levels of total coliform bacteria that are directly caused by biofilms in distribution lines rather than at water purification facilities.

Biofilms can cause your water to have a strong chlorine-like smell with very little effective free chlorine disinfectant in it. The chloramines, THMs, and other chlorination by-products have little disinfectant value but produce a strong smell that can easily fool an operator into thinking his water is biologically safe. Chlorination by-products also cause burning of eyes when bathing in such water. If you then try to reduce chlorine levels you will find your water taking on an unpleasant musty odor. Biofilms also cause differences in free chlorine levels at storage tank and at kitchen tap. I had a situation one time where the free chlorine level at the tank was 0.5 ppm but completely undetectable at the kitchen tap. I knew then that I had a biofilm problem in the distribution plumbing. Visual sediment or films, unpleasant odors, and elevated chlorine demand are indicators of a biofilm problem that can only be eliminated by tank cleaning, distribution system cleaning, and subsequent shock chlorination of your entire water system.

Although the cleaning maintenance of a rainwater cistern is usually much less than that of a swimming pool, the cleaning activities are very similar, including periodic brushing and vacuuming. Some dirt particles from your roof and atmosphere are fine enough to penetrate gutter screens and prefilter and can eventually build up over time inside tanks if not removed by circulation loop filters. Dissolved assimilable organic carbons from mold and biofilms in roof gutters become food sources for biofilm buildup inside tanks. Your raw water tank will have the heaviest buildup of sediment and biofilms, but the clean water tank is also susceptible to these problems. You must determine from experience how often tank cleaning is required. Scrubbing and vacuuming is easier when water levels are low because you can more easily see dirt and films on tank bottoms. Lower water levels also provide higher turnover rates through the vacuum filter and shortens the time required for tank cleaning. Unfortunately, low water levels usually occur when rain is scarce and you are trying to conserve water during these times. But a well-designed vacuum system can minimize water loss.

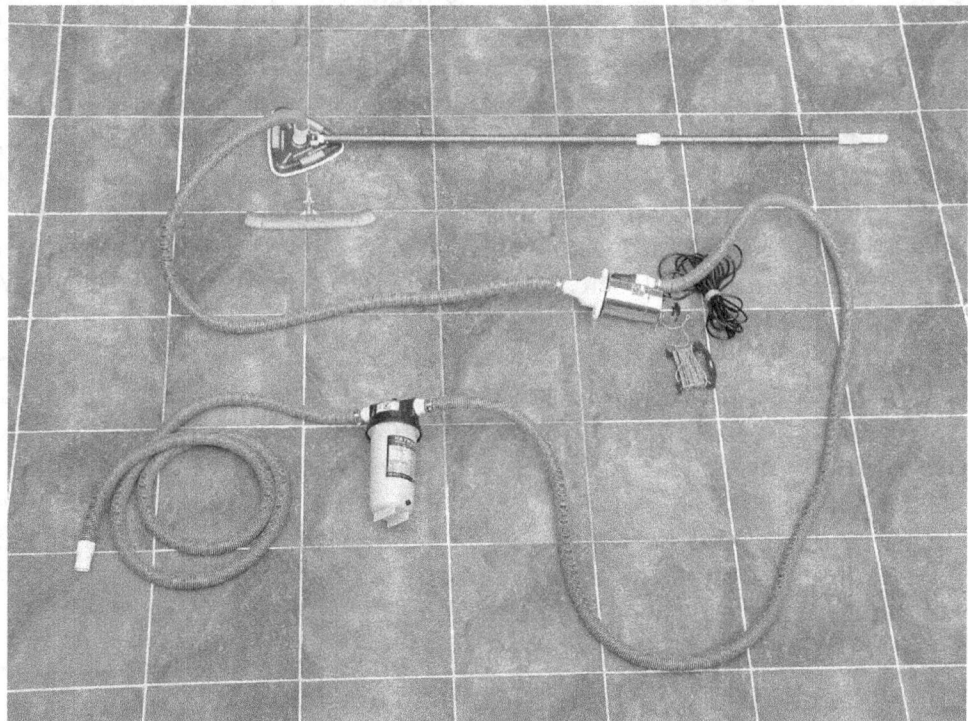

Fig. 6-1. Portable vacuum system for tank cleaning.

There are a number of low-cost portable hand-held pool vacuums on the market now that were not available when I first installed my rainwater system. You can find these at Lowe's or Pool & Spa suppliers. I have not personally tried any of these since the one I built several years ago is still working just fine. The vacuum system shown in Fig. 6-1 is the one I built from hardware purchased at Lowe's and on-line.

A good vacuum system requires a high flow rate pump of around 50 GPM in order to produce good suction. You also want flow rate to be high enough to get a good turnover rate, but not so high that you have to use a large heavy swimming pool sand filter to handle it. You can easily assemble a good portable vacuum system from a submersible waterfall pump, a cartridge filter used for spas or hot tubs, and a vacuum brush and hose used for swimming pool cleaning. I use a 50 GPM stainless steel submersible waterfall pump purchased from Lowes (Utilitech model no. 0435061) and a Hayward Micro Star Clear C225 cartridge filter, purchased online. This filter is a little small for the pump, but it works. The discharge from the submersible pump is connected to the cartridge filter sitting outside the tank with 1.5-inch flexible swimming pool vacuum hose. A return hose from the filter discharges filtered water back into the tank. You can either set the pump on the bottom of the tank and sweep dirt toward it with a brush, or you can do a little more assembly to connect the pump suction directly to a pool vacuum brush.

To do the latter (see Fig 6-1), remove the metal grill on the bottom of the pump and replace it with a coupler made from a PVC toilet flange. Drill holes in the toilet flange and attach it to the pump with four 1.5-inch long 8-32 screws. The plastic rotor housing must stay on the pump but the metal grill can be discarded. The four screws hold both the housing and PVC flange in place. Use a PVC reducer to connect the flange to a hose coupler. Hose couplers for pump and cartridge filter can be made from common PVC fittings. A 1.5-inch sink drain tailpiece fits nicely inside the 1.5-inch flexible pool sweep hose and can be secured with a hose clamp. The pipe thread on the Utilitech pump is 1.25-inch NPT and 1.5-inch NPT on the cartridge filter. Making a hose coupler with a male pipe thread on one end and a 1.5-inch sink tailpiece on the other is easy with common PVC fittings. Use silicone caulk on the high pressure line into the filter to help minimize leaking. The pump can then be suspended just below the water surface on a nylon rope, allowing you to vacuum all areas of the bottom of the tank, or simply lay it on the bottom and move it out of the way with your brush when necessary. Cut a length of flexible hose connecting the brush to the pump that is long enough to reach the farthest side of the tank, but not so long as to have hose coiled up in the tank getting in the way of sweeping. Figure 6-1 shows a picture of my portable vacuum system for tank cleaning.

You can also use an external pump for your vacuum system, but it may be difficult priming it and keeping it primed. With an external pump you must make sure there are no significant air leaks at suction couplings that could cause the pump to lose its prime. Both pump and suction line from the brush need to be filled with water before turning on the pump, otherwise you will get an air lock that prevents the pump from pumping water. If you turn on the pump and get no discharge then you probably have an air pocket in the suction line. If you have top discharge into the tank from the recirculation loop, you can simply remove the suction line from the brush and stick it on the circulation loop discharge to fill the line with water and drive out air. Otherwise fill it with a garden hose, reattach the brush and lower it into the water.

To minimize water loss and avoid recontaminating the clean water tank, start with the raw water tank first. We assume the water level is low enough to see the tank bottom, but the clean water tank may be topped off. Above normal disinfectant levels should be used in both tanks since tank cleaning necessarily exposes water to sources of bacteria (cleaning tools, loosened biofilms, body sweat, etc.). Hydrogen peroxide is preferable to chlorine in the raw water tank, especially if your system incorporates AOP, but chlorine is a valid less costly alternative. Hydrogen peroxide seems to work better on attacking biofilms. Temporarily disable or turn off the transfer pump controller and raw water circulation pump. You want to minimize sucking up loosened biofilms into circulation plumbing until manual cleaning is complete. Brush and vacuum tank to loosen and suck up as much visible sediment and biofilms as possible. If brushing makes the water so cloudy that you cannot see the bottom, leave the vacuum pump running and wait for the water to clear. Continue vacuuming until brushing no longer makes the water cloudy. If water flow through the vacuum slows, remove the cartridge filter and clean it with a garden hose. After vacuuming is complete turn the circulation pump back on. After the water begins to clear give the walls and bottom one final brushing with a pool brush (no vacuum) to stir up dirt that did not get removed with vacuuming so it can be removed with the circulation loop filters.

Now siphon or pump water from the clean water tank to the raw water tank to lower clean water level sufficiently. Depending on how much biofilm and dirt is in your clean water tank, you may want to raise chlorine to a "shock level" of about 10 ppm after the water level has been lowered. It will be diluted back down once you start refilling the tank after cleaning. You can also use hydrogen peroxide to lower chlorine back down after cleaning. If your toilet tank shows evidence of biofilm in distribution lines, you can pump some of this highly chlorinated water into distribution plumbing and let it shock chlorinate the system while you are cleaning your tanks. Run each household faucet until it discharges highly chlorinated water then shut it off. Avoid dumping too much of this strongly chlorinated water into your septic system to prevent killing anaerobic bacteria in septic tank. You can later flush this water out of distribution plumbing back into raw water storage to minimize water loss.

Before beginning brushing and vacuuming, turn off the clean water circulation pump and the house pressure pump. Repeat the same cleaning process as for the raw water tank. Turn circulation back on after finished vacuuming and brushing. After both raw water and clean water have become clear through circulation, turn the transfer pump controller back on and start refilling the clean water tank. Before energizing the household pressure pump, wait for the tank to substantially fill and chlorine levels are reduced back down and water clarity is completely restored. After energizing the household pressure pump you can flush the highly chlorinated water out of your household plumbing back into raw water storage through a garden hose sequentially connecting to each outside faucet. This will not flush it all, but it will substantially reduce the amount of chlorine dumped into your septic tank. If you find dilution does not reduce chlorine level to your satisfaction, add some hydrogen peroxide to neutralize excess chlorine. Add only small amounts at a time and test chlorine level.

After circulation loop filters have restored clarity in both tanks, clean or replace all filters. This is also a good time to replace the UV bulb if it is due to be changed, and test the check valve following the clean water circulator pump to make sure it is still working properly. The goal of this tank cleaning operation is to remove biofilms and sediment from your tanks with minimal wasted water and to prevent those particulates from getting inside household distribution plumbing. If heavy rains are forecast then you need not be so concerned with losing some water and the above procedure can be simplified. Tank cleaning should help reduce chlorine demand and reduce chlorination by-products that give your water a strong unpleasant smell. Similar to a swimming pool, you want crystal clear turbulent water continuously circulating in clean tanks with no sediment or biofilm. This paradigm is opposite to that promoted by most rainwater harvesting manuals (and the UPC) which advocate calming inlets to allow sedimentation in storage tanks. Encouraging sedimentation in storage tanks may be acceptable for non-potable rainwater harvesting, but not for potable rainwater harvesting. Over a hundred years of history in maintaining swimming pools teaches us that allowing water to stagnate is asking for serious problems. Keep water circulating and turbulent, and do not let sediment and biofilms build up. An example written procedure for tank cleaning is provided below.

Storage Tank Cleaning Procedure

Storage tank cleaning is best done during summer months when water level is low in at least one of the tanks so that the bottom of the tank can be clearly viewed. Start with the raw water tank, which will usually be the dirtiest tank. Tank cleaning will necessarily loosen biofilms and may cause water to become cloudy, depending on how dirty tank is. Isolate this dirty water and prevent it from being transferred to clean water tank or entering distribution system. Turn off transfer pump when cleaning raw water tank and turn off pressure pump when cleaning the clean water tank.

Assemble Vacuum System

There are three sections of flexible hose; two long sections and one short section. The short section connects between the brush and the pump. Connect end with the hose clamp to the pump on the end of the white cone-shaped flange on the end of the pump. Push the other end of the short hose onto the brush on the end of the vacuuming pole. One long flexible hose has hose clamps on both ends. This connects between the pump outlet and filter. Connect one end of hose to outlet pipe on side of pump and connect the other end to filter inlet. Connect the remaining long hose to the filter outlet. The other end of this hose discharges back into tank being cleaned. When finished using vacuum system, disassemble and hang hoses up to dry. Wash down filter cartridge with a hose sprayer. Wait for filter cartridge to thoroughly dry before reassembling filter.

Raw water tank cleaning

1. Depending on how dirty the tank is and how much water is in the tank, add hydrogen peroxide to help oxidize biofilms as they are loosened up. If hydrogen peroxide is not available use chlorine.
2. Assemble tank vacuum system. Turn off transfer pump. If the water is quite low and visibility is not a problem, you can leave circulation pump running during the cleaning process, otherwise turn it off. If the circulation pump is turned off, close valves at bottom of tank and in return line to tank to prevent siphoning from filters back into tank.
3. Lower pump motor into water first and lay it on its side on the bottom of the tank. With the tank circulation pump running, fill the suction hose with the upper discharge from the circulation loop. Disconnect the hose from the brush and push it onto the discharge outlet at the top of the tank. When the hose is filled reconnect the hose to the brush and lower brush into the water.
4. Assemble the filter housing and tie off on top of the tank to prevent it from falling. Lower the discharge end of the vacuum system into the water.
5. Turn on the vacuum pump briefly and confirm adequate suction. If so, turn off tank circulation pump and begin sweeping all areas of the bottom of the tank. Clean vacuum filter periodically by removing it from the housing and spraying down with a hose.
6. If the water becomes too cloudy to see the bottom, stop sweeping and let the vacuum continue running until the water clears. Cover the manhole with a clean towel to prevent leaves and insects from entering tank while waiting for the water to clear sufficiently.
7. Spray sides and top insides of tank with peroxide (or bleach) and water mixture. Brush sides of tank with a pool brush to loosen biofilm.
8. Once the water has become sufficiently clear from circulating through the vacuum, turn off the vacuum pump and remove the vacuum system from the tank. Restore normal circulation to the raw water tank, but leave the transfer pump turned off while cleaning the clean water tank.

Clean water tank cleaning

1. Siphon or pump the clean water into the raw water tank to lower the water level in the clean water tank. Raise the chlorine level to at least 2 ppm in the lowered clean water. You may wish to shock the distribution system with some of this highly chlorinated water while cleaning the tank. Make sure the transfer pump is turned off.
2. Assemble the tank vacuum system.
3. Repeat the same process as above to fill the suction hose of the vacuum with water.
4. Turn off the clean water circulation pump. Turn off the household pressure pump.
5. Repeat the same cleaning procedure on clean water tank as was performed on raw water tank.
6. After finished cleaning, restore circulation but leave the transfer pump and household pressure pump off until clarity is completely restored to the water in both tanks.
7. Replace cartridge filters in both circulation loops. Turn transfer pump back to "automatic" and begin refilling clean water tank.
8. Re-energize household pressure pump.
9. If water in distribution system is highly chlorinated, do not use until it is flushed out of the system back into the raw water tank. Once the clean water tank is full, test the chlorine level to make sure it is under 2 ppm. Use hydrogen peroxide to lower chlorine level if necessary. Flush the house distribution system with reduced chlorine clean water by discharging back into the raw water tank from outdoor faucets. The UV lamp will remove the remaining excess chlorine from the raw water tank.
10. You should notice an overall reduction in TDS and conductivity in the water after tank cleaning.

6.5 Gutters and Drain Line Cleaning

Cleaning gutters, gutter screens, and drain lines from gutters to raw water storage may become necessary if the prefilter is constantly getting clogged with clumps of mold and biofilm after every rainfall. If gutter screens are clogged, as indicated by water running off them to ground rather than penetrating through screens, first clean gutter screens with chlorine and pressure washing as previously mentioned. The process of cleaning screens will often clean out gutters and drain lines as well. If gutter screens are clean but you have reason to believe that drain lines are not, you can plug the drain pipe at the prefilter and fill the pipe with a concentrated solution of bleach and water. Use an air-inflatable expansion plug, such as McMaster-Carr item number 3058K5 or similar to plug drain line leading into first-flush valve. This is inflated with a hand tire pump. Leave the valve open to prevent wash water from getting into storage tank. Then pour a mixture of bleach and water, approximately 1:4 ratio, into the roof gutters and fill the drain pipes. Let the bleach sit in the pipes for several hours then flush it out to ground by deflating the plug. Rinse out drain lines and prefilter housing with fresh water from a garden hose and restore the system back to normal.

Another way to do this that uses less bleach is to scrub the pipes with a bleach-soaked rag. Even long drain pipes buried under ground can easily be scrubbed with a rag and liquid bleach using an electrician's trick for pulling electrical cable through buried conduit. Purchase a roll of light-weight poly line long enough for the run, which can usually be found in the electrical section of Lowes or Home Depot, and tie a 1-inch or 2-inch foam ball to it. Attach your shop vacuum to the far end of the drain pipe and turn it on, and then feed the poly line in at the other end. You may have to plug all the other drain pipes with a rag at the downspout to get sufficient vacuum. Having clean-outs at every downspout makes this easy. Once the poly line comes out at the shop vacuum, tie it to a heavier line that is at least twice the length of the drain pipe and pull it through for the cleaning operation. Tie a rag to the middle of the cleaning line, soak it with bleach and water, and pull it back and forth as if you were cleaning a rifle bore. Rinse off the rag and put fresh bleach on it each time it comes out one end of the drain pipe and then pull it back through again. When the rag comes out clean you know your drain pipes are clean. This operation is easier and quicker with two people, but one person can do it. An example written procedure for cleaning gutters and drain lines is provided below.

Gutters and Drain Lines Cleaning Procedure

Mold and organic matter buildup inside gutters and drain lines can cause frequent prefilter clogging, re-ducing collection efficiency and increasing decay by-products being added to collected water. If pollen season has ended and pre-filter is continuing to get clogged after every rainfall, gutters and drain lines may need cleaning. If too much roof water is lost, as indicated by large discrepancy between spreadsheet and actual water level after a rainfall, the micro-mesh gutter screens may be clogged with pollen and in need of cleaning. After cleaning gutter screens, gutters, and drain lines, you should observe an overall re-duction in TDS or conductivity of stored water. Clean gutters and gutter screens first, then drain lines, then storage tanks, and lastly the distribution system to minimize recontaminating a previously cleaned part of your water system.

Gutter screens and gutters

1. Mix a 1-to-4 solution of bleach and water. Turn valve control mode switch to "Open" position to prevent controller from closing first-flush valve.
2. Spray micro-mesh gutter screens with bleach and water. Let stand for about an hour, then rinse with high pressure water from hose or pressure washer.
3. Inspect screens for cleanliness. You should be able to barely see through them in bright sunlight. Repeat steps 1. and 2. above if necessary.
4. Remove one or two screws on screens and lift screens to inspect inside of gutter. Gutter should be clean and free of slime mold and algae. If not, first try dumping in full strength bleach at high end of gutters to loosen mold. If that does not work then remove screens and pressure wash gutters.
5. When done, reattach screens with screws.
6. Restore valve control mode switch to "Auto" position.

Drain lines from gutters to raw water tank

This method of drain line cleaning uses a strong oxidizing chemical, either hydrogen peroxide or chlorine, to oxidize and loosen mold and dirt in drain lines. This requires plugging the line in such a way that the chemical does not leak into your storage tank. Then the lines are filled with a concentrated solution of the oxidizing agent and allowed to soak overnight. Wear rubber gloves and eye protection when doing this cleaning procedure.

1. Open the first-flush diverter valve and remove the screen around the valve. Leave the valve open for this cleaning operation so that chemicals do not get into your storage tank.
2. Insert an air-inflatable rubber expansion plug into the pipe leading into the valve and inflate the plug with a hand operated tire pump.
3. Mix 1-to-4 solution of bleach and water (or hydrogen peroxide and water) and dump into all downspouts. Continue to add concentrate until all drain lines are full. Water leaking at downspout couplings indicates they are full.
4. Allow concentrate to sit in drain lines for 24 hours.
5. After soaking for 24 hours, slowly deflate rubber plug. Wear rubber gloves and eye protection when you do this. Let the wash water drain to ground.
6. Rinse oxidized sediment out of drain lines by spraying clean water in upper end of gutters.
7. Remove inflatable plug and re-assemble screen around valve.

6.6 Distribution System Cleaning

Distribution system should be a very rare necessity. Indeed it almost never occurs on houses connected to public water supplies, mainly because it is next to impossible to accomplish. Public water utilities keep chlorine levels high, close to those of swimming pools to battle biofilm build-up in water mains and household distribution plumbing. But potable rainwater harvesters can do better. Not only can we easily clean our distribution systems when necessary, but we can also maintain much lower chlorine levels than typical public utilities because we keep our distribution systems much cleaner.

A few clues indicating that a biofilm problem may exist in distribution plumbing include the following: 1) rust-colored biofilm forming in toilet tanks or around sink drains, 2) free chlorine level at kitchen faucet is significantly less than that at clean water storage, and 3) bad water taste or odor at kitchen faucet but not at the clean water tank. If a biofilm problem is indicated, first try raising chlorine levels for a few weeks to see if that will eliminate the problem. You can also shut off the water supply, drain all pipes, and then turn water back on with all faucets wide open. Be sure to turn off breaker power to water heater before draining it. I ruined a perfectly good plastic Marathon water heater one time by forgetting to turn off the power to it! The water hammer effect and rushing air/water through the pipes are often sufficient to strip biofilms off pipe walls, causing water discharge from shower heads and sink faucets to have a rusty appearance. Repeat this process a few times until water runs clear from faucets. Be careful not to overdo the water hammer and rupture a pipe!

Many swimming pool companies are aware that biofilms can build up a resistance to the current disinfectant being used in a pool, making biofilm removal extremely difficult for an operator. They sometimes recommend pool operators switch to a completely different disinfectant after a few years. Usually this means switching between chlorine-based disinfectants and hydrogen peroxide-based disinfectants. As with pools, biofilms in household plumbing can also build up resistance to disinfectant in potable water. If elevated shock levels of the normal disinfectant does not eliminate biofilms in distribution plumbing you may have to resort to shocking with a different disinfectant. For example, if you are currently using chlorine, then shocking with elevated levels of hydrogen peroxide may be required, or vice versa. Alternatively, you can try changing your system over to a completely different disinfectant at normal safe levels to see if that eliminates the biofilm. Remember that hypochlorite and hydrogen peroxide neutralize each other and cannot be used simultaneously. Before using hydrogen peroxide on a chlorine-based system, reduce chlorine levels to as low as possible in the clean water tank to minimize consumption of hydrogen peroxide in neutralizing residual chlorine. A UV lamp on the circulation loop can help reduce chlorine levels down in a matter of days. Alternatively, if you use only ozone in your raw water, you can turn off chlorine metering and wait several days for transferred raw water to dilute chlorine levels in the clean tank.

Use food-grade 35% hydrogen peroxide for this cleaning operation. You will probably need about a quart or more of this stuff. Shipping is expensive since hydrogen peroxide at this concentration is considered hazardous. If you can find a local supplier of food-grade hydrogen peroxide you can save money on shipping. This cleaning operation will take 24 hours or more, so stock up some drinking water in your refrigerator. You will raise hydrogen peroxide in your clean water to shock levels in excess of 100 ppm and soak your distribution plumbing in it for 24 hours. Lower the clean tank water level to just a few inches above the bottom outlet to minimize required amounts of 35% hydrogen peroxide. In ppm units 35% hydrogen peroxide equals 350,000 ppm. If residual chlorine did not consume hydrogen peroxide then the volume of concentrate can be determined from:

$$C1 \, x \, V1 = C2 \, x \, V2 \qquad\qquad (6\text{-}2)$$

where C1 and V1 are hydrogen peroxide concentrate ppm and volume respectively, and C2 and V2 are the desired concentration and water volume in the clean water tank. For example, if you have 300 gallons remaining in your clean water tank and you want it to have 100 ppm hydrogen peroxide, then you would need (100 ppm)(300 gal)/(350,000 ppm) = 0.09 gal of concentrate. This is the minimum amount you need. Residual chlorine remaining in water will demand more. Hydrogen peroxide test strips are useful for determining when you have added enough hydrogen peroxide to clean water storage.

After lowering the clean water level dump in the estimated amount of concentrated hydrogen peroxide into the clean water tank. Leave the circulation pump running. Test the clean water with hydrogen peroxide test strips to make sure it is at least 100 ppm. Turn off power to your hot water heater and pressure pump. Drain the chlorinated water from your distribution system. Turn pressure pump power back on (but not hot water heater power) and re-pressurize your distribution system with the hydrogen peroxide water. Remove all air from the system and use test strips at each faucet to determine when hydrogen peroxide comes out. When every faucet tests positive for hydrogen peroxide let the system soak for 24 hours. You can use this water for flushing toilets but try to minimize the amount of this water going into your septic tank. Do not drink this water; this disinfectant is at shock levels. Safe drinking concentrations of hydrogen peroxide are under 50 ppm.

After a 24-hour soak drain the distribution system again and re-pressurize using the water hammer effect to loosen dirt and biofilm. You will probably need to remove shower head and faucet screens when flushing the distribution system since hydrogen peroxide strips dirt and biofilms from pipe insides that can clog faucet screens. Repeat this process until all faucets run clear. Then pump the remaining clean water into the raw water tank until the water level is just barely above the bottom outlet on tank. This will dilute the hydrogen peroxide and also do some cleaning action on the raw water circulation system. Turn the transfer pump back on, turn on chlorine metering, and begin refilling the clean water tank. You may have to manually add some chlorine, depending on how you have the metering pump set up and how much hydrogen peroxide remained in the clean water tank when you started refilling. Purge the distribution system of remaining hydrogen peroxide, dumping it into the raw water tank through outside faucets.

Alternatively, you can just use the water as is and wait several days for the hydrogen peroxide to naturally break down, before turning chlorine metering back on. Safe levels for drinking hydrogen peroxide are around 25 to 30 ppm. I found that boiling water will drive out hydrogen peroxide and reduce its concentration to safe levels. Using this method allows running higher shock levels, of 80 ppm or more hydrogen peroxide, in a distribution system for several days to give it a thorough cleaning. Just make sure your family knows not to drink the water from the faucet during this time. An activated charcoal filter will not remove it. Distribution system cleaning takes advanced planning and some time to accomplish. To minimize impact on your family you might want to have on-hand a case of bottled water for drinking and cooking and a few 5-gallon buckets of water for toilet flushing while water is shut off during this hydrogen peroxide cleaning. An example written procedure for your log book is provided below.

Water Distribution System Cleaning Procedure

A potential biofilm problem in the distribution system is indicated by pinkish or rust-colored biofilms in toilet tank, around sink drains, and by a persistent significant difference in chlorine level and TDS between the clean water tank and kitchen faucet. A musty smell or taste to the water also indicates a biofilm problem. Dirty storage tanks are usually the primary cause of biofilm problems in a distribution system. First clean storage tanks and shock chlorinate distribution plumbing to see if that will solve the biofilm problem. If it does not, follow the procedure below to clean the distribution system using hydrogen peroxide. **Before handling concentrated hydrogen peroxide, a potentially dangerous chemical, read the Material Safety Data Sheet (MSDS) for 35% Hydrogen Peroxide.**

As with storage tank cleaning, distribution system cleaning is best done when water storage levels are low. This cleaning procedure uses shock levels of hydrogen peroxide in excess of 100 ppm in the clean water tank and the water hammer effect to clean distribution plumbing. Before adding hydrogen peroxide, reduce chlorine in clean water tank to bare minimum to avoid consuming too much hydrogen peroxide since chlorine neutralizes hydrogen peroxide. This can be accomplished by turning off power to peristaltic metering pump for several days.

1. Turn off power to transfer pump and lower water level in clean tank by filling raw water tank from outdoor faucet. Lower water level to a few inches above bottom outlet in clean water tank. This will permit using less hydrogen peroxide concentrate to obtain the required dilution level.
2. Turn off power to pressure pump and hot water heater. Drain water from hot water tank and distribution system. Drain into raw water tank if possible to minimize water loss.
3. Add food grade 35% hydrogen peroxide to clean water tank at a rate of 1 to 2 cups per 250 gallons. Use hydrogen peroxide test strips to insure at least 100 ppm hydrogen peroxide in clean tank. If no hydrogen peroxide is consumed by residual chlorine or other impurities, two cups of 35% hydrogen peroxide in 250 gallons of water should yield a concentration of 175 ppm. $C_1 \times V_1 = C_2 \times V_2$; $(350,000\ ppm)$ $(16\ oz) = C_2 \times (32,000\ oz)$; $C_2 = 175\ ppm$.
4. Turn pressure pump power back on and re-pressurize system. Bleed air from all faucets, including shower heads and outdoor faucets. Air traveling through plumbing will produce a small "water hammer" effect that helps loosen biofilms. When water runs clean through the faucet use a test strip to insure a high level of hydrogen peroxide from faucet and then turn off faucet. Repeat for all faucets until air is removed and distribution system is filled with shock level hydrogen peroxide.
5. Let system remain pressurized with hydrogen peroxide for 24 hours. You can use the system for bathing, cleaning, etc. but do not drink it. Use bottled water or other normal stored water for drinking.
6. After 24 hours of soaking in hydrogen peroxide, turn off pressure pump power and completely drain distribution system again. Turn pump power back on and purge air from system. Repeat steps 5. and 6. until water runs clear from faucets. When water runs clear from all faucets your system is clean.

7. Pump remaining clean water to raw water storage, if possible, but not so low that the circulation pump looses prime. Hydrogen peroxide is compatible with AOP and will help clean the raw water system.

8. Neutralize remaining hydrogen peroxide in clean water tank by adding a few ounces of chlorine until testing shows a detectable level of chlorine. Restore power to metering pump and transfer pump and begin refilling the clean water tank. If your metering pump is set to SECONDS mode you will need to manually add chlorine after tank is filled to restore proper chlorine levels. If metering pump is set to FLOW SWITCH mode, so that metering pump runs entire time that transfer pump is running, then you should not need to manually add any chlorine.

9. When clean water level is sufficiently above bottom tank outlet, restore power to pressure pump and begin flushing hydrogen peroxide from distribution system into raw water storage through outside faucets.

6.7 Electronic Controllers

There are three electronic controllers on the basic rainwater system; the transfer pump controller, the pressure pump runtime monitor, and the first-flush valve controller. A wireless system may have four controllers. You should have written procedures in your maintenance log book explaining how to operate these controllers and where they are located. Include an electronic schematic of the controller in your procedure in case it needs to be serviced. You should also provide contact information for someone who can repair the controller if it fails. Procedures for controllers that incorporate Arduino microcontrollers should also include sketch printouts. Example operating procedures for the two Arduino-based controllers are provided below (minus schematics and sketch listings).

Pump Runtime Monitor Operating Procedure

The runtime monitor is a safety device that prevents the pressure pump from emptying storage tanks in case of pipe rupture or major leak in the pressurized distribution plumbing. It counts the number of minutes the pressure pump runs and shuts off 120VAC power to pump if pump runs for longer than 20 minutes in a 24-hour period, which amounts to a loss of about 100 gallons. Runtime monitor displays total minutes of runtime during the current 24-hour period, and automatically resets to zero after 24 hours (assuming no alarm condition) and begins counting again. If runtime exceeds 20 minutes the display begins flashing total runtime, indicating the 20-minute limit has been exceeded. If this happens the monitor can be reset by pushing the Reset button to begin a new 24-hour monitoring period. Do this only after making sure there are no major leaks in your system.

The three-position toggle switch on the monitor allows the operator to select between; "on," "off," and "auto." Normally the switch is placed in the "auto" (automatic) position, which provides protection from major system leaks. On rare occasions you may need to consume more than 100 gallons in a 24-hour period, such as multiple clothes washer loads, tank cleaning, outdoor pressure washing, etc. There are two ways to handle this. First, you can put the toggle switch in "on" position which bypasses the monitor's protective feature. The monitor still records and displays pump runtime, flashing after 20 minutes, but it does not shut off power to the pump. Secondly, you can leave the monitor in "auto" and simply reset the monitor if the alarm condition is triggered, which is immediately noticed with a loss of water pressure and flashing of runtime. The "off" position shuts off power to pressure pump without having to trip the breaker in the breaker panel. This is useful when going on a long vacation and you want maximum protection against leaks.

Note that if the alarm condition is triggered (indicated by flashing runtime) the only way to remove the alarm condition is is to press the Reset button. A non-flashing continuous display of minutes runtime indicates no alarm condition has been triggered.

Changing Maximum Allowed Runtime

If 20 minutes maximum runtime is unacceptable you can easily change it to a different value. You need the Arduino IDE installed on a laptop computer and a proper USB cable. Remove the front panel from the Runtime Monitor Controller to access its Arduino board. Be aware that 120VAC voltages exist in this box on the two relays. Load the RuntimeMonitor sketch and change the number 20 in the first line in the code that reads:
const unsigned long maxMinutes = 20;
to whatever value you wish. Then recompile and upload the program to the Arduino board in your Runtime Monitor Controller. Remove the USB cable and reattach the front panel.

First Flush Controller Operation

The first flush controller is located in master bedroom closet. The control panel has LEDs at the top indicating status of first flush valve. A lit green LED indicates the first flush valve is closed and rainwater is going into storage tanks. A lit red LED indicates the valve is open and rainwater is being dumped to ground. The panel has two 4-digit numerical displays. The upper display shows current rainfall amount in inches. Lower display shows total water storage level in gallons. Displayed rainfall amount returns to zero and first-flush diverter valve is opened when Reset button is pressed. Record rainfall amount before pressing Reset button and enter measured rainfall in your WaterStorage2 spreadsheet.

The 3-position rotary switch just under LEDs controls operating mode of first-flush valve. Turning switch to "Close" position closes first-flush valve, regardless of rainfall, and keeps it closed unless raw water tank is filled to capacity, then it opens. This position allows collecting rainwater with zero first-flush amount. Turning switch to "Open" opens first-flush valve and keeps it open regardless of rainfall amount. Use this position when cleaning gutters or tanks. Turning switch to "Auto" allows the Arduino micro-controller to automatically control opening and closing of first-flush valve depending on rainfall and raw tank water level. For normal operation this switch should remain on "Auto." The other two positions (Close and Open) are generally only used for maintenance inspections or special circumstances (e.g. drought conditions or tank/gutter cleaning). If you suspect the first-flush valve is not working properly then turn this switch to "Close" and/or "Open" and check operation of valve is consistent with LED status lights.

The 6-position rotary switch adjusts amount of initial rainfall that is diverted to ground before closing first-flush valve. This initial rainfall washes dirt and contaminants from roof before sending captured rainfall to storage tanks. First-flush divert amount may be set to 0.02, 0.04, 0.06, 0.08, 0.10, and 0.12 inches with this switch. Normally this is set to an appropriate amount and left in that position. However, it can be changed to better adapt to local weather or seasonal conditions and stored water levels. For example, if tank is low and rain is frequent but only occurs in small amounts (e.g. less than 0.5 inches), then you may wish to move this switch to its lowest setting to capture more rainfall. Contrarily, if storage level is near full and heavy rain is predicted, then you may wish to move switch to its highest setting to divert maximum initial rainfall to ground and thoroughly clean gutters and drain lines before collecting rainfall. If you change first-flush amount be sure to enter new setting in your WaterStorage2 spreadsheet.

The first-flush amount switch is only considered if 3-position valve control mode switch is in "Auto" position. If that switch is in either "Open" or "Close" positions then first-flush amount setting is ignored. Also note that first-flush amount only takes effect with additional rainfall. If for example 0.08" first-flush was selected and rainfall ceased after 0.05". Turning this switch down to 0.04" does not immediately close the valve. The valve closes only after the rain gauge receives an additional 0.01" of rainfall, assuming the Reset button was not pushed. If you press the Reset button after reducing first-flush amount then counting begins at zero and valve will close when displayed rainfall equals first-flush amount setting. If tanks are not full and rainfall extends over multiple days, do not press Reset button in order to keep valve closed until rainfall has ceased. Roof is already clean and there is no need to repeat a first-flush cycle. Record rainfall and Reset after rainfall ceases.

Under severe drought conditions the first-flush amount can be reduced to zero by setting the 3-position valve control switch to "Close." Be aware that this could mean more frequent cleaning of the bag pre-filter and raw water tank filters since no roof washing is occurring in this position. But circumstances justify it.

6.8 Decommissioning

There may be rare times when a potable rainwater system must be shut down for an extended period of time due to a house being vacated for several months and no one available or willing to continue maintaining its rainwater system in the operator's absence. Temporarily decommissioning a system in such a way that it can be easily restarted at some indefinite date in the distant future is our objective in such a situation, especially if the house and rainwater system may be subjected to freezing temperatures. The goal is to drain all water from tanks, filters, circulation loops, and household plumbing and prevent rainwater from refilling storage tanks. Preferably, one should first flush household plumbing with hydrogen peroxide or elevated chlorine levels prior to draining all plumbing lines, to minimize bacterial growth during a long dormant period. Follow procedures for distribution system cleaning (section 6.6) to accomplish this.

To begin decommissioning, turn off power to the pressure pump, water heater, and all electrical hardware on the rainwater system. Open all faucets in and outside the house and drain all household plumbing, including hot water heater and pressure tank. Drain all rainwater storage tanks. Use the vacuum system to quickly drain these tanks. Open up and drain all filters and circulation loops on system. Dispose of old filter cartridges. Open and drain water out of UV lamp or AOP purifier. Make sure there is no water anywhere in the system that could freeze and crack pipes or pumps. Open diverter valve and secure it in open position so that it always dumps rainwater to ground. Unplug battery backup to first-flush valve controller and remove AC power to controller. Remove prefilter bag and plug inlet to raw water tank with a rubber stopper to prevent rainwater from entering tank and also to keep out insects and rodents. Make sure the screen is in place on the discharge from the gutters to prevent insects and rodents from entering the discharge lines. Cover the digital rain gauge so that it does not become clogged with leaves and mold or subject to freeze damage.

Once all the pipes are drained, close all faucets to keep insects out of water pipes. Lastly, secure the pump house and tank manhole openings to discourage tampering or vandalism. Securing of tank manhole openings is easily accomplished with a padlock hasp purchased from any hardware store. Bolt one end to the lid and the other to the rim of the tank opening. Since completely drying the tanks is impractical, pouring in a few ounces of concentrated bleach before securing tank openings helps reduce the potential for mold growth inside the tanks. Everything you do to decommission the hardware should be recorded in section 3 of the system log book. This information is extremely important for the next person who restores system operation at some later date. Do not skip on recording the details of decommissioning, even if you are the one who will recommission the system after a long vacation.

Lastly, if decommissioning was the result of death of original occupants, you must be sure to leave detailed instructions for the next occupants on how to recommission the system. Everyone knows how to turn on electrical power to common appliances, but a potable rainwater system is relatively unfamiliar hardware for most people. We wish that it were not so, but that is the case today. So make sure the next owners have all necessary instructions to get the system running again. Indeed, this potable rainwater system is a valuable capital asset on the property the next owners just purchased or inherited!

Chapter 7.
Alternative Concepts

This final chapter addresses issues not covered in the previous chapters. This book specifically covers potable rainwater harvesting as an everyday household water supply. We assume the house is tied to the electric grid, but may or may not be tied to a public water supply. After reading this book, one might ask, "How could this system be modified for an emergency water supply or an off-grid water supply?" This question is addressed in the first section below. The second section discusses chlorine venturi injectors as a lower cost alternative to peristaltic metering pumps. The third section discusses top suction from storage tanks since this is often promoted in some rainwater harvesting books. The fourth section discusses the numerous benefits of rainwater harvesting and my philosophy of decentralization, as opposed to centralization of supply and authority. I personally believe that a serious consideration of the benefits of decentralized supply versus centralized supply of essential resources is desperately needed in our nation today. Potable rainwater harvesting is, most definitely, a decentralized solution.

7.1 Emergency and Off-Grid Applications

A rainwater harvesting system that is only used as an emergency water supply in case of widespread failure of utilities and is otherwise simply used for garden watering, can be significantly simpler in design than that described in this book. In an emergency situation where public water supplies fail, either due to power failure or pathogenic contamination, a household is most likely not concerned about taking long hot showers or washing loads of clothes. Under water emergencies water consumption for drinking and personal hygiene quickly reduces to that of a wilderness backpacker, or at least a RV camper. Emergency consumption levels are on the order of a few gallons per day rather than tens or hundreds of gallons per day. Filtering and purifying water on demand at these consumption levels is much simpler and many different choices are available to a rainwater harvester. Various filters are available for recreational boating, backpacking, and other outdoor activities that can turn highly contaminated water into clean potable water. Obviously, the more contaminated your source water is, the more often your filters will need cleaning or replacement. So the question for an emergency rainwater harvester is, how clean do you want to maintain your emergency water source, assuming you will use slow filters and disinfectant to make it safe to drink?

For emergency-only applications there is no need to go through all the expense of a dual-tank design as discussed in this book; a single-tank design will work just fine as long as you realize that stored water needs further purification before drinking it. At a minimum stored water should be passed through a very small porosity filter, such as Doulton ceramic filter specifically designed to handle low quality water. Adding sodium hypochlorite or calcium hypochlorite to filtered water to provide a free chlorine level of 0.5 ppm provides a good margin of safety against viruses that might pass through the filter. Chlorine also helps reduce odor and improve taste by oxidizing natural organic matter in the water. Alternatively, a low level of chlorine could be maintained in a storage tank to help reduce mold and bacterial growth, even if stored water is mostly used for plant or animal watering. If a residual level of chlorine is maintained avoid overflowing the tank for reasons discussed earlier. Either a mechanical first-flush diverter valve like that discussed in this book's first edition or an electronic diverter valve as discussed in this edition would accomplish this. If you do not want to maintain a residual chlorine level in the tank, then overflowing your storage tank during rainfalls will help flush out old stagnant water with fresh rainwater and help control biofilms. Bring the inlet from the roof all the way to within an inch or two of the bottom of the tank to help stir up sediment and flush it out when the tank overflows. Prefiltering and micro-mesh gutter screens are much more important on a system that does not utilize continuous circulation through filters. No doubt periodic tank cleaning will also be more frequent.

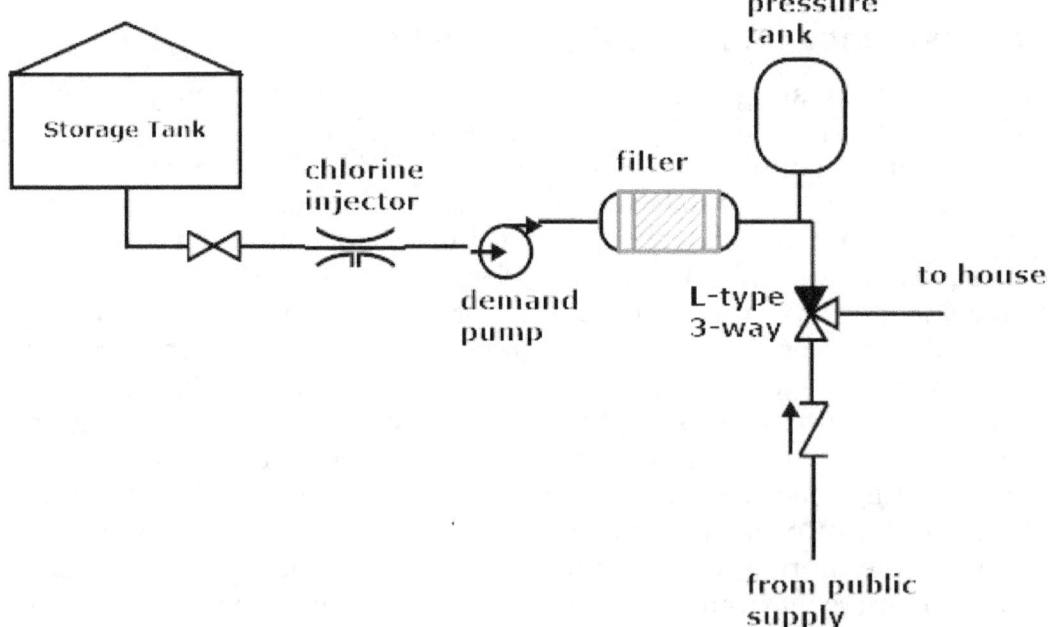

Fig. 7-1. Rainwater system for emergency-only water supply.

If you want to automate chlorine injection and plumb emergency water directly into house plumbing you will most likely need to get a permit from appropriate local water authorities to do so and prove that there is no possibility of cistern water getting into the public water supply. A check valve in the public water supply line can help prevent this from happening. An L-type three-way valve that automatically isolates public water supply when the house is connected to cistern water supply and vise versa will prevent back-flow contamination issues. These are sometimes called back-flow prevention valves and are discussed in sections 602 and 603 of the 2015 UPC.

A low-wattage pressure-actuated diaphragm pump, often called a demand pump on recreational vehicles, can push cistern water through a set of filters and a venturi chlorine injector to provide water pressure for the whole house. A small bladder-type pressure tank can be added to keep the pressure pump from rapidly cycling on and off. If the pump motor requires 12 VDC it can be powered off a car battery in an emergency. A possible emergency rainwater system is illustrated in Fig. 7-1.

An off-grid rainwater harvester has more options, assuming he has a better source of electrical power than an emergency-only rainwater harvester. If electrical power is not an issue then a complete dual-tank system could be implemented. If electrical power must be conserved then either the above emergency water supply or a slight modification of it could be implemented, depending on how much power is available for purification and pressurization. A small circulator pump could be turned on for a few hours each day with a timer to help keep stored water relatively clean. An off-grid rainwater harvester will presumably use his system as an everyday potable water supply and will want to minimize maintenance of filters. So keeping stored water as clean as practicable is a higher priority than with an emergency-only rainwater harvester. As before, chlorine can be added to the tank or injected when water leaves storage and enters the distribution system. To minimize electrical power consumption, ozone and UV can be eliminated and simply rely on chlorination and filtration. If THMs prove to be a problem then an activated charcoal filter can be added for drinking purposes only. Figure 7-2 shows a potential off-grid system design with minimal power consumption.

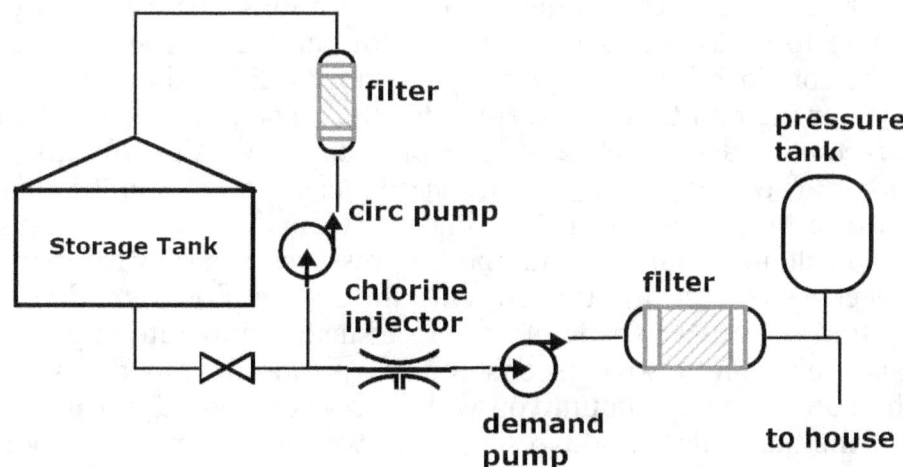

Fig. 7-2. Rainwater system for off-grid household water supply.

For those interested in implementing a UV purifier and/or ozone generator in an off-grid application, but still need to be somewhat concerned about electrical power usage, the table below provides power consumption levels as a function of maximum flow rate for the various Sterilight models manufactured by Viqua of Canada.

Table 7-1. Basic specifications of Sterilight UV purifiers.

Model	S1Q	S2Q	S5Q	S8Q	S12Q
Max flow rate	2 GPM	3 GPM	6 GPM	10 GPM	15 GPM
Power consumption	19W	22W	30W	46W	48W

Power consumption for my ozone generator is 14 watts. So a small circulating system could be implemented that consumes very little electrical energy. If the head pressure is under about 3 feet then a 1/200 HP magnetic drive circulator pump could probably be used, which only consumes 19 watts. As long as the off-grid home has a source of electrical energy (e.g. solar, wind, biomass, etc.) potable rainwater harvesting and clean pressurized running water are certainly feasible. It may be a scaled-down version of the system discussed in this book or a modified version using some of the disinfection techniques discussed. Most likely, an off-grid home will become even more conservation-minded than we have become due to the fact that electricity is also limited with off-grid living. Rainwater harvesting, whether on-grid or off-grid, does not mean doing without modern conveniences; it simply means knowing when and how to utilize those conveniences. Conservation increases when resources become limited.

7.2 Venturi Disinfectant Injectors

Venturi injectors are a low cost alternative to peristaltic metering pumps. Mazzei is a popular brand of disinfectant injectors that has a long history. A word of caution about sizing injectors: do not assume an injector salesman is an engineer and will recommend the correct injector for your application. I did and ended up purchasing the wrong injector the first time (Mazzei model 484). The low flow was insufficient to activate the recommended injector. Mazzei provides a website to help you select a correct injector (http://injectorselector.mazzei.net). If in doubt discuss your application with a Mazzei engineer. The injector I finally used on my system was a model 287. You will need to input a number of pressures and flow rates on the Mazzei website to select the correct injector. For chlorine suction flow rate, divide the pump flow rate (in GPM) by 213.3 to estimate chlorine suction flow rate in gallons per hour (GPH). This is equivalent to two cups of sodium hypochlorite added to 1600 gallons of water, which is about the maximum you will likely ever need. Typically, you will use even less chlorine than this. All Mazzei injectors will provide more suction flow than the above estimate, so you will need to add a metering valve on the chlorine suction line. Use a plastic metering valve that can handle concentrated sodium hypochlorite and use only plastic fittings in the concentrate feed (e.g. PVC, CPVC, nylon, polyethylene, etc.). The chlorine metering valve I chose for my system was a PVDF needle valve purchased from Ryan Herco Flow Solutions (www.rhfs.com). Placing a medical IV drip chamber in the chlorine concentrate line helped to properly set the metering valve.

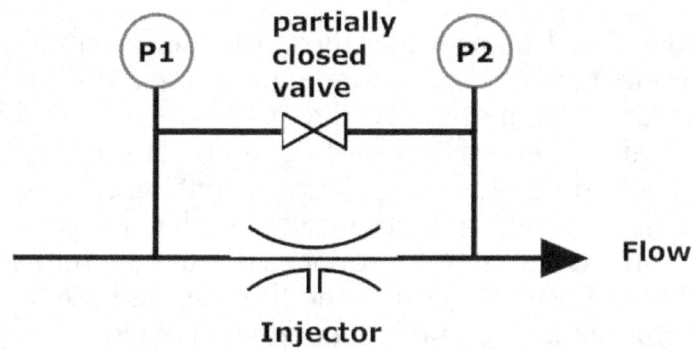

Fig. 7-3. Bypass valve to adjust injector pressure differential

Proper operation of these injectors requires a specific back-pressure on the outlet side and a higher pressure on the inlet side of the injector. Plumbing the injector as shown in Fig. 7-3 with two pressure gauges and a bypass valve allows setting correct pressures for the injector to work properly. Partially close the bypass valve just enough until the pressure difference causes the injector to begin sucking disinfectant. A medical IV drip chamber on the disinfectant suction line provides immediate feedback of disinfectant flow. With flow to the right as indicated in the figure, gauge pressure P1 will be greater than pressure P2 by about 10 psi. When flow is zero, P2 will equal P1 and disinfectant suction will stop.

The main problem with venturi injectors is that they all provide greater disinfectant suction flow than is needed. Throttling down disinfectant flow with a metering valve on the suction line can be problematic due to the injector's ball check valve on the suction line input. A metering valve adds back pressure on this check valve that can prevent it from opening. Check valves often need more pressure to get them open than to keep them open. At the low flows required for full strength liquid bleach this check valve is always right at the edge of fully closing. One way to avoid this is to dilute the liquid bleach and increase disinfectant flow to more normal rates expected by the injector. Another way to solve this problem (which I have not yet tried) is to dispense with the metering valve altogether and put an inexpensive plastic solenoid valve on the water flow input to the venturi. Then use an Arduino to control the solenoid valve, holding it open for a set amount of time. With a chlorine sensor input, the Arduino could automatically adjust metering time.

After messing with a Mazzei injector for 2 or 3 years I finally abandoned it for a more stable peristaltic metering pump. Nevertheless, venturi injectors are a low-cost alternative for injecting disinfectant. They work well when pressures and flows are in accord with injector specifications. Other than the small ball check valve they have no moving parts. Low cost chlorine metering is a subject that needs more research.

7.3 Top Suction from Storage Tanks

Low wattage centrifugal pumps, like those we use for circulation loops, cannot pump air and must remain primed with water at all times. For this reason these pumps are located at or near the bottom of tanks so that there is no possibility of their losing suction or becoming air-locked. If top suction and bottom discharge are utilized, special precautions need to be taken to prevent a centrifugal circulation pump from becoming air-locked and causing circulation to stop. All connections must be air-tight on suction side of pump. Furthermore, if ozone is discharged into the tank through its lower port those bubbles must not get into the top suction hose. Otherwise the pump could eventually become air-locked and circulation will cease.

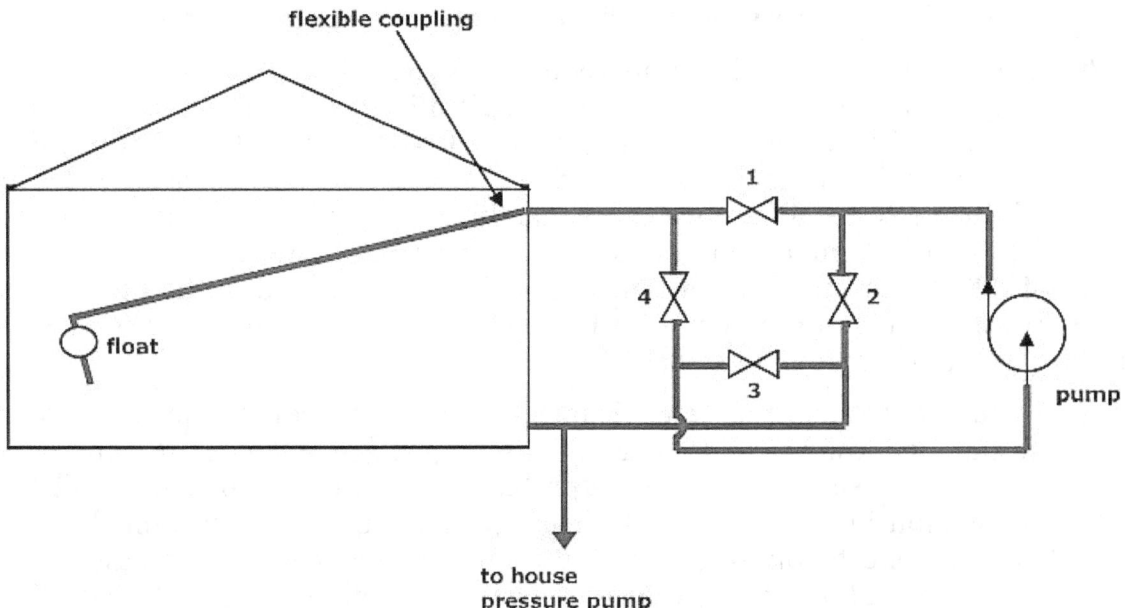

Fig. 7-4. Flow reversing 4-way valve for use with top suction.

If you incorporate top suction you should include a 4-way valve that allows easy flow reversal from top suction to bottom suction in order to purge air from a circulation loop when necessary. Finding a low-cost PVC 4-way valve is nearly impossible. Little demand for such valves exists. However, you can easily build a low-cost 4-way valve with four T's and four common ball valves, as shown in Fig. 7-4. (Electronics types will note the similarity to a full-wave bridge rectifier.)

When valves 1 and 3 are open and the other two are closed, the pump sucks from the bottom of the tank and discharges to the top. When valves 2 and 4 are open and the other two are closed, the pump sucks from the top and discharges to the bottom of the tank. The top and bottom ports on plastic tanks are usually on the same side of the tank. To laterally displace the suction to the opposite side of the tank and prevent ozone bubbles from the bottom port getting into top suction, fabricate a stiff L-shaped PVC suction line with a short section of flexible tubing to connect it to the top port and allow the L to pivot at this point. Use a plastic swimming pool rope float on the other end to float the suction just below the surface.

Figure 7-4 only shows the pump and 4-way valves, not the filters and purification equipment that would also exist in this circulation loop. We have also shown pressure pump suction as coming from bottom of tank, which is the best place to put it if the pressure pump is also a centrifugal type pump. However, note that this is also the discharge line of the circulation pump immediately after circulating water passes through filters and purifier. So the pressure pump is sucking from the cleanest water in storage. This is the one advantage I see of top suction versus bottom suction. However, I think it can also be problematic in terms of the circulation pump loosing prime. If this system is implemented then a flow sensor should be included that shuts off circulation pump power if an air lock should occur since some circulator pumps can never be run dry (e.g. Laing Thermotech pumps) without permanent damage to the pump.

7.4 Shared Potable Rainwater System

Rainwater harvesting is primarily a private solution, not a centralized government or public utility solution. The reason for this is that people cannot truly learn to conserve water when the only consequence of their actions is a utility bill at the end of every month. Public water is cheap and even a doubling or tripling of cost is not sufficient to force people to learn to conserve water. When people pay for something they expect their vendor to supply it, regardless of cost. The only way that people can learn to conserve water is when they see immediate and direct consequences of their actions, and those consequences include running out of water. No amount of money can overcome the problem of limited supply, but a public water utility always appears to have an unlimited supply of water. If a public utility tries to raise the cost of water in order to encourage conservation, voters will bring political pressure upon that utility to bring prices back down. So a strong motivation to learn water conservation is simply not available with public water supply. Centralization has removed the need for self control. In fact, all environmental conservation must start with the private individual, not government.

Assuming adequate collection area and average rainfall, a rainwater harvesting system can potentially be implemented on a shared basis among multiple family dwellings (e.g. single story townhouse or apartment) in order to reduce installation and maintenance costs, if that community maintains it as a privately-owned decentralized water supply, rather than centralized public supply. If management of that system degenerates into a centralized socialist form of governance, then that water supply will fail. All socialist forms of government, in spite of claims to the contrary, are inherently centralized with benefits only going to the elitist tyrants at the top. Everywhere that socialism has been practiced in the world, including in the early American colonies, it has proven to become a dismal failure. Shared rainwater harvesting systems must be managed in a decentralized way that punishes irresponsibility. In other words, self control and individual responsibility must be encouraged, not eliminated. There can be no possibility of manipulation, borrowing, redistribution, or abuse of neighbors. There must be individual consequences for individual actions, including running out of water even when your neighbors have plenty. The maintenance costs and labor of that system must also be shared, not just the benefits. And when anyone in that community fails to do his duty, he alone should suffer the consequences without expecting others to pay the penalty for him.

The technological problem here with a shared rainwater system is that each dwelling must clearly understand that their allotted portion of water storage, not the whole storage tank, is in fact their own limited supply, regardless of the number of persons in their own household, and regardless of what is available to their neighbors. That allotment must be based on roof area, not on how many persons are living under the roof in a particular family. (This eliminates government system control and wealth redistribution!) Each dwelling must see direct and immediate feedback of their own daily water consumption and its impact on their portion of remaining water storage. This can be done with an LED display, computer log-in, or some other means. Electronic flow meters and solenoid valves on each dwelling, monitored by a computer (maintained by entire community), would automatically shut off water supply to a dwelling if it used up its allotted portion of water storage, regardless of actual amount in total storage. Only additional rainfall would restore that careless dwelling's portion of water storage. The computer algorithm (under oversight of entire community) could adjust daily recommended consumption, depending on expected rainfall. Remaining allotted storage, usage rate, and when they run out of water all need to be clearly visible to each and every dwelling in a decentralized connected community. This provides both visibility of limited supply and immediate feedback on consumption that is so essential for learning to live within one's limited resources (as opposed to living within the resources of others).

Centralized public water utilities cannot maintain personal responsibility because there is no immediate feedback on limited resources and usage rates. Everyone thinks the public water resource is unlimited when in fact it is not. There are no serious consequences for wastefulness and slothfulness, other than a little higher water bill. Centralization makes human society pay for the sins of individuals.

A shared rainwater system obviously has unique challenges that a completely private system does not have. The primary challenge is maintaining the private ownership and individual responsibility that are so vital with potable rainwater harvesting. The small local community of a shared system might choose to elect a system caretaker and even pay him to perform maintenance activities. But that community must be extra careful that they do not turn their shared system into a small public utility that facilitates water waste. Otherwise their system will fail. The consequence of running out of water, even when others have plenty, must be very real to each and every person sharing a common rainwater system. No one can expect the community to be penalized for their own water wasteful practices.

Liberal socialist governments always end up abusing their people and creating shortages of essential resources. For example, consider the current water shortage in Cape Town, South Africa where residents are now restricted to only 13.2 gallons per day per person or face stiff fines from government. Although this is still greater than the 9.1 gallons per day per person that we voluntarily achieved during our 2016 drought, most would consider this abusive. Several articles on the Internet show that this water shortage in Cape Town is mostly due to political incompetence and corruption. This year the socialist South African parliament overwhelmingly voted to begin confiscating white-owned land without compensation, which is nothing more than government theft. Many are blaming global warming (i.e. more advanced nations) on their own water shortage. Now consider the facts. Cape Town has an average annual rain fall of 31 inches per year, which is 44% more than Boulder, Colorado's average annual rainfall of 21.6 inches. We showed in section 3.3 that potable rainwater harvesting was indeed feasible for Boulder with a consumption rate of 50 to 75 gal/day. Obviously, Cape Town should be able to do potable rainwater harvesting even more easily than Boulder. But who in their right mind would install a rainwater harvesting system in Cape Town when their government will most likely confiscate it? So the people suffer water shortages when God gives them more than enough rain!

7.5 Benefits of Rainwater Harvesting

This book shows that modern potable rainwater harvesting incorporates a vast amount of advanced modern technology, some of which is more advanced than that currently used by municipal water suppliers. Modern rainwater harvesting, even in areas where it is not essential, actually promotes the advancement of society by making it more conservation-minded; more aware of how thoughtless and wasteful activities can destroy precious natural resources; less dependent on centralized government to supply our most basic needs for survival; and more knowledgeable of basic mechanical and electrical skills. No government program can come close to conserving water like private rainwater harvesting.

A widespread practice of rainwater harvesting would stimulate technological advances in water treatment, building materials, and practices that are less polluting of groundwater resources, such as metal roofs versus asphalt shingle roofs. Septic tanks and sewage processing facilities become less polluting with rainwater harvesting than with municipal water supply simply due to reduced water consumption and reduced amount of chlorine dumped into the environment. Rainwater harvesting motivates landscaping with native plants, rather than exotic water-consuming plants. It can help extend the life of existing aging municipal water supplies and reduce risks of costly municipal bonds for rebuilding water supplies. Rainwater is naturally soft, pleasant to bathe in, pleasant to drink, and eliminates the need for water softeners and their salt-laden waste water. Rainwater is absolutely wonderful in the kitchen: it leaves no spots on washed dishes or mineral deposits in tea kettles.

Designing, building, and maintaining a potable rainwater harvesting system provides parents tremendous opportunities to teach their children a multitude of basic practical skills and experience in mathematics, microbiology, chemistry, electricity and electronics, plumbing, pumps and motors, filters, the physics of pressure and water flow, water conservation, masonry, and much more. There is no better teacher than actual hands-on experience. The advantages and benefits of rainwater harvesting are so many that it is truly amazing that its practice is not more widespread than it is today. I hope this book contributes, at least in some small way, to help change that. For those that do choose to design and build your own system, I hope this book helps you avoid some of the mistakes I made along the way in building my own potable rainwater system.

Rainwater harvesting is obviously a so-called "green technology." Clean potable water is among the most basic of essential resources needed for survival, and nothing encourages water conservation more than rainwater harvesting. We are disappointed that more local and city governments do not strongly promote rainwater harvesting. Public dependence on municipalities to always supply abundant cheap potable water has actually encouraged water wasteful practices, increased pollution, and increased strain on aging municipal water purification plants. In fact, even centralized electric power generation (as opposed to decentralized power) is highly wasteful of resources and environmentally polluting, but that is another story for another time. The naïve faith and trust in centralized government-regulated public utilities to always make available an infinite supply of clean potable water has become a dangerous phenomenon in our modern society. It has resulted in a dumbing-down effect on society as a whole, causing society to think they cannot survive without centralized government-controlled water supplies. Society is continually at risk to disruption of clean potable water supplies, and our environment is at risk to wasteful and polluting human activities as a result of the idolatrous worship of centralized government.

Taking personal ownership and responsibility for potable water supply reduces risks to both human society and environment. Decentralized modern potable rainwater harvesting can be implemented practically anywhere in the world that has rainfall, if people are first willing to learn how to conserve water, and then properly match their water consumption to their collection area, storage tank size, and average rainfall. Self-reliant people learn to adapt to available resources; government-dependent people do not. Public utilities are not truly interested in decentralization, conservation, and self-reliance because that eats into their income streams. Maximizing waste helps maximize utility profits and government tax revenue. When demand exceeds production public utilities can always go to local government to obtain taxpayer-backed bonds to build bigger facilities. For this reason, real conservation of resources will not come from government nor public utilities. Real conservation and environmental protection only come from a society that refuses to be dependent on centralized government to supply their daily needs.

A large amount of information has been published on rainwater harvesting, but much of it seems to be written by theoreticians rather than practitioners. This becomes evident when you see the same unworkable or highly risky concepts showing up in different publications. In addition to more experimental research on modern potable rainwater harvesting, a great need exists for actual practitioners of potable rainwater harvesting to write on this subject and share their experiences with others. Modern potable rainwater harvesting is not difficult, it just seems difficult to most people because so very few people practice it. We have become dependent on government to supply us safe and healthy water. Yet, incredible technology exists today to cheaply provide the cleanest drinking water that human society has ever seen! Our ancestors successfully practiced rainwater harvesting without the technology we have today. We should be even more successful rainwater harvesters.

Building a modern potable rainwater harvesting system is not difficult. Any reasonably intelligent person with a willingness to study and learn a few new skills can build a rainwater harvesting system that provides cleaner potable water than practically all public water supplies. Needed tools are not expensive and you probably already have them in your garage or shop. If you are unfamiliar with basic plumbing and electrical skills, I recommend Plumbing A House by Peter Hemp[49] and Wiring A House by Rex Cauldwell.[50] Your local homebuilder supplier usually has several excellent educational books. Numerous educational books on basic electronics exist, including Getting Started in Electronics[51] and the Engineer's Mini Notebook series[52] authored by Forrest M. Mims. I highly recommend the Arduino Cookbook by Michael Margolis[53] for learning to electronically interface with the world using

---

49 Peter Hemp, Plumbing a House, The Taunton Press, 1998.

50 Rex Cauldwell, Wiring a House, The Taunton Press, 2002.

51 Forrest M. Mims, III, Getting Started in Electronics, Tandy Corporation, 1988.

52 Forrest M. Mims, III, Engineer's Mini Notebook, Vols. I, II, and III, Master Publishing, Inc. 2007.

53 Michael Margolis, Arduino Cookbook, O'Reilly Media, Inc. 2012.

microcontrollers. And of course a wealth of free information exists on the Internet on all subjects and skills needed to build a potable rainwater harvesting system.

Even if you never build a potable rainwater harvesting system, you should at least acquire the knowledge and skills needed to do so. Overspecialization in modern society is actually a dangerous stupefying of human society that reduces innovation. In thermodynamic terms, increasing specialization and division of labor is an increasing entropy process that ultimately ends when useful work can no longer be accomplished. I prove this fact in another book, but suffice it here to say that entropy in the engine of human society increases as specialization increases. Innovation usually comes from those who are familiar with multiple disciplines. But most people today cannot even perform basic mechanical and electrical servicing of the systems they depend on daily in their homes due to overspecialization, much less create innovative alternatives to those systems during widespread disaster. Many have become too obsessed with recreation and entertainment to be concerned about acquiring basic skills that might help them and their neighbors survive natural or politically-induced disasters.

In spite of all the benefits of rainwater harvesting, demand for centralized government-regulated water supplies will continue to serve those without capacity or desire to manage their own potable water supply. So utilities need not fear that rainwater harvesters will put them out of business. However, it is naïve to think of rainwater harvesting as a low-technology reversion to the past, suitable only for back-woods living. On the contrary, society's increasing dependence on centralized governmental supply of essential resources is a move backwards in societal advancement. Societal dependence on centralized supply of essential resources make societies less secure from human error, natural disasters, terrorism, wars, and environmental pollution. Such dependence on centralized government degrades and stupefies society, encouraging government to expand its tyranny. This puts human society dangerously at risk to unexpected disasters, which we all witnessed with hurricane Katrina in 2005. When human society looks to government to meet all its needs and wants, it gets the kind of government it deserves: an ever-increasing draconian bureaucracy that sucks innovation and productive life out of human society! As the prophet Samuel warned in 1 Samuel 8 of the Bible, the fault is not with the government, but with a populace that demands tyrants rule over them. Instead of learning self-control, they wish to be slaves of other men.

As government becomes larger and more bureaucratic, it becomes less able to innovate and make technological advancements. This is simply the nature of increasing entropy. When I worked in government in Washington DC, many of my fellow engineers complained how much government lagged private industry in technology. We wrote engineering documents using a pair of scissors and a jar of glue to do cutting and pasting! We hand-printed reports and a secretary typed them out. Some frustrated engineers went out and bought their own personal computers just to work more efficiently. Even military systems were obsolete by the time they were deployed. When President Kennedy told Wernher von Braun, "I want you to put a man on the moon," von Braun replied in so many words, "I'll do it on one condition: you leave me alone and let me do it my way!" Von Braun understood that if government interfered he could never make the technological leap to put a man on the moon.

Why is government such a huge roadblock to technological advancement? The answer is actually quite simple. Government is run by politicians whose job security depends on pleasing most of the people the majority of the time. Government is, therefore, the ultimate consensus organization. By definition, innovation **never** comes from consensus thinking! For this reason, government has an inherent hindrance to innovation and technological advancement. Therefore, human society is foolish to look to its government to solve environmental and technological problems, including problems of the medical industry. President Eisenhower understood this when he warned the nation about the "military-industrial-congressional complex" in his farewell speech on January 17, 1961, although his political advisers forced him to remove the word "congressional" from his actual speech. Very few heeded his warnings and today we have government spending money on useless enterprises that even its own technical experts say are huge wastes of time and resources. Unfortunately, political job security is their primary objective in all this spending, not advancing technology, productivity, or the self-reliance of human society.

I close this book with one final reason for pursuing rainwater harvesting, which is actually more important than all other reasons stated above. I mention this reason for theological readers of this book, who will appreciate this reason more than other reasons given. The fundamental philosophy behind open source and the so-called "maker movement" is decentralization, as opposed to empire building. Makers build stuff and then show others how to do it. Mentoring or discipleship is the primary means by which that is accomplished. Makers despise obfuscation and hiding essential details under under an excuse of "proprietary" for the purpose of empire building. This book has been written in the spirit of showing others how to do it. The world is big enough for many more profitable businesses involved in designing and building potable rainwater systems. If you find a potential business opportunity in this book that interests you, then I wholeheartedly encourage you to pursue it.

When enough of us are doing potable rainwater harvesting the right way, then we will see government follow with good, beneficial, and informative regulations. Not all government regulatory codes are bad. In fact, many of those codes are quite informative, especially when codes are written by intelligent engineers and scientists who want to help human society do things safely and correctly. Indeed, all conscientious makers and tradesmen consult applicable code books to understand how to build things correctly. The problem occurs when businesses (both big and small) try to use government regulation as a means of empire building and gaining an advantage over competitors. This is sometimes called, "using the sword of government to eliminate your competition," and is an evil practice that has continued for centuries.

There are countless examples of this practice in every industry. It produces bad regulation which causes many businesses to simply relocate overseas or to another state. It increases the entropy of doing business and reduces human productivity. The only beneficiaries of bad regulation are the tyrants at the top.

One can show that decentralization is in fact a Biblical mandate originating in God's institution of marriage in Genesis 2:24. This topic is too extensive to develop here, but I have developed it in another book. Suffice it to say that sinful man's natural inclination is to establish centralized control and build empires at all levels of human society; be it Mafia-style families, big church denominations, big corporations, or big bureaucratic governments. Centralization at any level of human society always leads to tyranny, abuse, waste, and destructiveness. Specialization leads to professionalization, which then leads to centralized control and protection of those professions. A humorous example of this is that one can now become a certified professional rainwater harvester, in spite of the fact that mankind has successfully practiced rainwater harvesting for thousands of years! In thermodynamic terms, specialization, professionalization, and centralization are increasing entropy processes that ultimately lead to "heat death" in the engine of human society when profitable productivity can no longer overcome the inherent increasing waste (or increasing entropy).

Very few economists and businessmen today consider (or even understand) the problem of entropy in human productivity. Most all of them believe in the myth of perpetual motion in the engine of human productivity, and are then surprised when their empires finally implode. Every engine, including the engine of human society, must have an entropy-reducing step in its cycle in order to keep on running. Work output always accompanies an increase in entropy. A portion of that work must be spent on an entropy-reducing step in that engine. In human society this step is called "decentralization." But this step is fundamentally absent in all empire building.

The first historical record of centralized empire building is recorded in Genesis 11 concerning the Tower of Babel. When God said, "Now nothing that they propose to do will be withheld from them," He was not making a statement about all those grand and wonderful things that centralized empire might accomplish. On the contrary, it was a statement about the tyranny and abuse that top-level elitist governors would inflict upon those below them. Even human sacrifice or state-sanctioned murder would not be withheld from them, as the rest of the Bible goes on to show. Empires waste and abuse human resources; they increase the entropy of human labor.

Some reviewers of this book asked me why I don't try to patent some of the hardware discussed here. Even though I am an inventor on ten U.S. patents I have never made a single dime on any of those patents! And I know of very few inventors who have made money on their patents. I learned long ago that the primary beneficiaries of patents are patent lawyers, big corporations that hire them, and the U.S. Patent Office. However, we inventors and "makers" have far more ambitious goals than just making a lot of money. And those goals extend long after our own short lifetimes. Our goals are nothing short of helping to decentralize productive human society, making it more self-reliant, independent, and able to enjoy the fruits of its own labor. We want to be a part of helping put productive human society beyond the reach of government tyranny and big corporate tyranny. For these reasons I now consider U.S. patents to be a high entropy waste of time and effort!

The so-called "anti-federalist" founding fathers of our nation, men like Patrick Henry, Samuel Adams, Melancton Smith, and George Mason, understood the Biblical mandate for decentralization and strongly argued against a centralized federal government in the constitutional debates of 1787, using scriptures like 1 Samuel 8 in their arguments. Just as the Tower of Babel building project never would have begun if those at the bottom had not agreed to it, so Samuel Adams blamed the "stupid servility" of common people in his private letter to Richard Henry Lee in 1787. That staunch Puritan wrote, "The body of the people tamely consent and submit to be their slaves." Of course these "anti-federalists" lost those constitutional debates, but they have been vindicated because practically everything that they warned would happen has now come to pass in our day. The blame for tyranny in government rests not so much on government itself, but on the governed who consent to be tyrannized. Therefore, whether a rainwater harvester realizes it or not, he is actually acting upon a Biblical mandate and serving his Creator by helping to decentralize human society!

Appendix A
Structured Flowcharting

A.1 Introduction

This appendix provides greater detail on the discipline of structured flowcharting for those programming systems with microcontrollers. A professor in graduate school first introduced me to structured flowcharting while studying computer science and I have used it ever since. I have modified the technique only slightly to more rigorously follow a fixed set of rules, but it is basically the same as I learned it many years ago. The occasional reference to "SmartDraw" is due to the fact that this document came out of a request to the developers of the SmartDraw software to include the discipline of structured flowcharting. Structured flowcharts could be drawn with SmartDraw, but not easily. The user had to toggle between two different libraries. There was also no set of rules in SmartDraw that prevented the user from drawing unstructured flowcharts. Rather than completely rewrite this document, I have kept the references to SmartDraw because they are instructive. They show that even modern software developers do not truly understand this discipline of structured flowcharting. Once you have mastered and internalized this discipline it will revolutionize the way you think about problems and look for solutions. It will help you write better code for the electronic controls on your rainwater system and any other system.

A.2 The Problem

Figure A-1 shows an example of a "modern flowchart" produced in SmartDraw. The flowchart is for a pump runtime monitor that prevents a pump from running longer than a specified predefined limit in a 24-hour period. The logic is clear but the flowchart does not help the software programmer develop good code that can be easily tested. Indeed a sloppy coder might use GOTO statements to implement this algorithm and the flowchart does not prevent that from happening. But the infamous GOTO statement is the primary cause of "spaghetti code" with multiple entry points and multiple exit points in a section of code. Allowing multiple entries and exits can quickly raise the complexity of large software systems to the point that the software cannot be fully tested in a reasonable amount of time. For example, simply following the arrows, one can show that there are eight different paths from Start to End in Fig. A-1 (including only one loop back to the branch, "Is pump running?"), all of which must be tested. When we see buggy software that has not been fully tested, or cannot be tested on time and within budget, the product is usually the result of a bunch of spaghetti coders. Good software only has one entry point and only one exit point. Furthermore, it is not immediately clear to the reviewer of Fig. A-1 whether or not this algorithm was developed in a structured manner, due to the fact that the flowchart does not rigorously follow a set of rules for structured flowcharting. Indeed, arrows still point left and right; rotate Fig. A-1 by 90 degrees and they would point up and down, one of key features of unstructured spaghetti code. The only good thing that can be said of the modern flowchart in SmartDraw is that it uses a reduced set of only two symbols, the oval and the box.

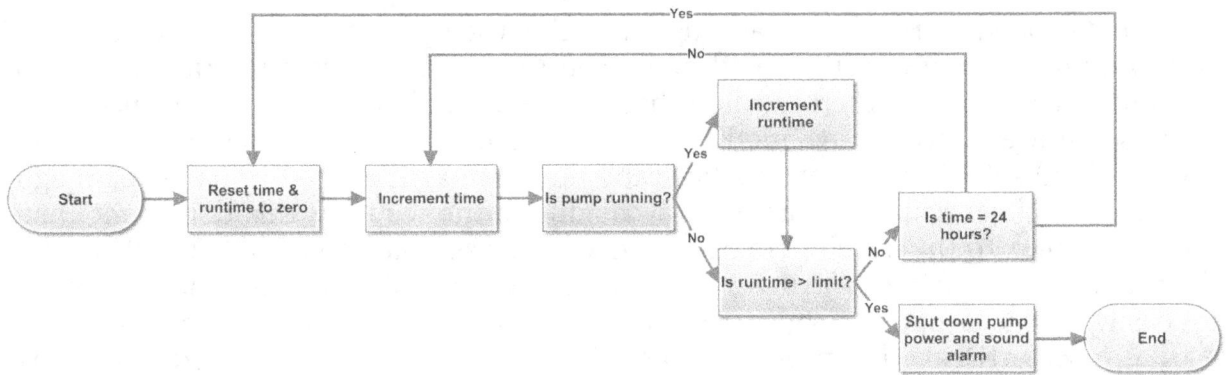

Fig. A-1. Example modern flowchart in SmartDraw

A.3 The Solution

The solution to the problem above is to force the programmer to develop algorithms with structured flowcharts rather than the modern flowcharts currently in SmartDraw. Figure A-2 is an example of a structured flowchart that performs the same task as Fig. A-1. Unlike Fig. A-1, the structured flowchart does not permit the use of GOTO statements. Flow is from top to bottom with one entry and one exit everywhere in the code. Every section of code that includes decisions can be collapsed into a single box with only one entry at the top and one exit at the bottom of the box. All the statements inside the WHILE loop in Fig. A-2 can be collapsed into a single box that reads, "Increment clocks and check for alarm or reset conditions." All the statements inside the REPEAT-UNTIL loop can be collapsed into a singe box that reads, "Monitor pump runtime for alarm condition." Connector lines never cross and never go back up. Lines never go to the left except as a return to the exit point of a decision branch. Indeed arrows are not even needed with structured flowcharts because the flow is clearly understood to be from top to bottom. The arrows on connectors always point down, and sometimes to the left at tie-in points on diamonds, but never point up. Connecting lines to decision and loop branches only move to the right, never to the left. The only time that a connector moves to the left is as a return from the TRUE branch of a decision diamond symbol.

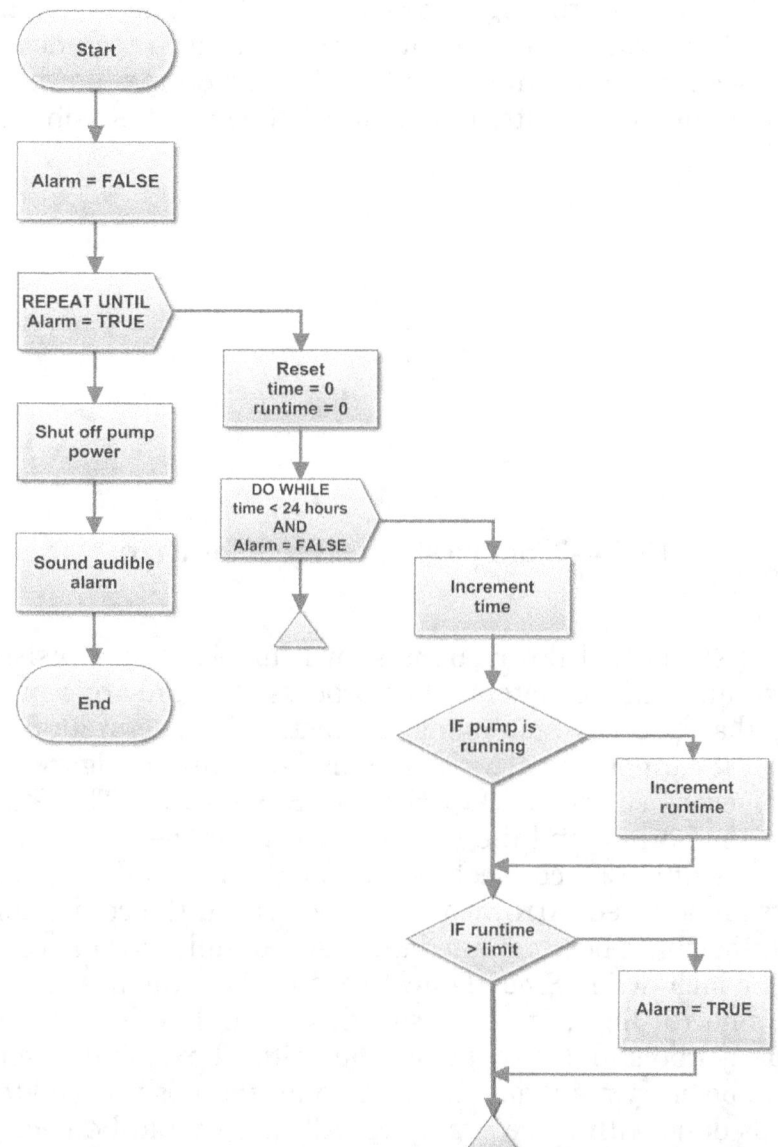

Fig. A-2. Structured flowchart of algorithm in Fig. A-1.

Only five symbols are used: 1.) ovals marking the algorithm start and end points, 2.) boxes with flow always entering in at the top and exiting at the bottom of the box, 3.) right-pointing boxes for DO loops, FOR loops, WHILE loops, REPEAT-UNTIL loops, etc. 4.) a small triangle marking the end of the loop, 5.) diamonds for conditional branches such as IF-THEN, IF-THEN-ELSE, and CASE statements. Figure A-2 uses all five of these symbols. As with boxes, flow always enters both the pointed box and diamond symbols from the top and exits out the bottom. Temporary conditional flow is always to the right, never to the left. The flow and indentation is somewhat akin to the structure of a table of contents in a technical document with sections and subsections. Most importantly, the structured flowchart immediately shows how the code should be written to maintain the lowest possible complexity.

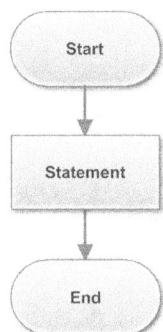

Fig. A-3. Simplest structured flowchart

The simplest structured flowchart is shown in Fig. A-3, consisting of one start point, one end point, and one statement of process with only one input and only one exit connecting that process statement. In fact, all structured flowcharts can be reduced to this. If it cannot, the algorithm is not structured. Figure A-3 would be the starting point for every structured flowchart in SmartDraw. The Start oval has only one connector at the bottom and the End oval has only one connector at the top. The rectangular box has only two connectors, one at the top and one at the bottom. No other connectors are allowed. Arrows are not needed on the connectors and generally are not used, but they have been retained in these examples for instructional purposes. A button would be included in SmartDraw to easily allow the user to turn on or off the arrows on the connector lines. Starting with this basic flowchart, the user would then insert additional symbols above or below the initial box. Since every symbol to be entered has only one entry point and one exit point, there is no need to manually draw connectors as was done with these examples. All that should be necessary is to insert or delete symbols from the existing flowchart, forcing the user to develop a structured algorithm and allowing SmartDraw to automatically format and space the symbols.

Figure A-4 shows a loop symbol entered immediately after the box in Fig. A-3. Loops are used when a set of instructions must be repeatedly executed either a fixed number of times or an indefinite number of times until certain conditions are met. This loop consists of at least one process statement which hangs off to the right of the pointed box. The loop symbol could be either a WHILE loop, FOR loop, REPEAT-UNTIL loop, or DO loop, which the user specifies by typing text in the pointed box. The conditions governing loop execution are also typed in the pointed box. The small triangle marks the end of the loop, causing the flow to jump immediately back to the pointed box that governs the loop. When a loop is executed, program flow repeatedly flows down through the symbols hanging off to the right of the pointed box until the looping is finished. After the looping is finished, flow resumes back out through the bottom of the pointed box. There is only one entry into the loop and only one exit out of it. Thus the entire loop can be collapsed into a single rectangular box that says, "Repeat a set of instructions until done."

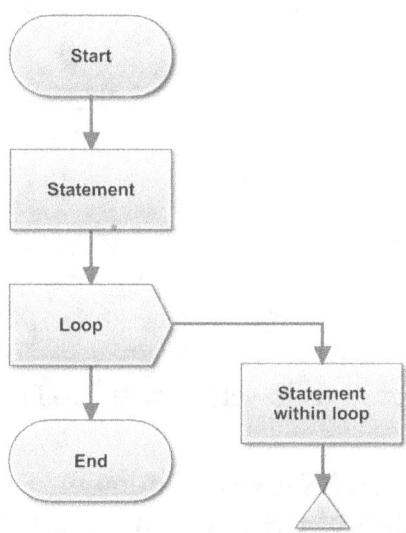

Fig. A-4. Structured flowchart with loop symbol

Figure A-5 shows a simple IF-THEN statement using a diamond symbol. Diamonds have only three connection points: top, bottom, and right side. Connections are never made to the left side of a diamond. Program flow always enters at the top of the diamond. If the condition is TRUE, flow continues to the statements hanging off the right side of the diamond. If the condition is FALSE, flow continues out the bottom of the diamond. The right branch always ties back into the connector leaving the bottom of the diamond. This tie-in point marks the end of an IF-THEN statement, thus providing only one entry and only one exit for this conditional branching section of code. As additional symbols are entered into the TRUE branch the tie-in point moves further down from the diamond since flow can never go upward. Since there is only one way into and only one way out of a decision diamond, the diamond can be reduced to a rectangular statement block that reads, "execute a sequence of statements if the condition is TRUE."

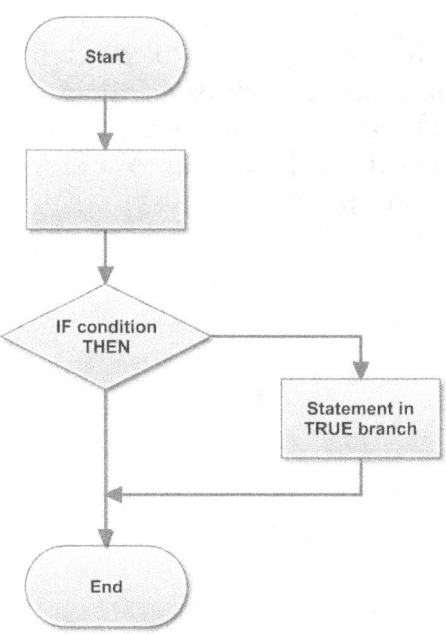

Fig. A-5. Structured flowchart with IF-THEN statement.

Figure A-6 shows an IF-THEN-ELSE statement. Symbols can be entered in both the TRUE and FALSE branches of the diamond. Again, the two branches must tie in below the diamond, which marks the end of an IF-THEN-ELSE statement and the single exit point of this conditional block of code. An ELSEIF statement in some programming languages actually tests for two different conditions and would be represented with a second nested diamond in the ELSE branch of the original diamond. Nevertheless, connector lines never cross, never go up or to the left out of a diamond, and the entire structure can be reduced to a single rectangular box with one input and one output that reads, "Execute a sequence of instructions depending on whether the condition is TRUE or FALSE." The IF-THEN and IF-THEN-ELSE decisions may have very complex conditions including one or more logical operators (OR, AND, NOT, XOR, etc.) but the condition always resolves into one of two answers: "yes" or "no," or more precisely, TRUE or FALSE. The TRUE branch always hangs off the right of the diamond and the FALSE branch always hangs directly below it. The two branches tie back in together at some point below the diamond, allowing the entire conditional statement to be reduced to a single box with one entry and one exit. Flow lines never cross, never go to the left out of a diamond, and never go up, even with multiple nested diamonds. If this rule is violated the code is likely not structured.

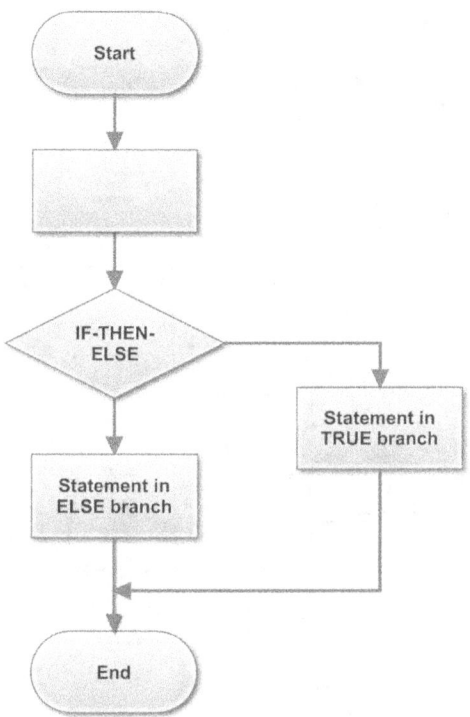

Fig. A-6. Structured flowchart with IF-THEN-ELSE statement.

Some conditionals do not have only two answers: yes or no. An example of a conditional with more than two answers is "What is the color of the crayon?" Of course the correct answer could eventually be arrived at with multiple diamonds asking, "Is it red?", "Is it blue?", or "Is it yellow?" But it is much simpler to ask what the color is and execute instructions depending on that color. Figure A-7 shows a multi-branch conditional, such as a CASE or Switch statement. The CASE values are shown as labels above each of the multiple branches. All of the branches tie back in to a single point below the diamond, allowing the entire CASE statement to be reduced to a single box with one entry and one exit. As with the IF-THEN-ELSE, the CASE statement could include code in the ELSE branch below the diamond and above the tie-in point when the condition value does not match any of the specified values to the right of the diamond, as shown in Fig. A-7. Alternatively, the CASE statement could have a straight line from the bottom of the diamond to the tie-in point, like the IF-THEN statement. Again, even with multiple nested CASE statements and other conditional statements, lines can never cross, never go upward, and never exit to the left out of a diamond symbol. The only flow line that goes to the left is the final tie-in at the very end of the conditional.

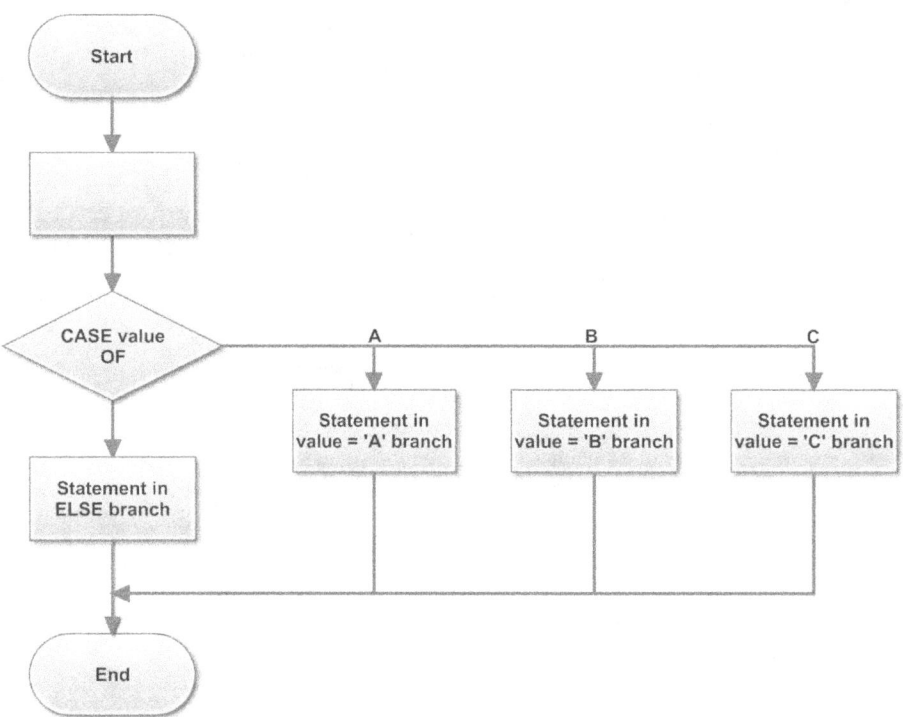

Fig. A-7. Structured flowchart with CASE statement with more than two branches.

A consideration of code complexity clearly illustrates the advantages of structured flowcharts over conventional unstructured flowcharts. Consider figures A-8 and A-9 that perform the exact same function. Figure A-8 is an unstructured flowchart because it prevents the decision branches (diamonds) from being reduced to single rectangular boxes with one input and one output. The output of process "1" enters a point inside one of the branches of the first diamond. The flowchart violates a couple of rules of structured flowcharts. First, it has a branch off to the left of the second decision diamond. This is not allowed in structured flowcharts. If the branch for process "1" were redrawn to the right of the second diamond, flow lines would have to cross, which is also not allowed with structured flowcharts. Secondly, the flowchart has flow lines that go up, which is not allowed with structured flowcharts. Structured flowcharts only allow flow lines that go down or to the right and left at the end of a process branch, but never up.

Any unstructured flowchart can be redrawn into a structured flowchart equivalent, and the process of doing this usually results in simpler and more reliable code. We first list all the unique one-pass paths through the unstructured flowchart, excluding repetitive loops through the same block of code. Figure A-8 has four unique one-pass paths through it. Using the process numbers to list the paths from "Start" to "End", these paths are; 2, 3, 2-1-2, and 3-1-2. Since neither of the decision diamonds can be reduced to a single-input-single-output box, all four of these paths count toward the complexity value of this block of code. The irreducible complexity equals the number of one-pass paths through the simplest flowchart representation of the software, which in this case is a complexity of four. Reducing code to a simpler flowchart form does not mean automatically leaping to a one-box representation of the entire code, but rather sequentially reducing each loop and conditional block into their one-box representations.

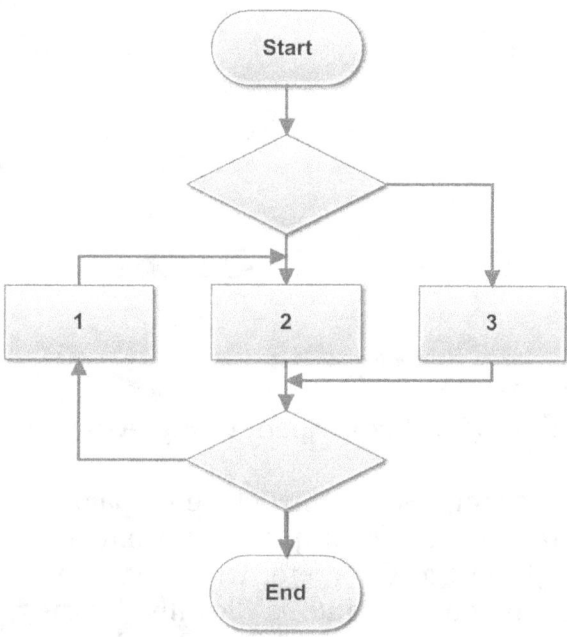

Fig. A-8. Unstructured flowchart with two intertwined decision branches.

Note that the last two paths listed above consist of process "2" or "3" with a process "1-2" added to it. This shows that the basis set for this block of code is actually only; 2, 3, and 1-2, which gives us a clue how to redraw the flowchart into a simpler structured flowchart. The code needs to first select between process "2" or "3," then decide whether or not to add on a process "1-2." The structured equivalent of the above flowchart is shown in Fig. A-9. Enumerating the paths through Fig. A-9 will show that it has exactly the same unique one-pass paths as Fig. A-8. However the difference is that both decision blocks can now be reduced to two rectangular boxes that read, "select between either process 2 or 3," and "select either straight through or process 1-2." Therefore, this code can be further reduced to an even simpler flowchart with only two boxes in it giving a complexity value of one. All structured flowcharts have an irreducible complexity value of unity. Anything greater than one is unstructured.

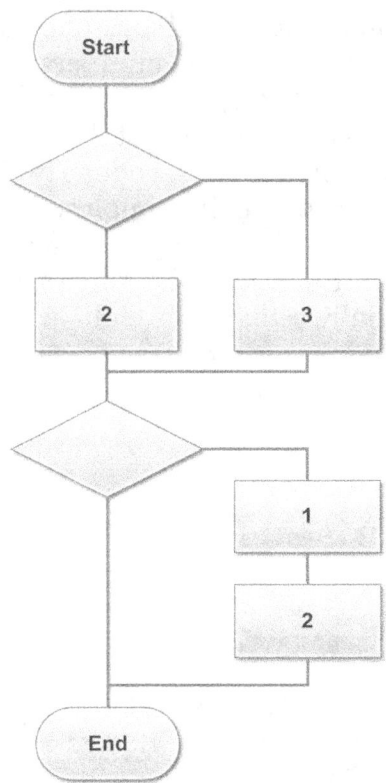

Fig. A-9. Structured flowchart equivalent of Fig. A-8 with complexity of one.

There are numerous examples of unstructured "spaghetti code" flowcharts on the Internet. One of the more notorious examples is in Chapter 5 of ACI-318-95, <u>Building Code Requirements for Structural Concrete</u>, which can be found in pdf form on the Internet. This code was made law and is "legally binding upon all citizens and residents of the United States of America." The flowchart specifies the process for a concrete production facility to obtain approval for a specific concrete mix. This same flowchart has been propagated into subsequent years of the code book for almost two decades. The flowchart consists of nine intertwined decision diamonds, five boxes with more than one entry point (three of which have 3 or 4 entries), with an irreducible complexity of 24. This is nothing but bone-headed obfuscation! Government bureaucrats and regulators may get away with obfuscation since they do not produce anything of marketable value at a competitive price in society and obfuscation helps them with their self-esteem! But do not go down that path. You are building hardware and software that you want to be maintained long after you have gone to the grave. Intelligent hardware and software design follows the "KISS" principle (keep it simple stupid). And that is the goal of structured flowcharting: designing reliable processes that can be easily understood and maintained by others when you are unavailable. If the irreducible one-pass path complexity of your flowchart is greater than one then you need to work harder on the process design. If you enforce the discipline of structured flowcharts upon your process design, you will automatically have a process with the lowest possible complexity of unity.

A.4 Conclusion

Structured flowcharts lead directly to structured code and processes that follow clear logical sequences. It leads to code that can be fully tested because every section of code has only one entry point and one exit point. Some might argue that structured flowcharts are not conducive to event-driven multi-processor systems. That is not true. Every process within a multi-tasking environment should be written with well-structured, fully-tested code. The fact that the process may be temporarily interrupted so the computer can go work on some other task is irrelevant. When the computer returns and picks up where it left off in the original process, the program flow still follows a structured logical sequence of instructions. Furthermore, the operating system that manages the interrupting of processes and multi-tasking should be written in a highly structured, fully tested manner, not as buggy spaghetti code. Good reliable code has an irreducible complexity of one, which means that it has one fully tested high level path through it that anyone can understand.

Now that you understand how to draw structured flowcharts, we want to show a simple flowchart for managing a potable rainwater harvesting system. Figure A-10 illustrates two facts. One is that you do not need arrows on connector lines in a structured flowchart because the flow direction is understood to be down and sometimes temporarily to the right. Second, there are some processes that never stop repeating, and they should not unless something drastic happens. Waking up in the morning after a good night's sleep is one example. That does not mean that the process has no exit point. But the exit may be extremely rare, such as dying in the middle of the night or entering into a coma. Figure A-10 shows what is to be done in the extremely rare event that proper water system management ceases: you either decommission the water system or you find someone else who can properly manage the system. There are no other alternatives.

Of course the weekly management activities could be expanded to include more detail. And the flowchart does not show longer term (annual) activities such as laboratory testing or tank cleaning. But the flowchart illustrates the clean, simple, and clear logic of a structured flowchart. And it emphasizes the repetitive nature of certain essential tasks in maintaining a potable rainwater harvesting system. The tasks are not necessarily difficult, but they cannot be ignored. A structured flowchart is a good way to clearly communicate to others what needs to be done to maintain your system. You should consider making structured flowcharts a part of your written procedures notebook. Most of life is just simple routine requiring only faithful execution. No doubt unexpected interruptions occur along the way requiring flexibility and development of new solutions. But if you can reduce those solutions to a structured flowchart, you will find execution is a whole lot easier and much more understandable for others following in your footsteps.

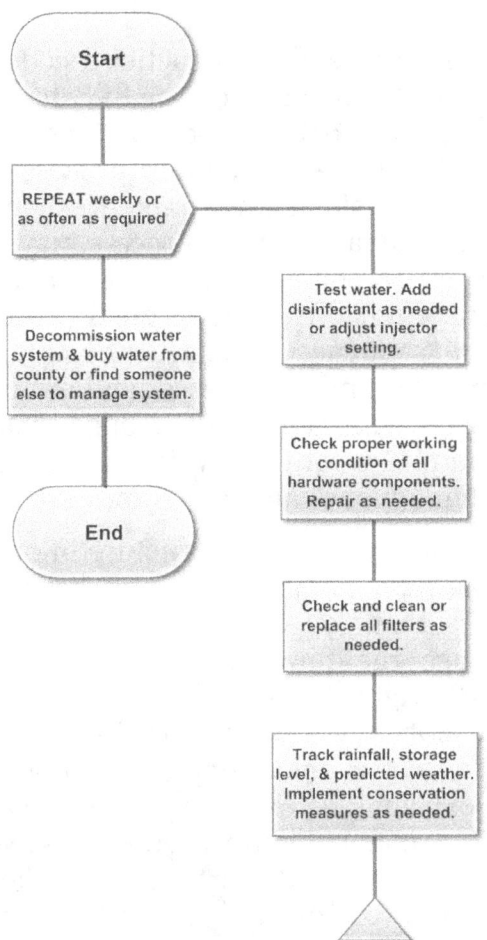

Fig. A-10. Structured flowchart for rainwater system management.

Appendix B.
References and Resources

The list of resources below is not meant to be exhaustive. They are simply the specialized resources I used to design and build my own system, and may or may not be useful in your own search for resources. Much of the miscellaneous hardware can be purchased at your local home builders supply (e.g. Lowe's or Home Depot) or your local hardware or plumbing store, so they are not listed here. There may be other competitors not listed here that provide products of equal or better quality. Failure to list them implies nothing about their ability to do so. This information is provided to help you get started in your own search for local suppliers of hardware and information for building a modern potable rainwater harvesting system. You can also search NSF International's web site for other suppliers of NSF certified products (http://www.nsf.org/certified-products-systems). The two NSF standards we are mainly concerned with are NSF/ANSI 61 for drinking water system components and NSF/ANSI 60 for drinking water treatment chemicals. I am also in the process of building a web site (www.PotableRainwater.com) to provide a resource for unique rainwater harvesting hardware and information.

B.1 Publications

Alternative Disinfectants and Oxidants Guidance Manual, EPA 815-R-99-014, United States Environmental Protection Agency, April 1999.

Richard Blum and Christine Bresnahan, Python Programming for Raspberry Pi, Second Edition, Pearson Education, 2016.

Bruce Carter and Ron Mancini, Op Amps for Everyone, 3rd Ed., Texas Instruments, 2009.

Rex Cauldwell, Wiring A House, The Taunton Press, 2002.

Christopher C. Burt, Extreme Weather: A Guide and Record Book, W.W. Norton & Company Ltd., London, 2004.

J.A. Davis and G.P. Curtis, Consideration of Geochemical Issues in Groudwater Restoration at Uranium In-Situ Leach Mining Facilities, NUREG/CR-6870, U.S. Geological Survey, Menlo Park, CA, January 2007.

Disinfectants and Disinfectant By-Products, Environmental Health Criteria 216, World Health Organization, 2000.

Robert Faludi, Building Wireless Sensor Networks, O'Reilly Media, Inc. 2011.

Georgia Rainwater Harvesting Guidelines, Georgia Department of Community

Affairs, 2009.

Guidelines for Canadian Drinking Water Quality, Guideline Technical Document, Chlorine, HC Pub: 4188, Publications Health Canada, Ottawa, Ontario, 2009.

Guidelines for Drinking-water Quality, 4th Edition, World Health Organization, 2011.

Susan Hall, Groundwater Restoration at Uranium In-Situ Recovery Mines, South Texas Coastal Plain, Open-File Report 2009-1143, U.S. Geological Survey, Central Energy Resources Science Center, Denver, CO, 2009.

Harvesting, Storing, and Treating Rainwater for Domestic Indoor Use, Texas Commission on Environmental Quality (TCEQ), January 2007.

Health Risks from Microbial Growth and Biofilms in Drinking Water Distribution Systems, Office of Ground Water and Drinking Water, U.S. Environmental Protection Agency, June 17, 2002.

Peter Hemp, Plumbing A House, The Taunton Press, 1998.

Holistic Approach to Sustainable Water Management in Northwest Douglas County, prepared for: Colorado Water Conservation Board, Dominion Water and Sanitation District, Castle Ines North Metropolitan District, Douglas County, Thunderbird Water and Sanitation District, and Plum Valley Heights HOA. Prepared by: Leonard Rice Engineers, Inc., Meurer & Associates, and Ryley Carlock & Applewhite Professional Association. January 2007.

Dr. Hari J. Krishna, et. al. The Texas Manual on Rainwater Harvesting, 3rd Edition, Texas Water Development Board, 2005.

Mark W LeChevallier, Kwok-Keung Au, Water Treatment and Pathogen Control, World Health Organization, 2004.

Mark Lutz, Programming Python, 4th Edition, O'Reilly Media, 2011.

Patricia S. H. Macomber, Guidelines on Rainwater Catchment Systems for Hawaii, College of Tropical Agriculture and Human Resources, University of Hawaii at Manoa, 2010.

Michael Margolis, Arduino Cookbook, O'Reilly Media, 2012.

Simon Monk, Programming the Raspberry Pi, Second Edition, McGraw Hill Education, 2016.

Robert J. Potwora, Trihalomethane Removal with Activated Carbon, Water Contioning & Purification, pp. 22-24, June 2006.

Rainwater Harvesting: System Planning, AgriLIFE Extension Texas A&M System, 2009.

Julie Rogers, A.B. Dowsett, P.J. Dennis, J.V. Lee, C.W. Keevil, <u>Influence of Plumbing Materials on Biofilm Formation and Growth of Legionella pneumophila in Potable Water Systems</u>, Applied and Environmental Microbiology, 1994, 60(6):1842.

Paul Scherz, <u>Practical Electronics for Inventors</u>, McGraw-Hill, 2000.

Jacques M Steininger, <u>PPM or ORP: Which Should Be Used?</u>, Swimming Pool Age & Spa Merchandiser, November 1985.

J. Steininger, <u>ORP Control in Pools and Spas</u>, Santa Barbara Control Systems, 1998.

Jacques M. Steininger, Catherine Pareja, <u>ORP Sensor Response in Chlorinated Water</u>, NSPI Water Chemistry Symposium, Vol 1, 1996.

Richard James Spahl, <u>Groundbreaking Measurement of Free Chlorine Disinfecting Power in a Handheld Instrument</u>, Myron L Company White Paper, January 2012.

<u>Uniform Plumbing Code 2015</u>, International Association of Plumbing and Mechanical Officials, Ontario, CA.

<u>Virginia Rainwater Harvesting Manual</u>, 2nd Edition 2009, The Cabell Brand Center, Salem, Virginia.

B.2 Industrial Mechanical Supplies

You may not be able to find some specialized hardware items at your local hardware store or home building supply. Online industrial suppliers are a good source for these hard-to-find items. A few that I use are listed here:

MSC Industrial Supply Co. (www.mscdirect.com)

McMaster-Carr (www.mcmaster.com)

Grainger (www.grainger.com)

80/20 Inc. (www.8020.net)

FlexPVC (www.flexpvc.com)

B.3 Food-Grade Adhesives and Coatings

Any adhesives and coatings that come in contact with the rainwater should be NSF-approved for potable water or safe to use in the food industry.

Polymer Composites, Inc. (www.theepoxyexperts.com)

Smooth-On, Inc. (www.smooth-on.com)

Reynolds Advanced Materials (www.reynoldsam.com)

EMI Supply (www.emisupply.com)

Sherwin-Williams (http://protective.sherwin-williams.com/)

The Epoxy Experts (www.theepoxyexperts.com)

B.4 Plastic Storage Tanks

There are a number of manufacturers of rotationally molded plastic storage tanks NSF-approved for potable water. Online drop shippers usually have arrangements with several of these manufacturers and provide huge selections of tank sizes and prices. Shop around for the best price and delivery to your local area. A few drop shippers of plastic tanks are listed here:

National Tank Outlet (www.ntotank.com)

TanksForLess (www.tanksforless.com)

Plastic-Mart (www.plastic-mart.com)

The Tank Depot (www.tank-depot.com)

B.5 Filters, Purifiers, Disinfectant, and Injectors

Fresh Water Systems, Greenville, South Carolina (www.freshwatersystems.com)

Microfiltration Gutter Products, Paramus, New Jersey (www.mastershield.com)

Klean Gutter, retail affiliate of MasterShield (www.micromeshgutterguards.com)

Prozone Water Products, Huntsville, Alabama (www.prozoneint.com)

Rain Harvest Systems, Cumming, Georgia (www.rainharvest.com)

Global Water Treatment Chemicals, Weatherford, Texas (www.gwtcinc.com)

Pure Water Products, Denton, Texas (www.purewaterproducts.com)

Mazzei Injector Company, LLC, Bakersfield, California (www.mazzei.net)

Ryan Herco Flow Solutions, Burbank, California (www.rhfs.com)

Big Brand Water Filter, Inc., Newbury Park, California(www.bigbrandwater.com)

B.6 Electronic Parts

Sparkfun Electronics (www.sparkfun.com)

Digi-Key Corporation (www.digikey.com)

Jameco Electronics (www.jameco.com)

Mouser Electronics (www.mouser.com)

Newark Electronics (www.newark.com)

Arduino microcontroller reference (http://arduino.cc/en/Reference/HomePage)

Ambient Weather (www.ambientweather.com)

RainWise Inc (www.rainwise.com)

SJE Rhombus (www.sjerhombus.com)

INDEX

www.ingramcontent.com/pod-product-compliance
Lightning Source LLC
Chambersburg PA
CBHW081717220526
45468CB00008B/1881